W. T. KOITER'S ELASTIC STABILITY OF SOLIDS AND STRUCTURES

This book deals with the elastic stability of solids and structures, for which Warner Koiter was the world's leading expert of his time. It begins with fundamental aspects of stability, relating the basic notions of dynamic stability to more traditional quasi-static approaches. The book is concerned not only with buckling, or linear instability, but most importantly with nonlinear postbuckling behavior and imperfection sensitivity. After laying out the general theory, Koiter applies the theory to a number of applications, with a chapter devoted to each. These include a variety of beam, plate, and shell structural problems and some basic continuum elasticity problems. Koiter's classic results on the nonlinear buckling and imperfection sensitivity of cylindrical and spherical shells are included. The treatments of both the fundamental aspects and the applications are completely self-contained. This book was recorded as a detailed set of notes by Arnold van der Heijden from W. T. Koiter's last set of lectures on stability theory at TU Delft.

Arnold M. A. van der Heijden has his own consultancy, HESTOCON Consultancy B.V. He received his master's and Ph.D. degrees, with honors, in mechanical engineering and applied mechanics under Professor Koiter. He has been a technical and scientific staff member in the Applied Mechanics Laboratory at Delft University of Technology, an honorary research Fellow at Harvard University, a professor at Eindhoven Technical University, a board member of the Department of Applied Mechanics of the Royal Dutch Institute for Engineers, and co-editor (with J. F. Besseling) of the Koiter symposium book *Trends in Solid Mechanics*. Dr. van der Heijden has worked on staff and consulted for many corporations, including Royal Dutch Shell (Pernis and The Hague), ABB Lummus Global, and as a project leader of ATEX at General Electric Advanced Materials, SABIC, and Essent. He has done gas explosion calculations for offshore platforms, including structural analysis. He is currently working on improvements in safety management for ProRail.

W. T. Koiter's Elastic Stability of Solids and Structures

Edited by

Arnold M. A. van der Heijden

Technische Universiteit Delft

CAMBRIDGE
UNIVERSITY PRESS

CAMBRIDGE UNIVERSITY PRESS
Cambridge, New York, Melbourne, Madrid, Cape Town,
Singapore, São Paulo, Delhi, Mexico City

Cambridge University Press
The Edinburgh Building, Cambridge CB2 8RU, UK

Published in the United States of America by Cambridge University Press, New York

www.cambridge.org
Information on this title: www.cambridge.org/9781107407015

First published 2009
First paperback edition 2012

A catalogue record for this publication is available from the British Library

Library of Congress Cataloguing in Publication Data

Heijden, Arnold van der.
W. T. Koiter's elastic stability of solids and structures / Arnold van der Heijden.
p. cm.
Includes bibliographical references and index.
ISBN 978-0-521-51528-3 (hardback)
1. Elastic analysis (Engineering). 2. Structural stability.
I. Heijden, Arnold van der. II. Title.
TA653.H44 2008
620.1′1232 – dc22 2008017910

ISBN 978-0-521-51528-3 Hardback
ISBN 978-1-107-40701-5 Paperback

Contents

Preface

These lecture notes were made after Professor Koiter's last official course at Delft's University of Technology, in the academic year 1978–79. Although these notes were prepared in close collaboration with Professor Koiter, they are written in the author's style. The author is therefore fully responsible for possible errors.

This course covers the entire field of elastic stability, although recent developments in the field of stiffened plates and shells are not included. Hopefully, these lecture notes reflect some of the atmosphere of Dr. Koiter's unique lectures.

Delft, June 10, 2008 A. M. A. v. d. Heijden

1

Stability

1.1 Discrete systems

Consider a system with a finite number of degrees of freedom. The position of this system is represented by a position vector $\mathbf{q}(q^1, q^2 \ldots q^n)$, where q^i $(i \in 1 \ldots n)$ are n independent coordinates. It is assumed that the system is *holonomic*, i.e., no relations exist between the derivatives of the coordinates, and *scleronomic*, i.e., the factor time is not explicitly needed in the description of the system.† Let $\dot{q}^i$ be the generalized velocities. The kinetic energy T is then a homogeneous quadratic function of the generalized velocities, and hence T can be written as

$$T = \frac{1}{2} a_{ij} (\mathbf{q}) \dot{q}^i \dot{q}^j . \tag{1.1.1}$$

When the system is non-sclerononic, terms linear in the velocities and a term independent of $\dot{q}^i$ must be added. The coefficient a_{ij} $(\mathbf{q})$ is called the *inertia matrix*. The forces acting upon the system can be expressed by a generalized force vector $\mathbf{Q}$ defined by

$$Q_i \delta q^i = \text{v.w.}, \tag{1.1.2}$$

where the right-hand side stands for the virtual work of all the forces acting upon the system. In general, this expression is not a total differential. However, for an important class of problems, it is. Systems for which (1.1.2) is a total differential are called *conservative systems*. In that case we have

$$Q_i \delta q^i = -\delta P (\mathbf{q}) , \tag{1.1.3}$$

where δP $(\mathbf{q})$ is a total differential and P $(\mathbf{q})$ is called the potential energy. In the following, we mainly restrict our attention to conservative systems because for elastic systems, conservative forces play an important role.

Introducing a kinetic potential L defined by

$$\mathcal{L} = T - P, \tag{1.1.4}$$

† This implies that $d\mathbf{q} = \mathbf{q},_k \, dq^k$.

the Lagrangian equations for a conservative system are

$$\frac{d}{dt}\frac{\partial \mathcal{L}}{\partial \dot{q}^i} - \frac{\partial \mathcal{L}}{\partial \dot{q}^i} = 0, \quad (\mathrm{i} \in 1, \ldots, \mathrm{n}). \tag{1.1.5}$$

Using the expression (1.1.1), we may rewrite this equation to yield

$$\frac{d}{dt}(a_{ij}\dot{q}^i) - \frac{1}{2}a_{hk,i}\dot{q}^h\dot{q}^k + P_{,i} = 0, \tag{1.1.6}$$

where

$$a_{hk,i} = \frac{\partial a_{hk}(\mathbf{q})}{\partial q^i}, \quad P_{,i} = \frac{\partial P(\mathbf{q})}{\partial q^i}$$

However, it often happens that non-conservative forces are present (e.g., damping forces). It is then advantageous to take these into account separately as follows:

$$Q_i\delta q^i = \delta P(\mathbf{q}) + Q_i^*\delta q^i, \tag{1.1.7}$$

where $\mathbf{Q}^*$ is the vector of non-conservative forces. The equations of motion then read

$$\frac{d}{dt}\frac{\partial \mathcal{L}}{\partial \dot{q}^i} - \frac{\partial \mathcal{L}}{\partial \dot{q}^i} = Q_i^*, \quad (\mathrm{i} \in 1 \cdots \mathrm{n}). \tag{1.1.8}$$

These are n second-order ordinary differential equations.

Let us now consider the stability of discrete systems. For a system to be in equilibrium, the velocities (and hence the kinetic energy) have to vanish. This implies that for a conservative system, we have

$$P_{,i} = 0. \tag{1.1.9}$$

In words: The potential energy has a *stationary* value.

By *stability* we mean that a small disturbance from the state of equilibrium does not cause large deviations from this state of equilibrium. A disturbance from the state of equilibrium implies that the velocities are nonzero or that the position differs from the equilibrium position. We can always choose our coordinate system such that the equilibrium position is given by $\mathbf{q} = \mathbf{0}$. Furthermore, we can always choose the potential energy in such a way that it vanishes in the equilibrium position. Doing so, we may write

$$P = \frac{1}{2}P_{,ij}(\mathbf{0})\, q^i q^j + \cdots. \tag{1.1.10}$$

To be able to give a more exact definition of stability, we need a measure to denote the deviation from the state of equilibrium. Remembering that in equilibrium we have $\mathbf{q} = \dot{\mathbf{q}} = \mathbf{0}$, a number $\rho(\mathbf{q}, \dot{\mathbf{q}})$ is introduced with the following properties:

$$\begin{aligned}
&1)\quad \rho(\mathbf{q}, \dot{\mathbf{q}}) \leq 0 \quad \text{for} \quad \mathbf{q} \neq \mathbf{0} \quad \text{or} \quad \dot{\mathbf{q}} \neq \mathbf{0},\\
&2)\quad \rho(\mathbf{q}_1, \mathbf{q}_2, \dot{\mathbf{q}}_1 + \dot{\mathbf{q}}_2) \leq \rho(\mathbf{q}_1, \dot{\mathbf{q}}_1)\,\rho(\mathbf{q}_2, \dot{\mathbf{q}}_2)\\
&\qquad \text{(triangle inequality)},\\
&3)\quad \rho(\alpha\mathbf{q}, \alpha\dot{\mathbf{q}}) = |\alpha|\,\rho(\mathbf{q}, \dot{\mathbf{q}}) \quad \alpha \in \mathbb{R}.
\end{aligned} \tag{1.1.11}$$

We are now in a position to define the following stability criterion.

An equilibrium position is *stable* if and only if for *each* positive number ε there exists a positive number $\delta(\varepsilon)$ such that for *all* disturbances of the equilibrium at the time $t > 0$, with $\rho[\mathbf{q}(0), \dot{\mathbf{q}}(0)] < \delta$, the motion for $t > 0$ satisfies $\rho[\mathbf{q}(t), \dot{\mathbf{q}}(t)] < \varepsilon$.

Notice that the statement about stability depends on the measure that is used. Different measures yield different criteria for stability. Notice further that different measures may be used for $t = 0$ and $t > 0$. This freedom is of great importance for applications. For example, suitable choices for ρ are

$$\rho = \left[\sum_{i=1}^{n} (q^i)^2 + \sum_{i=1}^{n} (\dot{q}^i)^2\right]^{1/2},$$

$$\rho = \max\left|q^i\right| + \max\left|\dot{q}^i\right|.$$

For a conservative system, $T + P =$ constant. This well-known result can easily be derived from Lagrange's equations for a conservative, holonomic, and scleronomic system. Multiplying the equations by $\dot{q}^i$, we obtain

$$\dot{q}^i \frac{d}{dt}\frac{\partial\mathcal{L}}{\partial\dot{q}^i} - \dot{q}^i\frac{\partial\mathcal{L}}{\partial q^i} = 0$$

or

$$\frac{d}{dt}\left(\dot{q}^i\frac{\partial\mathcal{L}}{\partial\dot{q}^i}\right) - \ddot{q}^i\frac{\partial\mathcal{L}}{\partial\dot{q}^i} - \dot{q}^i\frac{\partial\mathcal{L}}{\partial q^i} = 0$$

or

$$\frac{d}{dt}\left(\dot{q}^i\frac{\partial\mathcal{L}}{\partial\dot{q}^i}\right) - \frac{d}{dt}\mathcal{L} = 0.$$

Using Euler's theorem for homogeneous quadratic functions, we readily obtain

$$\frac{d}{dt}(2T) - \frac{d}{dt}(T - P) = 0,$$

from which follows

$$T + P = E, \tag{1.1.12}$$

where $E = T(t = 0) + P(t = 0)$. This equation enables us to make the following statement about stability.

Theorem. *The equilibrium is stable provided the potential energy is positive-definite.*

To prove this theorem, we introduce the following norms:

$$\|\mathbf{q}\|^2 = \sum_{i=1}^{n}(q^i)^2$$

$$\|\dot{\mathbf{q}}\|^2 = \sum_{i=1}^{n}(\dot{q}^i)^2.$$

Let $d(c)$ denote the minimum of $P(\mathbf{q})$ on the hypersphere $\| \mathbf{q} \| = c$. $P(\mathbf{q})$ is positive-definite when $d(c)$ is a monotonically increasing function of c on the sphere $\delta \leq c < R$.

Proof. $T + P = \text{constant} = E$, T is positive or zero, and P is positive-definite. Restrict the initial disturbance so that

$$\| \mathbf{q}\,(0) \| < c_1 \quad \text{and} \quad E < d\,(c_1)\,.$$

This means that $T\,(t = 0) < d\,(c_1)$. Because $T + P < d\,(c_1)$ and P is positive-definite, it follows that $T < d\,(c_1)$ for all t. On the other hand, because T is positive or zero, it follows that $P < d\,(c_1)$ for all t. A similar argument holds for a disturbance $\| \dot{\mathbf{q}}(0) \| < c_2$. Hence, we may choose an arbitrary (small) disturbance and the displacements and velocities will always remain within definable bounds. ■

The converse of this theorem has not yet been proven in all generality. To see some of the difficulties that are encountered, we consider the following example (one degree of freedom):

$$P(q) = e^{-q^{-2}} \cos q^{-2}.$$

For $q = 0$, all the derivatives vanish. However, in the immediate vicinity of the origin there are always negative values of P. In spite of this, the system is stable for sufficiently small disturbances.

Actual physical systems are never exactly conservative, i.e., there is always some dissipation. The approximation by a conservative system is often a very good approximation. In the presence of damping forces, we need the Lagrangian equations with an additional term for the non-conservative forces. Multiplying by $\dot{\mathbf{q}}^i$,

$$\dot{q}^i \frac{d}{dt} \frac{\partial \mathcal{L}}{\partial \dot{q}^i} - \dot{q}^i \frac{\partial \mathcal{L}}{\partial q^i} = Q_i^* \dot{q}^i, \quad (\mathrm{i} \in 1, \ldots, \mathrm{n})$$

from which follows

$$\frac{d}{dt}\,(T + P) = Q_i^* \dot{q}^i \equiv -D\,(\mathbf{q}, \dot{\mathbf{q}})\,. \tag{1.1.13}$$

Damping implies that the dissipation function $D > 0$ for $\dot{\mathbf{q}} \neq 0$. We now make the following assumptions:

1) The damping forces have the property that $Q_i^* \to 0$ for $\|\dot{\mathbf{q}}\| \to 0$.
2) $D(\mathbf{q}, \dot{\mathbf{q}}^i) > 0$ for $\|\dot{\mathbf{q}}\| \neq 0$.
3) $P\,(\mathbf{q})$ does not possess stationary values for $\| \dot{\mathbf{q}} \| < c$except at $\mathbf{q} = \mathbf{0}$.

Systems satisfying these conditions are called *pseudo-conservative.* Notice that because of restriction (1), dry friction forces are excluded.

Theorem. *In the presence of (positive) damping forces, a system with an indefinite potential energy is unstable.*

Proof. If P is indefinite, consider a disturbance of the equilibrium configuration with zero velocity and negative potential energy. The initial total energy is thus negative and, as this configuration cannot be in equilibrium, motion must result, as a result of which energy is dissipated. The total energy must decrease, so the system cannot stay in the vicinity of the origin, which means that the equilibrium configuration is unstable. ∎

The great advantage of this stability theorem is that it does not involve the kinetic energy, and hence the inertia matrix $a_{ij}(\mathbf{q})$. For a conservative or pseudo-conservative system, the stability criterion only depends on the potential energy (a quasi-static criterion). In general, the stability problem is a dynamic problem, and the kinetic energy plays an essential role. An example of such a problem is the behavior of the wings of an airplane in an airflow. In this case, the forces do not depend on only the geometry but also on the velocities.

For static loads, it is often sufficient to restrict oneself to conservative loads (e.g., deadweight loads). A more severe restriction for continuous systems is that we must restrict ourselves to elastic systems, i.e., to systems where there is a potential for the internal energy. Such a potential does not exist when plasticity occurs.

Let us now have a closer look at the stability problem. As mentioned previously, the stability criterion is fully determined by the potential energy $P(\mathbf{q})$. In the equilibrium position, we have chosen $P(\mathbf{0}) = 0$ and $\mathbf{q} = \mathbf{0}$ so that we may write

$$P(\mathbf{q}) = \frac{1}{2}P_{,ij}(\mathbf{0})\,q^i q^j + \cdots \equiv \frac{1}{2}c_{ij}q^i q^j + \cdots, \tag{1.1.14}$$

where c_{ij} denotes the stiffness matrix in the equilibrium position. It follows that when the stiffness matrix is positive-definite, $P(\mathbf{q})$ is positive-definite and the system is stable. If c_{ij} is indefinite (or negative-definite) then the system is unstable. If c_{ij} is semi-definite-positive (i.e., non-negative and zero for at least a deflection in one direction), then we must consider higher-order terms in the expansion for P. This case is called a *critical case of neutral equilibrium*. We shall consider this case in more detail.

It is convenient to transform the quadratic form (1.1.14) to a sum of quadratic terms. If the form is positive-definite, then the coefficients in the transformed form are all positive. Applying this transform to (1.1.14) and denoting the transformed coordinates again by q^i, we may write

$$P(\mathbf{q}) = \frac{1}{2}\sum_{i=1}^{n} c_i (q^i)^2 + \cdots + ()\,(q^1)^3 + \cdots. \tag{1.1.15}$$

Further, we order the coefficients c_i such that

$$c_1 \le c_2 \le c_3 \le \;\cdots\; \le c_n.$$

We now consider the case $c_1 = 0, c_2 > 0$. Taking all $q^i = 0\,(i > 1)$, the dominant term will be $(q^1)^3$. This term can attain negative values, and hence the system will be unstable. A necessary condition for stability is that the coefficient of $(q^1)^3$ is equal

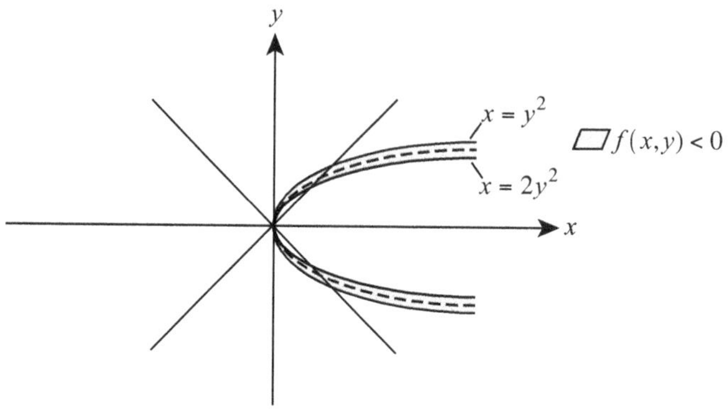

Figure 1.1.1

to zero. A further necessary condition for stability is that the coefficient of $(q^1)^4$ is positive. However, this condition is insufficient, as will be shown in the following example. Consider the function

$$P = f(x, y) = (x - y^2)(x - 2y^2) = x^2 - 3xy^2 + 2y^4. \qquad (1.1.16)$$

The graphs of the functions $x - y^2 = 0$ and $x - 2y^2 = 0$ are given in Fig. 1.1.1.

The function $f(x, y)$ in an arbitrary small neighborhood of the origin takes on both positive and negative values. In this case, the quadratic form in y vanishes at the origin, and there is no cubic term, but the coefficient in the quartic term is positive. Hence, the necessary conditions for stability are satisfied. However, this system is unstable because in an arbitrarily small neighborhood of the origin, P takes on negative values.

The reason that the conditions mentioned here are not sufficient is that we have restricted our investigation to straight lines through the origin (see Fig. 1.1.1). Following these straight lines, we always find only positive values in a sufficiently small neighborhood of the origin. However, if we follow curved lines through the origin (see the dashed lines), we easily find negative values. Once we have recognized the reason why the conditions imposed are insufficient, it is easy to find a remedy. To this end, we consider a line $y =$ constant in the neighborhood of the origin, and we minimize $f(x, y)$ with respect to x, i.e.,

$$\operatorname*{Min}_{y=y_1} f(x, y) = x^2 - 3xy_1^2 + 2y_1^4. \qquad (1.1.17)$$

This yields $2x - 3y_1^2 = 0$, and hence $x = 3/2\, y_1^2$. Substitution of this value into $f(x, y)$ yields $\min f(x, y) = -1/4\, y_1^4$, which means that the function is indefinite.

In general, the function P is minimized with respect to q^i $(i > 1)$ for fixed q^1. When the coefficient of $(q^1)^4$ is positive-definite, the system is stable.

2

Continuous Elastic Systems

2.1 Thermodynamic background

Consider a body that is in a state of equilibrium under conservative loads. Our aim is to investigate this equilibrium state.

For an *elastic* body, the internal energy per unit mass may be represented by $U(s, \boldsymbol{\gamma})$, where s denotes the specific entropy and $\boldsymbol{\gamma}$ is the deformation tensor. Let x_i $(i = 1, 2, 3)$ be the components of the position vector $\mathbf{x}$, which describe the position of a material point in the "fundamental state" I, which is to be investigated. Let $\mathbf{u}(\mathbf{x})$ be the displacement vector from the fundamental state ($\mathbf{u}$ is a small but finite displacement). The corresponding position in the "adjacent state" II is then $\mathbf{x} + \mathbf{u}$. The (additional) deformation tensor is now defined by

$$\gamma_{ij} = \frac{1}{2}(u_{i,j} + u_{j,i}) + \frac{1}{2}u_{h,i}u_{h,j}. \tag{2.1.1}$$

The fact that the body has undergone deformations to arrive in the fundamental state is unimportant because the state I is kept fixed.

The temperature T is now defined by

$$T = \frac{\partial U}{\partial s} \tag{2.1.2}$$

($\boldsymbol{\gamma}$ is kept constant).

From (2.1.2) we obtain

$$\frac{\partial T}{\partial s} = \frac{\partial^2 U}{\partial s^2}.$$

The specific heat of the material is now defined by

$$T\frac{\partial s}{\partial T} = C_\gamma, \tag{2.1.3}$$

where $C_\gamma > 0$ for a thermodynamically stable material. As $\partial^2 U/\partial s^2$ is positive (nonzero) we may solve (2.1.2) for s, which yields $s = s(T, \boldsymbol{\gamma})$.

We now introduce the function $F(T, \boldsymbol{\gamma})$, defined by

$$F(T, \boldsymbol{\gamma}) = U - Ts. \tag{2.1.4}$$

$F(T, \boldsymbol{\gamma})$ is called the *free energy*.

Writing (2.1.4) as a total differential, we find (for fixed $\boldsymbol{\gamma}$)

$$\frac{\partial F}{\partial T}\delta T = \frac{\partial U}{\partial s}\delta s - T\delta s - s\delta T. \tag{2.1.5}$$

Using (2.1.2) we find

$$s = -\frac{\partial F}{\partial T}. \tag{2.1.6}$$

Let us denote the temperature in the fundamental state (which by virtue of the equilibrium state is equal to the temperature of its surroundings) by T_{I}. A disturbance of the equilibrium state will cause a heat flux in the body. In the following, we will assume that the temperature of the surrounding medium is constant (T_{I}). Denoting the heat flux by $\mathbf{q}$, the heat flux per unit time through a closed surface is given by $\int_A \mathbf{q} \cdot \mathbf{n}\, dA$, where $\mathbf{n}$ denotes the unit normal vector on the surface, positive in the outward direction. According to the second law of thermodynamics,† vheat will flow out of the body when its surface temperature is higher than that of the surrounding medium, i.e.,

$$(T - T_{\mathrm{I}})\mathbf{q} \cdot \mathbf{n} \geq 0 \quad \text{(on the surface)}. \tag{2.1.7}$$

The heat flux will cause an entropy flux. The entropy flux vector $\mathbf{h}$ is given by

$$\mathbf{h} = \frac{1}{T}\mathbf{q} \quad \text{(per unit time and per unit area)}. \tag{2.1.8}$$

For an arbitrary part of the body, the entropy balance is given by

$$\int_V \rho \dot{s}\, dV = \int_A \mathbf{h} \cdot \mathbf{n}\, dA, \tag{2.1.9}$$

where ρ is the specific mass. This equation only holds in the absence of irreversible processes in the body. When the state of the body also depends on the deformation rates, irreversible processes will occur, which implies entropy production. In that case, the entropy balance reads

$$\int_V \rho \dot{s}\, dV = \int_A \mathbf{h} \cdot \mathbf{n}\, dA + \int_V \rho\sigma\, dV, \quad \sigma \geq 0, \tag{2.1.10}$$

where σ denotes the entropy production per unit time and mass. This is the more general formulation of the second law of thermodynamics (CLAUSIUS-DUHEM). The first law of thermodynamics states that the total amount of heat that flows into a body is transformed into internal energy.

Let $P_L[\mathbf{u}(\mathbf{x}(t))]$ be the potential energy of the external loads and let

$$K[\dot{\mathbf{u}}(\mathbf{x}(t))] = \frac{1}{2}\int_V \rho\dot{\mathbf{u}} \cdot \dot{\mathbf{u}}\, dV$$

† This is an early formulation by CLAUSIUS (1854).

be the kinetic energy. The total energy balance is then given by

$$\frac{d}{dt}\left\{\int_V \rho U(s,\gamma)\, dV + K[\dot{\mathbf{u}}(\mathbf{x}(t))] + P_L[\mathbf{u}(\mathbf{x}(t))]\right\} = -\int_A \mathbf{q}(\mathbf{x}(t))\cdot\mathbf{n}\, dA, \quad \text{(1st law)} \tag{2.1.11}$$

where we have a negative sign on the right-hand side of this equation because the heat flux is regarded as positive in the outward direction.

To draw conclusions from the first and the second laws, we subtract (2.1.11) from (2.1.10) multiplied by T_I. This yields

$$\frac{d}{dt}\left\{\int_V \rho[U(s,\gamma) - T_\mathrm{I} s]\, dV + K[\dot{\mathbf{u}}(\mathbf{x}(t))] + P_L[\mathbf{u}(\mathbf{x}(t))]\right\} = \int_V \left(\frac{T_\mathrm{I}}{T} - 1\right)\mathbf{q}\cdot\mathbf{n}\, dA - T_\mathrm{I}\int_V \rho\sigma\, dV \le 0 \tag{2.1.12}$$

(Duhem, 1911).
Here we have made use of the relation

$$\int_V \rho\dot{s}\, dV = \frac{d}{dt}\int_V \rho s\, dV \tag{2.1.13}$$

The first term on the right-hand side of (2.1.12) is negative because the heat flux is in the outward direction when $T > T_\mathrm{I}$, and the second term is negative because the entropy production is always positive. The integral on the left-hand side of (2.1.12) may be expressed in terms of the free energy. Using the relation

$$U(s,\gamma) - T_\mathrm{I} s = U(s,\gamma) - Ts + (T - T_\mathrm{I})s = F(T,\gamma) + (T - T_\mathrm{I})\frac{\partial F}{\partial T}, \tag{2.1.14}$$

we obtain

$$\frac{d}{dt}\left\{\int_V \rho\left[F(T,\gamma) + (T_\mathrm{I} - T)\frac{\partial F}{\partial T}\right] dV + P_L + K\right\} \le 0. \tag{2.1.15}$$

Duhem (1911) already discussed the stability of a system on the basis of this equation and came to the conclusion that a system is stable when the form between the braces is positive-definite. In this form, K is a positive-definite function. However, the terms between the square brackets depend on the deformation tensor and the temperature, whereas P_L depends on the displacement field. The problem is to separate the influence of the temperature and the displacement field. A straightforward expansion

$$F(T,\gamma) = F(T_\mathrm{I},\gamma) + \left(\frac{\partial F}{\partial T}\right)_{T_\mathrm{I}}(T - T_\mathrm{I}) + \frac{1}{2}\left(\frac{\partial^2 F}{\partial T^2}\right)_{T_\mathrm{I}}(T - T_\mathrm{I})^2 + \cdots$$

does not solve the problem. Following ERICKSEN (1965), we may write the Taylor expansion of the free energy at constant deformation $\boldsymbol{\gamma}$ in the form

$$F(T_{\mathrm{I}}, \boldsymbol{\gamma}) = F(T, \boldsymbol{\gamma}) + \left(\frac{\partial F}{\partial T}\right)_T (T_{\mathrm{I}} - T) + \frac{1}{2}\left(\frac{\partial^2 F}{\partial T^2}\right)_* (T_{\mathrm{I}} - T)^2, \quad (2.1.16)$$

where the first derivative is evaluated at the deformation $\boldsymbol{\gamma}$ and temperature T, and the second (starred) derivative at the deformation $\boldsymbol{\gamma}$ and an intermediate temperature $T^* = T + \theta(T_{\mathrm{I}} - T)$, where $0 < \theta < 1$. Using (2.1.16) we may rewrite the term between the square brackets in (2.1.15) as follows:

$$\begin{aligned} F(T, \boldsymbol{\gamma}) + (T_{\mathrm{I}} - T)\frac{\partial F}{\partial T} &= F(T_{\mathrm{I}}, \boldsymbol{\gamma}) - \frac{1}{2}\left(\frac{\partial^2 F}{\partial T^2}\right)_* (T_{\mathrm{I}} - T)^2 \\ &= F(T_{\mathrm{I}}, \boldsymbol{\gamma}) + \frac{1}{2}\left(\frac{c_\gamma}{T}\right)_* (T_{\mathrm{I}} - T)^2, \end{aligned} \quad (2.1.17)$$

where we have used the relation

$$c_\gamma \equiv T\frac{\partial s}{\partial T} = T\frac{\partial^2 F}{\partial T^2}.$$

The first term on the right-hand side of (2.1.17) depends only on the displacement field. The second term is positive-definite. The energy balance may now be written in the form

$$\frac{d}{dt}\left\{\int_V \rho F(T_{\mathrm{I}}, \boldsymbol{\gamma})\, dV + P_L[\mathbf{u}(\mathbf{x}(t))] + K[\dot{\mathbf{u}}(\mathbf{x}(t))] + \frac{1}{2}\int_V \rho\left(\frac{c_\gamma}{T}\right)_* (T_{\mathrm{I}} - T)^2\, dV\right\} \le 0. \quad (2.1.18)$$

The last two terms in the left-hand member are positive-definite, and the remaining terms depend only on the displacement field. Our energy balance is not affected when we subtract from the expression between the braces a time-independent quantity,

$$\int_V \rho F(T_{\mathrm{I}}, \mathbf{0})\, dV.$$

Further, we introduce the notation

$$W(\boldsymbol{\gamma}) \equiv \rho[F(T_{\mathrm{I}}, \boldsymbol{\gamma}) - F(T_{\mathrm{I}}, \mathbf{0})], \quad (2.1.19)$$

where $W(\boldsymbol{\gamma})$ is the (additional) stored elastic energy in the isothermal (additional) deformation $\boldsymbol{\gamma}$ at constant temperature T_{I}, from the fundamental state I to the current state. The potential energy functional P is now defined by

$$P[\mathbf{u}(\mathbf{x}(t))] = \int_V W(\boldsymbol{\gamma})\, dV + P_L[\mathbf{u}(\mathbf{x}(t))]. \quad (2.1.20)$$

In words: The potential energy is equal to the sum of the increase of the elastic energy for isothermal deformations and the potential energy of the external loads.

Hence, stability for the class of problems discussed depends on *isothermal* constants.[†] The question of stabilty when T_I is not constant is still unsolved (which is important, for example, in problems with thermal stresses).

2.2 Theorems on stability and instability

In our discussion on the stability of discrete systems, we have seen that we need measures to be able to specify expressions like "small disturbance" and "not large deviations." In our discussion of the stability of continuous systems, which is governed by the character of the potential energy, we need a measure for the displacements. A suitable measure is the L_2-norm of the displacement field, defined by

$$\left\|\mathbf{u}\left(\mathbf{x}\left(t\right)\right)\right\|^2 = \frac{1}{M}\int_V \rho\,\mathbf{u}\left(\mathbf{x}\left(t\right)\right)\cdot\mathbf{u}\left(\mathbf{x}\left(t\right)\right)\,dV, \tag{2.2.1}$$

where M is the total mass of the elastic body. We shall employ the same measure for the initial disturbance. We shall assume that the potential energy functional is regular in the following sense. On every ball $\|\mathbf{u}\| = c$ in the function space of kinematically admissible displacement fields $\mathbf{u}\left(\mathbf{x}\left(t\right)\right)$ where the radius c is sufficiently small, the energy functional $P\left[\mathbf{u}\left(\mathbf{x}\left(t\right)\right)\right]$ has a proper minimum that is a continuous function $d(c)$ in the range of c under consideration. The potential energy functional is called *positive-definite* if the function $d(c)$ is a (positive) increasing function in a range $0 \leq c < c_1$. The functional is called *indefinite* if the function $d(c)$ is a (negative) decreasing function for $0 \leq c < c_1$.

We are now in a position to formulate the *stability criterion*: The equilibrium in the fundamental state is stable if the potential energy functional $P\left[\mathbf{u}\left(\mathbf{x}\left(t\right)\right)\right]$ is positive-definite.

To show this, we introduce the notation

$$\begin{aligned} V\left[\mathbf{u}\left(\mathbf{x}\left(t\right)\right), \dot{\mathbf{u}}\left(\mathbf{x}\left(t\right)\right), T\left(x\left(t\right)\right)\right] &= P\left[\mathbf{u}\left(\mathbf{x}\left(t\right)\right)\right] \\ &+K\left[\dot{\mathbf{u}}\left(\mathbf{x}\left(t\right)\right)\right] + \frac{1}{2}\int_V \rho\left(\frac{c_\gamma}{T}\right)_* \left(T_I - T\right)^2\,dV, \end{aligned} \tag{2.2.2}$$

where V is the total energy.[‡]

[†] In the literature, one frequently encounters vague and loose statements to the effect that buckling is "rapid" and that it is therefore "reasonable" to assume that the motion is adiabatic. This would imply that elastic stability would be governed by the adiabatic elastic constants rather than by the isothermal elastic properties. This reasoning is erroneous as follows from the foregoing analysis. A simple example is a strut with pinned ends under a compressive load N. For sufficiently small values of the compressive load the straight configuration is stable, and this stability is manifested by a non-vanishing fundamental frequency. This frequency decreases when the critical Euler load N_1 is approached, and it vanishes for N_1. The motion at the critical load is thus infinitely slow, and hence isothermal.

[‡] DUHEM called it the "ballistic energy."

According to (2.1.18) we have $dV/dt \leq 0$. Let the initial disturbance $\mathbf{u}(\mathbf{x}(0))$ satisfy the condition

$$\|\mathbf{u}(\mathbf{x}(0))\| < \alpha_1 \tag{2.2.3}$$

and let the total energy satisfy the condition

$$0 < V_0 < d(\alpha_1), \tag{2.2.4}$$

where $V_0 = V[\mathbf{u}(\mathbf{x}(0)), \dot{\mathbf{u}}(\mathbf{x}(0)), T(\mathbf{x}(0))]$.

Because $dV/dt < 0$, it follows that $P[\mathbf{u}(\mathbf{x}(t))] < d(\alpha_1)$, and hence $\|\mathbf{u}(\mathbf{x}(t))\| < \alpha_1$.

In words: For a given, sufficiently small initial disturbance and a given total energy, the displacements at $t > 0$ are bounded by the value of $\|\mathbf{u}(\mathbf{x}(0))\|$.

For an instability criterion, we need the following assumptions:

1. For nonvanishing deformation rates, the entropy production is positive.
2. In a sufficiently small neighborhood of the fundamental state, the potential energy has no stationary values at a different energy level.

Under these additional conditions, the following theorem holds.

Theorem. *If the potential energy functional is indefinite, the system is unstable.*

To show this, we consider an initial disturbance $\mathbf{u}(\mathbf{x}(0))$, as small as we please, in the region in function space where $P[\mathbf{u}(\mathbf{x})]$ is negative, and we select initial velocities and temperature variations that are both identically zero. The initial value V_0 of the total energy is then negative, and by (2.1.18) this energy will decrease until the motion comes to a final stop. Because no equilibrium will be possible in the range $0 < c < c_1$ of $\|\mathbf{u}\|$, it follows that this norm at some time must approach or exceed the value c_1, no matter how small the initial disturbance has been chosen. It follows that the equilibrium in the fundamental state is unstable.

The only case that is not covered by our discussion so far is the case $d(c) = 0$. However, this case is not important because this condition never occurs in practical problems.

The results obtained so far are not as useful as it may seem because so far it has proved to be impossible to show that the elastic energy functional is positive-definite, even for an elastic body without external loads. To state this problem more clearly, we consider the potential energy in the elastic body

$$P[\mathbf{u}(\mathbf{x}(t))] = P_L\mathbf{u}(\mathbf{x}(t)) + \int_V W(\boldsymbol{\gamma})\, dV, \tag{2.2.5}$$

where the elastic potential $W(\boldsymbol{\gamma})$ is expanded about its value in the fundamental state

$$W(\boldsymbol{\gamma}) = \left(\frac{\partial W}{\partial \gamma_{ij}}\right)_{\mathrm{I}} \gamma_{ij} + \frac{1}{2}\left(\frac{\partial^2 W}{\partial \gamma_{ij}\,\partial \gamma_{k\ell}}\right)_{\mathrm{I}} \gamma_{ij}\gamma_{k\ell} + \cdots \tag{2.2.6}$$

This expansion contains linear and quadratic terms in the strains, which in their turn contain linear and quadratic terms in the derivatives of the displacements. Consider now the displacement fields $\mathbf{u} = \alpha \mathbf{u}_1(\mathbf{x})$, where $\mathbf{u}_1(\mathbf{x})$ is a given (fixed) displacement field and α is a positive number. Stability is now determined by α for $\alpha \to 0$. Now consider all kinematically admissible displacement fields $\mathbf{u}_1(\mathbf{x})$. It has been demonstrated that it is impossible to show that W is positive-definite on this basis. We shall also need restrictions on the derivatives of the displacements. Strictly speaking, this condition implies that we cannot use the stability criterion. However, it can be shown that for an indefinite elastic energy functional, the system is unstable.

Let us now consider the first term in (2.2.6). By virtue of

$$\delta W = S_{ij}\,\delta\gamma_{ij}, \tag{2.2.7}$$

where S_{ij} is a symmetric stress tensor, we have

$$S_{ij} = \frac{1}{2}\left(\frac{\partial W}{\partial \gamma_{ij}} + \frac{\partial W}{\partial \gamma_{ji}}\right)_{\mathrm{I}} \equiv \left(\frac{\partial W}{\partial \gamma_{(ij)}}\right)_{\mathrm{I}}. \tag{2.2.8}$$

The potential energy may now be written as

$$\begin{aligned} P[\mathbf{u}(\mathbf{x}(t))] = \int_V \Bigg[& \frac{1}{2} S_{ij}\left(u_{i,j} + u_{j,i} + u_{h,i}u_{h,j}\right) \\ & + \frac{1}{2}\left(\frac{\partial^2 W}{\partial \gamma_{ij}\,\partial \gamma_{k\ell}}\right)_{\mathrm{I}} \gamma_{ij}\gamma_{k\ell} + \cdots \Bigg] dV + P_{L_1}[\mathbf{u}(\mathbf{x}(t))] + P_{L_2}[\mathbf{u}(\mathbf{x}(t))] + \cdots, \end{aligned} \tag{2.2.9}$$

where we have expanded the potential of the external loads,

$$P_L[\mathbf{u}(\mathbf{x}(t))] = P_{L_1}[\mathbf{u}(\mathbf{x}(t))] + P_{L_2}[\mathbf{u}(\mathbf{x}(t))] + \cdots, \tag{2.2.10}$$

where P_{L_1} is linear in $\mathbf{u}$, P_{L_2} is quadratic in $\mathbf{u}$, and so forth.

Because the fundamental state I is an equilibrium state, the first variation of P must vanish for all kinematically admissible displacement fields, which implies

$$P_1[\mathbf{u}(\mathbf{x}(t))] = \int_V \frac{1}{2} S_{ij}\left(u_{i,j} + u_{j,i}\right) dV + P_{L_1}[\mathbf{u}(\mathbf{x}(t))] = 0. \tag{2.2.11}$$

Hence (2.2.9) may be written as

$$P[\mathbf{u}(\mathbf{x}(t))] = \int_V \left[\frac{1}{2} S_{ij} u_{h,i}u_{h,j} + \frac{1}{2}\left(\frac{\partial^2 W}{\partial \gamma_{ij}\,\partial \gamma_{k\ell}}\right)_{\mathrm{I}} \gamma_{ij}\gamma_{k\ell} + \cdots\right] dV + P_{L_2}[\mathbf{u}(\mathbf{x}(t))] + \cdots \tag{2.2.12}$$

In the remaining part of these lectures, we shall restrict ourselves to dead-weight loads, unless mentioned otherwise. Thus,

$$P_{d.w.L}[\mathbf{u}(\mathbf{x}(t))] = P_{L_1}[\mathbf{u}(\mathbf{x}(t))], \tag{2.2.13}$$

so that the discussion of stability is focused on the integral in (2.2.12). We now define a tensor of elastic moduli

$$E^{\mathrm{I}}_{ijk\ell} \equiv \left(\frac{\partial^2 W}{\partial \gamma_{(ij)} \partial \gamma_{(k\ell)}} \right)_{\mathrm{I}}, \tag{2.2.14}$$

where W is written symmetrically with respect to γ_{ij} and $\gamma_{k\ell}$. Notice that this is not the tensor of elastic moduli that is usually used in the theory of elasticity because that tensor is defined by

$$E^{0}_{ijk\ell} \equiv \left(\frac{\partial^2 W}{\partial \gamma_{(ij)} \partial \gamma_{(k\ell)}} \right)_{0}, \tag{2.2.15}$$

where the index 0 indicates that the second derivatives of W must be evaluated in the undeformed state. Then the tensor of elastic moduli for a homogeneous isotropic material is given by

$$E^{0}_{ijk\ell} = G_\gamma \left[\delta_{ik}\delta_{j\ell} + \delta_{i\ell}\delta_{jk} + \frac{2\nu}{1-2\nu}\delta_{ij}\delta_{k\ell} \right]. \tag{2.2.16}$$

This tensor gives a complete description of the elastic material when the elastic potential is given as

$$W^0 = \frac{1}{2} E^{0}_{ijk\ell} \gamma^{0}_{ij} \gamma^{0}_{k\ell}, \tag{2.2.17}$$

i.e., when W^0 is a homogeneous quadratic form in the strain components. Notice that here we have used Cartesian coordinates in the undeformed state. The deformation tensor is

$$\gamma_{ij} = \frac{1}{2} \left(u_{i,j} + u_{j,i} + u_{h,i} u_{h,j} \right), \tag{2.2.18}$$

where $\mathbf{u}$ now denotes the displacements with respect to the undeformed configuration and $()_{,i} = \partial () / \partial x_i$, where x_i are Cartesian coordinates. The description of W^0 by (2.2.17) is in principle only valid for infinitesimally small strains, and even then only the linear terms in the strain tensor are important. The expression for the elastic potential may now be generalized by assuming that for finite strains (where the quadratic terms in γ_{ij} may become important), the elastic potential can still be represented by a quadratic function. The fact that quadratic terms in the strain tensor may become important even when the linearized strain tensor θ_{ij},

$$\theta_{ij} \equiv \frac{1}{2} \left(u_{i,j} + u_{j,i} \right), \tag{2.2.19}$$

is small is immediately clear from the fact that $|\theta_{ij}| \ll 1$ does not imply that the linearized rotation tensor

$$\omega_{ij} \equiv \frac{1}{2} \left(u_{i,j} - u_{j,i} \right) \tag{2.2.20}$$

is small.

The variation of W for a small disturbance from the equilibrium configuration is given by

$$\delta W = E^0_{ijk\ell}\gamma^0_{ij}\delta\gamma^0_{k\ell} + \frac{1}{2}E^0_{ijk\ell}\delta\gamma^0_{ij}\delta\gamma^0_{k\ell} \tag{2.2.21}$$

(with respect to Cartesian coordinates in the undeformed configuration).

Notice that Cartesian coordinates in the undeformed configuration become curvilinear coordinates in the fundamental state I, and vice versa. It is always possible, at least in principle, to find curvilinear coordinates in the undeformed configuration that become Cartesian coordinates in the fundamental state I. In a curvilinear system, the variation of W is given by

$$\delta W = \underset{0}{E}^{\alpha\beta\lambda\mu}\gamma^0_{\alpha\beta}\delta\gamma^0_{\lambda\mu} + \frac{1}{2}\underset{0}{E}^{\alpha\beta\lambda\mu}\delta\gamma^0_{\alpha\beta}\delta\gamma^0_{\lambda\mu}, \tag{2.2.22}$$

where the contravariant components of the tensor of elastic moduli are given by

$$E^{\alpha\beta\lambda\mu} = G\left(\underset{0}{g}^{\alpha\lambda}\underset{0}{g}^{\beta\mu} + \underset{0}{g}^{\alpha\mu}\underset{0}{g}^{\beta\lambda} + \frac{2\nu}{1-2\nu}\underset{0}{g}^{\alpha\beta}\underset{0}{g}^{\lambda\mu}\right). \tag{2.2.23a}$$

The difference of the metric tensors in the undeformed configuration and in the fundamental state is

$$\underset{0}{g}^{ij} - \delta_{ij} = O(\varepsilon),$$

where ε is largest principal extension in the fundamental state. Hence it follows that

$$E^{\alpha\beta\lambda\mu} = E^0_{ijk\ell}\left[1 + O(\varepsilon)\right] \tag{2.2.23b}$$

so that

$$\frac{1}{2}\underset{0}{E}^{\alpha\beta\lambda\mu}\gamma^0_{\alpha\beta}\gamma^0_{\lambda\mu} = \frac{1}{2}E^{\mathrm{I}}_{ijk\ell}\gamma_{ij}\gamma_{k\ell}\left[1 + O(\varepsilon)\right]$$

and hence

$$\frac{1}{2}E^{\mathrm{I}}_{ijk\ell}\gamma_{ij}\gamma_{k\ell} \approx \frac{1}{2}E_{ijk\ell}\gamma_{ij}\gamma_{k\ell} > 0, \tag{2.2.24}$$

where $E_{ijk\ell}$ is the tensor of elastic moduli that is used in the theory of elasticity. The fact that we have approximated the elastic energy with a relative error of $O(\varepsilon)$ does not affect the positive-definite character of the potential energy if the first term in the potential energy (2.2.12) is also multiplied by a factor $(1 \pm \varepsilon)$.

Remarks

1. For large elastic deformations, the approximation (2.2.23b) is not valid because then ε is large. This can occur, e.g., in rubber-like materials.
2. The fact that we have approximated the elastic energy by a quadratic function in the strains with the classical tensor of elastic moduli implies that we may also apply additional approximations used in the theory of elasticity, e.g., beam theory, plate, and shell theories.

Example. Consider a simply supported strut, loaded in compression by a force N.

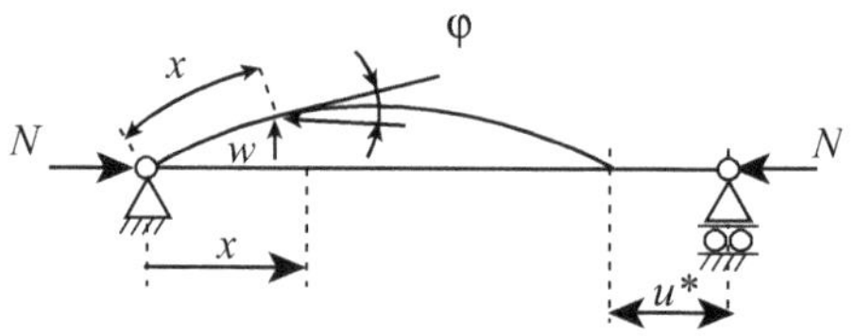

Figure 2.2.1

In the engineering approach, the strut is usually assumed to be inextensible. We have the following relations:

$$\sin\varphi = \frac{dw}{dx} \equiv w' \quad \cos\varphi\frac{d\varphi}{dx} \equiv w''$$

$$\cos\varphi = \sqrt{1-w'^2} \quad \frac{d\varphi}{dx} = \kappa = \frac{w''}{\sqrt{1-w'^2}}$$

$$u^* = \ell - \int_0^\ell \sqrt{1-w'^2}\,dx.$$

For an inextensible strut, the potential energy is

$$P\left[w\left(x\right)\right] = \int_0^\ell \frac{1}{2}EI\frac{w''^2}{1-w'^2}\,dx + \int_0^\ell N\left[\sqrt{1-w'^2}-1\right]dx. \tag{2.2.25}$$

Notice that now the potential energy of the external loads is a nonlinear function of the displacement, in contradistinction to what we have used in our theory. This is due to the fact that we have used an auxiliary condition; namely we have assumed that the strut is inextensible, which means

$$\left(1+u'^2\right) + w'^2 = 1 \quad \text{or} \quad u' = \sqrt{1-w'^2} - 1.$$

In this case, we may say that buckling occurs when the energy supplied by the loads is equal to the strain energy in bending. This may not be generalized. ■

Returning to the general theory and restricting ourselves to dead-weight loads and to materials that follow the generalized Hooke's Law, we may write

$$P\left[\mathbf{u}\left(\mathbf{x}\right)\right] = \int_V \left[\frac{1}{2}S_{ij}u_{h,i}u_{h,j} + \frac{1}{2}E_{ijk\ell}\gamma_{ij}\gamma_{k\ell}\right]dV, \tag{2.2.26}$$

where the strain tensor is given by

$$\gamma_{ij} = \frac{1}{2}\left[u_{i,j} + u_{j,i} + u_{h,i}u_{h,j}\right] \equiv \theta_{ij} + \frac{1}{2}u_{h,i}u_{h,j}. \tag{2.2.27}$$

The potential energy functional may now be written as (only for dead-weight loads)

$$P[\mathbf{u}(\mathbf{x})] = \int_V \left[\frac{1}{2}S_{ij}u_{h,i}u_{h,j} + \frac{1}{2}E_{ijk\ell}\theta_{ij}\theta_{k\ell}\right] dV$$
$$+\frac{1}{2}\int_V \left[E_{ijk\ell}\theta_{ij}u_{m,k}u_{m,\ell} + E_{ijk\ell}\theta_{k\ell}u_{h,i}u_{h,j}\right] dV + \frac{1}{8}\int_V E_{ijk\ell}u_{h,i}u_{h,j}u_{m,k}u_{m,\ell}\, dV, \tag{2.2.28}$$

where we have arranged the terms so that the integrants contain only terms of second, third, and fourth degree, respectively, in the displacements. Writing

$$P[\mathbf{u}(\mathbf{x})] = P_2[\mathbf{u}(\mathbf{x})] + P_3[\mathbf{u}(\mathbf{x})] + P_4[\mathbf{u}(\mathbf{x})] \tag{2.2.29}$$

and using the fact that $E_{ijk\ell} = E_{k\ell ij}$, we have

$$\begin{aligned} P_2[\mathbf{u}(\mathbf{x})] &= \frac{1}{2}\int_V \left[S_{ij}u_{h,i}u_{h,j} + E_{ijk\ell}\theta_{ij}\theta_{k\ell}\right] dV \\ P_3[\mathbf{u}(\mathbf{x})] &= \frac{1}{2}\int_V E_{ijk\ell}\theta_{ij}u_{m,k}u_{m,\ell}\, dV \\ P_4[\mathbf{u}(\mathbf{x})] &= \frac{1}{8}\int E_{ijk\ell}u_{h,i}u_{h,j}u_{m,k}u_{m,\ell}\, dV. \end{aligned} \tag{2.2.30}$$

A positive-definite energy functional means $P_2 + P_3 + P_4 \geq 0$. However, because P_3 and P_4 contain only higher-order terms, it is usually sufficient to consider only P_2. $P_2[\mathbf{u}(\mathbf{x})]$ must be positive-definite for all kinematically admissible displacement fields $\mathbf{u}(\mathbf{x})$. When u_i and $u_{i,j}$ are sufficiently small, then $P_2[\mathbf{u}(\mathbf{x})] > 0$ is a sufficient condition for stability. (No rigorous proof is offered here.)

The limiting case that $\min P_2[\mathbf{u}(\mathbf{x})] = 0$ for a nonzero displacement field $\mathbf{u}_1(\mathbf{x})$ is called a *critical case of neutral equilibrium*. P_2 is called the *second variation*, and $\mathbf{u}_1(\mathbf{x})$ is called the *buckling mode*.

In the following, we shall argue that this case is only possible in slender constructions. To see this, we first notice that according to (2.2.19) and (2.2.20) we may write

$$u_{i,j} = \theta_{ij} - \omega_{ij} \tag{2.2.31}$$

so that

$$S_{ij}u_{h,i}u_{h,j} = S_{ij}\theta_{hi}\theta_{hj} - 2S_{ij}\theta_{hi}\omega_{hj} + S_{ij}\omega_{hi}\omega_{hj}. \tag{2.2.32}$$

The components of the stress tensor S_{ij} are small compared to those of the tensor of elastic moduli $E_{ijk\ell}$, which are of $O(G)$ where G is the shear modulus. In the elastic range (for engineering materials) the strains must be small so that only terms involving the rotations might compete with terms with $E_{ijk\ell}$. This is only possible for large rotations, so the last term in (2.2.32) is the principle term. To show that the construction under consideration must be slender, we write

$$\begin{aligned}\omega_{ij,k} = \frac{1}{2}\left(u_{i,j} - u_{j,i}\right)_{,k} = \frac{1}{2}\left(u_{i,jk} - u_{j,ik}\right) = \frac{1}{2}\left(u_{j,k} - u_{k,j}\right)_{,i} \\ -\frac{1}{2}\left(u_{i,k} + u_{k,i}\right)_{,j} = \theta_{jk,i} - \theta_{ik,j}.\end{aligned} \tag{2.2.33}$$

For a sufficiently supported construction,

$$\omega_{ij,k} = O\left(\frac{\omega}{\ell}\right),$$

where ℓ is a characteristic length of the construction. Large values of one or more components of the rotation tensor are only possible if one or more strains are of order $(\omega/\ell)\ell^*$ where $\ell^*/\ell \ll 1$. This condition means that the construction has a second characteristic length ℓ^* that is considerably smaller than ℓ. This implies that the construction is slender (e.g., beams, plates, shells).

A rigorous mathematical proof of this scenario was given by FRITZ JOHN. He showed that when $\gamma_{ij} = O(\varepsilon)$, $\varepsilon \ll 1$, and the dimensions of a body are all of the same order of magnitude, the rotation vector is given by

$$\boldsymbol{\omega} = \boldsymbol{\omega}_0 + O(\varepsilon)$$

where $\boldsymbol{\omega}_0$ is a constant rotation vector. For an adequately supported construction, this means that

$$\boldsymbol{\omega} = \mathrm{O}(\varepsilon),$$

i.e., the rotations and the strains are of the same order of magnitude.

2.3 The stability limit

For conservative dead-weight loads, and under the assumption that the material follows the generalized Hooke's Law, the potential energy is given by

$$P[\mathbf{u}(\mathbf{x})] = \int \left[\frac{1}{2} S_{ij} u_{h,i} u_{h,j} + \frac{1}{2} E_{ijk\ell} \gamma_{ij} \gamma_{k\ell}\right] dV. \tag{2.3.1}$$

As discussed in Section 2.2, stability is primarily determined by the character of P_2, given by

$$P_2[\mathbf{u}(\mathbf{x})] = \int \left[\frac{1}{2} S_{ij} u_{h,i} u_{h,j} + \frac{1}{2} E_{ijk\ell} \theta_{ij} \theta_{k\ell}\right] dV. \tag{2.3.2}$$

If the second variation is positive-definite, the equilibrium is stable, and if the second variation is indefinite (or negative-definite), the equilibrium is unstable. The second variation is unable to give a valid decision on the stability of instability in the critical case that it is semi-definite positive. Let $\mathbf{u}(\mathbf{x})$ be a minimizing displacement field. Then

$$P_2[\mathbf{u}(\mathbf{x}) + \varepsilon\boldsymbol{\zeta}(\mathbf{x})] \geq P_2[\mathbf{u}(\mathbf{x})] \tag{2.3.3}$$

for all kinematically admissible displacement fields $\boldsymbol{\zeta}(\mathbf{x})$ and for sufficiently small values of $\varepsilon \in \mathbb{R}$. Here and in the following it will be assumed that $\boldsymbol{\zeta}$ is continuously differentiable. Expanding the left-hand side in (2.3.3), we obtain

$$P_2[\mathbf{u}] + P_{11}[\mathbf{u}, \varepsilon\boldsymbol{\zeta}] + P_2[\varepsilon\boldsymbol{\zeta}] \geq P_2[\mathbf{u}], \tag{2.3.4}$$

where we have written $\mathbf{u}$ instead of $\mathbf{u}(\mathbf{x})$ and so on, from which

$$\varepsilon P_{11}[\mathbf{u}, \boldsymbol{\zeta}] + \varepsilon^2 P_2[\boldsymbol{\zeta}] \geq 0 \quad \text{for } \forall\boldsymbol{\zeta} \quad \text{and} \quad \forall\varepsilon \in \mathbb{R}, \quad |\varepsilon| \ll 1. \tag{2.3.5}$$

Because this expression must hold for all sufficiently small values of $|\varepsilon|$, it follows that

$$P_{11}[\mathbf{u}, \boldsymbol{\zeta}] = 0. \tag{2.3.6}$$

This equation is the variational equation for neutral equilibrium. From the functional (2.3.2), we now obtain for the bilinear term

$$P_{11}[\mathbf{u}, \boldsymbol{\zeta}] = \int \left[S_{ij} u_{h,i} \zeta_{h,j} + E_{ijk\ell}\theta_{ij} \frac{1}{2} (\zeta_{k,\ell} + \zeta_{\ell,k}) \right] dV, \tag{2.3.7}$$

where ζ_k are the Cartesian components of $\boldsymbol{\zeta}$.

Due to the symmetry of $E_{ijk\ell}$ in the indices $k\ell$, we may write

$$E_{ijk\ell}\theta_{ij} \cdot \frac{1}{2} (\zeta_{k,\ell} + \zeta_{\ell,k}) = E_{ijk\ell}\theta_{ij}\, \zeta_{k,\ell}.$$

Further, we introduce the notation

$$E_{ijk\ell}\theta_{ij} = \sigma_{k\ell}. \tag{2.3.8}$$

Here $\sigma_{k\ell}$ are the stresses corresponding to the linearized strain tensor in the absence of prestresses. Thus (2.3.7) may now be rewritten as

$$\int (S_{ij} u_{h,j} + \sigma_{hj})\, \zeta_{h,j}\, dV = 0, \tag{2.3.9}$$

and using the divergence theorem we obtain

$$\int_{A_P} (S_{ij} u_{h,i} + \sigma_{hj})\, \zeta_h n_j\, dA - \int_V (S_{ij} u_{h,i} + \sigma_{hj})_{,j}\, \zeta_h\, dV = 0 \quad \text{for } \forall\boldsymbol{\zeta}. \tag{2.3.10}$$

According to the principal theorem in the calculus of variations, we then must have

$$(S_{ij} u_{h,i} + \sigma_{hj})_{,j} = 0 \quad \text{in } V,^{\dagger} \tag{2.3.11}$$

$$(S_{ij} u_{h,j} + \sigma_{hj})\, n_j = 0 \quad \text{on } A_P, \qquad u_i = 0, \quad \text{on } A_u. \tag{2.3.12}$$

† Suppose (2.3.11) does not hold in a point $\mathbf{x}^*$, say, $(S_{ij} u_{h,i} + \sigma_{hj})_j > 0$. Then choose

$$\zeta = \begin{cases} \left[(x_i - x_i^*)(x_i - x_i^*) - R^2\right]^2 & \text{for } (x_i - x_i^*)(x_i - x_i^*) \leq R^2 \\ 0 & \text{for } (x_i - x_i^*)(x_i - x_i^*) \geq R^2 \end{cases}$$

and thus, the surface integral in (2.3.10) vanishes. However, the volume integral is positive, so (2.3.10) is violated, and hence (2.3.11) must hold.

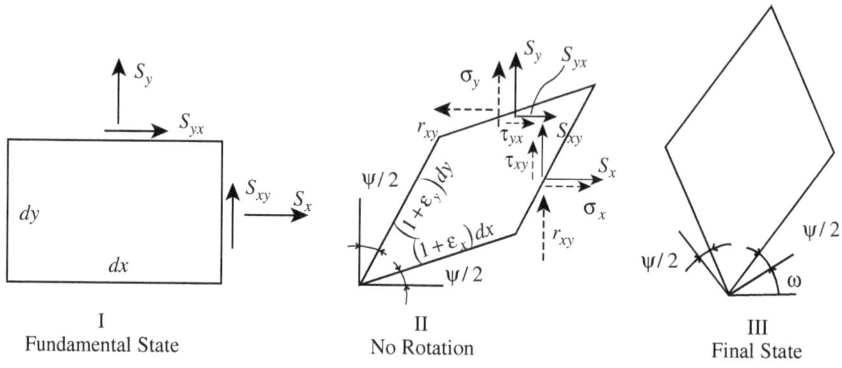

Figure 2.3.1

Notice that in the absence of prestresses S_{ij}, these equations reduce to the equations from the classical theory of elasticity. Performing differentiation by parts, we obtain from (2.3.11)

$$S_{ij,j}u_{h,i} + S_{ij}u_{h,ij} + \sigma_{hj,j} = 0. \tag{2.3.13}$$

The equilibrium equations and boundary conditions in the fundamental state I are given by

$$\begin{cases} S_{ij,j} + X_i = 0 & \text{in } V \\ S_{ij}n_j = P_i & \text{on } A_p \end{cases}, \tag{2.3.14}$$

where X_i are the mass forces and P_i are prescribed tractions on A_p. Using these expressions, we can rewrite (2.3.12) and (2.3.13) as

$$-X_i u_{h,i} + S_{ij}u_{h,ij} + \sigma_{hj,j} = 0 \quad \text{in } V \tag{2.3.15}$$

$$p_i u_{h,i} + \sigma_{hj}n_j = 0 \quad \text{on } A_p, \qquad u_i = 0 \quad \text{on } A_u. \tag{2.3.16}$$

These equations and boundary conditions were derived for the first time by TREFFTZ (1930, 1933).

Different but equivalent equations were derived earlier by BIEZENO and HENCKY (cf. C.B. BIEZENO and R. GRAMMEL, *Engineering Dynamics*, Vol. I). We shall reproduce here their derivation for the two-dimensional case (to simplify the analysis).

Consider a rectangular material element with dimensions dx and dy in the fundamental state, loaded by stresses S_x, S_y, and S_{xy} (see Figure 2.3.1).

The final state is reached in two steps: First, a deformation without a rotation of the deformed element (state II), and then a rotation of the deformed element (state III).

In state II, the element will not be in moment equilibrium under the forces S_x, S_y, and S_{xy} acting on the deformed element. To restore equilibrium (to a first approximation), we add additional (small) forces σ_x, σ_y, and τ_{xy} ($\tau_{xy} = \tau_{yx}$). These additional forces do not enable us to satisfy the equilibrium of moments exactly.

To reach this goal, skew-symmetric shear forces r_{xy} ($r_{yx} = -r_{xy}$) must be added. It is obvious that the final rotation does not disturb the equilibrium of moments.

The equilibrium of moments requires

$$S_x dy\, dx \frac{1}{2}\psi + S_{yx}\, dx\,(1+\varepsilon_y)\, dy - S_{xy}\, dy\,(1+\varepsilon_x)\, dx$$
$$- S_y\, dx \frac{1}{2}\psi dy - r_{xy}\, dy\, dx - r_{xy}\, dx\, dy = 0,$$

from which follows

$$2r_{xy} = \frac{1}{2}\psi(S_x + S_y) + S_{xy}\,(\varepsilon_y - \varepsilon_x)\,. \tag{2.3.17}$$

The equilibrium of forces yields

$$\begin{aligned}(S_x + \sigma_x - S_{xy}\omega)\,,_x + (S_{yx} + \tau_{yx} - r_{xy} - S_y\omega)\,,_y + X = 0\\ (S_y + \sigma_y - S_{yx}\omega)\,,_y + (S_{xy} + \tau_{xy} - r_{xy} - S_x\omega)\,,_x + Y = 0.\end{aligned} \tag{2.3.18}$$

In the fundamental state, we have

$$\begin{aligned}S_{x,x} + S_{yx,y} + X = 0\\ S_{y,y} + S_{xy,x} + Y = 0.\end{aligned} \tag{2.3.19}$$

Substitution of these equations into (2.3.18) and using the relation (2.3.17) yields

$$\begin{aligned}\sigma_{x,x} + \tau_{yx,y} - (S_{xy}\omega)\,,_x - (S_y\omega)\,,_y - \frac{1}{4}\,[\psi(S_x - S_y)]\,,_y - \frac{1}{2}\,[S_{xy}\,(\varepsilon_y - \varepsilon_x)]\,,_y = 0\\ \tau_{xy,x} + \sigma_{y,y} + (S_{yx}\omega)\,,_y + (S_x\omega)\,,_x + \frac{1}{4}\,[\psi(S_x - S_y)]\,,_x + \frac{1}{2}\,[S_{xy}\,(\varepsilon_y - \varepsilon_x)]\,,_x = 0.\end{aligned} \tag{2.3.20}$$

With $\omega = \frac{1}{2}\,(v,_x - u,_y)$ and $\psi = v,_x + u,_y$, for the first of the equations we finally obtain (2.3.20)

$$\begin{aligned}\sigma_{x,x} + \tau_{yx,y} - \left[\frac{1}{2}S_{xy}\,(v,_x - u,_y)\right]_{,x} - \left[S_y\frac{1}{2}\,(v,_x - u,_y)\right]_{,y}\\ - \frac{1}{4}\,[(v_{,x} + u_{,y})\,(S_x - S_y)]_{,y} - \frac{1}{2}\,[S_{xy}\,(v_{,y} - u_{,x})]_{,y} = 0.\end{aligned} \tag{2.3.21}$$

The general result from Biezeno and Hencky may be written in the form

$$\left(\sigma_{ij} + \frac{1}{2}S_{hi}\theta_{hj} - \frac{1}{2}S_{hj}\theta_{hi} + S_{ih}\omega_{hj}\right)_{,i} = 0 \tag{2.3.22}$$

or rewritten as

$$\left[\sigma_{ij} + S_{hi}\,(\theta_{hj} + \omega_{hj}) - \frac{1}{2}S_{hi}\theta_{hj} - \frac{1}{2}S_{hj}\theta_{hi}\right]_{,i} = 0$$

or

$$\left(\sigma_{ij} + S_{hi}u_{j,h} - \frac{1}{2}S_{hi}\theta_{hj} - \frac{1}{2}S_{hj}\theta_{hi}\right)_{,i} = 0. \tag{2.3.23}$$

Equation (2.3.21) is the two-dimensional form of (2.3.22) for $j = 1$. Our earlier result (2.3.11) was

$$(\sigma_{ij} + S_{hi} u_{j,h})_{,i} = 0. \tag{2.3.24}$$

The additional terms in (2.3.23) can easily be derived from the variational approach by adding to the energy density the term $\frac{1}{2} S_{hi} \theta_{hj} \theta_{ij}$. This is small, of order $O(\varepsilon)$ compared to $\frac{1}{2} E_{ijk\ell} \theta_{ij} \theta_{k\ell}$, in which we had already admitted a relative error of $O(\varepsilon)$. Hence it follows that the equations for neutral equilibrium derived by BIEZENO and HENCKY and those derived by TREFFTZ are equivalent within the scope of our theory.

We now continue with our general discussion of neutral equilibrium, and we consider the case that

$$P_2 [\mathbf{u}(\mathbf{x})] \geq 0, \qquad P_2 [\mathbf{u}_1(\mathbf{x})] = 0,$$

where $\mathbf{u}_1(\mathbf{x})$ is the buckling mode. The question now arises whether $\mathbf{u}_1$ is the only kinematically admissible displacement field for which $P_2 [\mathbf{u}(\mathbf{x})]$ vanishes. To investigate this condition we might look for additional solutions of the equations of neutral equilibrium, but it proves to be more useful to proceed as follows. Consider the set of orthogonal (kinematically admissible) displacement fields. To define orthogonality, we introduce the positive-definite auxiliary functional

$$T_2 [\mathbf{u}(\mathbf{x})] \geq 0. \tag{2.3.25}$$

This functional defines the measure in the energy space. Applying similar arguments as in the discussion of P_2, we find that the bilinear term must vanish,

$$T_{11} [\mathbf{u}(\mathbf{x}), \mathbf{v}(\mathbf{x})] = 0, \tag{2.3.26}$$

which defines the *orthogonality* of $\mathbf{u}$ and $\mathbf{v}$.

A possible choice for T_2 is

$$T_2 [\mathbf{u}] = \frac{1}{2} \int_V G u_{i,j} u_{i,j} \, dV, \tag{2.3.27}$$

where the integrant is a positive-definite quadratic function of the displacement gradients. The factor G has been added to give T_2 the dimension of energy. The vanishing of the bilinear term yields

$$G \int_V u_{i,j} v_{i,j} \, dV = 0. \tag{2.3.28}$$

Another suitable choice is

$$T_2 [\mathbf{u}] = \frac{1}{2} \int_V \rho u_i u_i \, dV, \tag{2.3.29}$$

where ρ is the mass density. This choice is motivated by Rayleigh's principle for the determination of the lowest eigenfrequency of small vibrations about an equilibrium configuration, which states

$$\omega^2 = \mathrm{Min}\frac{P_2[\mathbf{u}(\mathbf{x})]}{\int_V \frac{1}{2}\rho u_i u_i \, dV}. \qquad (2.3.30)$$

In the critical case of neutral equilibrium, ω vanishes, which means that the motion is infinitely slow. It will turn out that the results are independent of the particular choice for T_2.†

An arbitrary displacement field can always be written in the form

$$\mathbf{u}(\mathbf{x}) = a\,\mathbf{u}_1(\mathbf{x}) + \bar{\mathbf{u}}(x), \quad T_{11}[\mathbf{u}_1, \bar{\mathbf{u}}] = 0. \qquad (2.3.31)$$

Namely,

$$T_{11}[\mathbf{u}_1, \mathbf{u}] = a\,T_{11}[\mathbf{u}_1, \mathbf{u}_1] + T_{11}[\mathbf{u}_1, \bar{\mathbf{u}}],$$

where the last term vanishes by virtue of (2.3.31). It follows that

$$a = \frac{T_{11}[\mathbf{u}_1, \mathbf{u}]}{T_{11}[\mathbf{u}_1, \mathbf{u}_1]} = \frac{T_{11}[\mathbf{u}_1, \mathbf{u}]}{2T_2[\mathbf{u}_1]}, \qquad (2.3.32a)$$

where we have used the relation

$$T_2[\mathbf{u}_1 + \mathbf{u}_1] = T_2[2\mathbf{u}_1] = 4T_2[\mathbf{u}_1] = T_2[\mathbf{u}_1] + T_{11}[\mathbf{u}_1, \mathbf{u}_1] + T_2[\mathbf{u}_1].$$

The second variation for an arbitrary displacement field can now be written as

$$P_2[\mathbf{u}] = P_2[a\mathbf{u}_1 + \bar{\mathbf{u}}] = a^2 P_2[\mathbf{u}_1] + aP_{11}[\mathbf{u}_1, \bar{\mathbf{u}}] + P_2[\bar{\mathbf{u}}], \qquad (2.3.32b)$$

where the first term on the right-hand side vanishes because $\mathbf{u}_1$ is the buckling mode, and the second term vanishes because P_{11} vanishes for all displacement fields, and hence also for displacement fields orthogonal to $\mathbf{u}_1$. In other words, it follows that when $P_2[\mathbf{u}] = 0$ for $\mathbf{u} \neq \mathbf{u}_1$, $P_2[\mathbf{u}]$ also vanishes for a displacement field orthogonal to $\mathbf{u}_1$. We may thus restrict ourselves to displacement fields that are orthogonal to $\mathbf{u}_1$. As $P_2[\alpha\bar{\mathbf{u}}] \to 0$ for $\alpha \to 0$, it is more suitable to consider the following minimum problem:

$$\underset{T_{11}[\mathbf{u}_1, \bar{\mathbf{u}}]=0}{\mathrm{Min}} \frac{P_2[\bar{\mathbf{u}}(\mathbf{x})]}{T_2[\bar{\mathbf{u}}(\mathbf{x})]} = \lambda_2, \qquad (2.3.33)$$

where λ_2 is the minimum of the left-hand member. If $\lambda_2 = 0$, then we have a second buckling mode, and when $\lambda_2 > 0$, $\mathbf{u}_1$ is then the unique buckling mode. Let $\mathbf{u}_2$ be the second buckling mode; then

$$\frac{P_2[\mathbf{u}_2 + \varepsilon\eta]}{T_2[\mathbf{u}_2 + \varepsilon\eta]} \geq \frac{P_2[\mathbf{u}_2]}{T_2[\mathbf{u}_2]} = \lambda_2 \qquad (2.3.34)$$

and

$$T_{11}[\mathbf{u}_2, \mathbf{u}_1] = T_{11}[\eta, \mathbf{u}_1] = 0. \qquad (2.3.35)$$

† See following equation (3.3.55).

Rewriting (2.3.34), we find

$$P_2[\mathbf{u}_2] + \varepsilon P_{11}[\mathbf{u}_2, \eta] + \varepsilon^2 P_2[\eta] \geq \lambda_2 \left\{T_2[\mathbf{u}_2] + \varepsilon T_{11}[\mathbf{u}_2, \eta] + \varepsilon^2 T_2[\eta]\right\}$$

where $P_2[\mathbf{u}_2] - \lambda_2 T_2[\mathbf{u}_2] = 0$, so that

$$\varepsilon\left\{P_{11}[\mathbf{u}_2, \eta] - \lambda_2 T_2[\mathbf{u}_2, \eta]\right\} + \varepsilon^2\left\{P_2[\eta] - \lambda_2 T_2[\eta]\right\} \geq 0. \tag{2.3.36}$$

Because this equation must hold for arbitrary small values of ε, the term linear in ε must vanish, i.e.,

$$P_{11}[\mathbf{u}_2, \eta] - \lambda_2 T_2[\mathbf{u}_2, \eta] = 0 \tag{2.3.37}$$

for all displacement fields with $T_{11}[\mathbf{u}_1, \eta] = 0$. In this condition, the displacement field η is orthogonal to $\mathbf{u}_1$. In our original condition of $P_{11}[\mathbf{u}, \zeta] = 0$, ζ was not submitted to this requirement. We shall now show that (2.3.37) also holds for arbitrary displacement fields ζ. To show this, we write $\zeta = t\mathbf{u}_1 + \eta$, where η satisfies the orthogonality condition. Replacing η in (2.3.37) by ζ, we find

$$\begin{aligned}
&P_{11}[\mathbf{u}_2, t\mathbf{u}_1 + \eta] - \lambda_2 T_{11}[\mathbf{u}_2, t\mathbf{u}_1 + \eta] \\
&= tP_{11}[\mathbf{u}_2, \mathbf{u}_1] + P_{11}[\mathbf{u}_2, \eta] - \lambda_2 t T_{11}[\mathbf{u}_2, \mathbf{u}_1] - \lambda_2 T_{11}[\mathbf{u}_2, \eta] \\
&= t\left\{P_{11}[\mathbf{u}_2, \mathbf{u}_1] - \lambda_2 T_{11}[\mathbf{u}_2, \mathbf{u}_1]\right\} + P_{11}[\mathbf{u}_2, \eta] - \lambda_2 T_{11}[\mathbf{u}_2, \eta] \\
&= P_{11}[\mathbf{u}_2, \eta] - \lambda_2 T_{11}[\mathbf{u}_2, \eta] = 0
\end{aligned} \tag{2.3.38}$$

It follows that without a loss of generality we may restrict ourselves to displacement fields that are orthogonal to the ones that are already known.

Suppose we have found m linearly independent solutions. The solution of $P_{11}[\mathbf{u}, \zeta] = 0$ can then be written as

$$\mathbf{u}(\mathbf{x}) = a_h \mathbf{u}_h, \quad h \in (1, \ldots m) \quad \text{with } T_{11}[\mathbf{u}_h, \mathbf{u}_k] = 0 \quad \text{for } h \neq k.$$

Further, we may normalize the buckling modes by requiring

$$T_{11}[\mathbf{u}_1, \mathbf{u}_1] = \cdots = T_{11}[\mathbf{u}_m, \mathbf{u}_m] = 1. \tag{2.3.39}$$

To investigate the stability of the critical case of neutral equilibrium, we write for an arbitrary displacement field $\mathbf{u}(\mathbf{x})$

$$\mathbf{u}(\mathbf{x}) = a_h \mathbf{u}_h(\mathbf{x}) + \bar{\mathbf{u}}(\mathbf{x}), \quad T_{11}[\bar{\mathbf{u}}, \mathbf{u}_k] = 0 \quad k \in (1, 2, \ldots, m). \tag{2.3.40}$$

This is always possible, namely,

$$T_{11}[\mathbf{u}_k, \mathbf{u}] = T_{11}[\mathbf{u}_k, a_h, \mathbf{u}_h + \bar{\mathbf{u}}] = a_h T_{11}[\mathbf{u}_k, \mathbf{u}_h] + T_{11}[\mathbf{u}_k, \bar{\mathbf{u}}] = a_k. \tag{2.3.41}$$

A substitution of (2.3.40) into the energy functional yields

$$\begin{aligned}
P[\mathbf{u}(\mathbf{x})] &= P_2[a_h\mathbf{u}_h + \bar{\mathbf{u}}] + P_3[a_h\mathbf{u}_h + \bar{\mathbf{u}}] + P_4[a_h\mathbf{u}_h + \bar{\mathbf{u}}] + \cdots \\
&= P_2[a_h\mathbf{u}_h] + P_{11}[a_h\mathbf{u}_h, \bar{\mathbf{u}}] + P_2[\mathbf{u}_h] \\
&\quad + P_3[a_h\mathbf{u}_h] + P_{21}[a_h\mathbf{u}_h, \bar{\mathbf{u}}] + P_{12}[a_h\mathbf{u}_h, \bar{\mathbf{u}}] + P_3[\bar{\mathbf{u}}] \\
&\quad + P_4[a_h\mathbf{u}_h] + P_{31}[a_h\mathbf{u}_h, \bar{\mathbf{u}}] + P_{22}[a_h\mathbf{u}_h, \bar{\mathbf{u}}] + P_{13}[a_h\mathbf{u}_h, \bar{\mathbf{u}}] + P_4[\bar{\mathbf{u}}] + \cdots.
\end{aligned} \tag{2.3.42}$$

When the material follows the generalized Hooke's Law, the expansion terminates after $P_4[\overline{\mathbf{u}}]$; in other cases, higher-order terms follow. Because $\mathbf{u}_h$ are buckling modes, the first term on the right-hand side vanishes, $P_{11}[a_h\mathbf{u}_h, \zeta] = 0$, for all displacement fields ζ, and hence also for $\overline{\mathbf{u}}$. Because we have already found m buckling modes, $P_2[\overline{\mathbf{u}}]$ must satisfy the relation

$$P_2[\overline{\mathbf{u}}] \geq \lambda_{m+1} T_2[\overline{\mathbf{u}}], \quad \lambda_{m+1} > 0. \tag{2.3.43}$$

Let us now first consider the case that $\overline{\mathbf{u}} = \equiv \mathbf{0}$. We then only have to deal with the terms $P_3[a_h\mathbf{u}_h]$ and $P_4[a_h\mathbf{u}_h]$. A necessary condition for stability is that $P_3[a_h\mathbf{u}_h]$ vanishes and that $P_4[a_h\mathbf{u}_h] \geq 0$. Hence it follows that from the mathematical point of view, systems will generally be unstable because functions for which $P_2 = P_3 = 0$ and $P_4 \geq 0$ are exceptions. However, in applications this often happens due to the symmetry of the structure. The conditions mentioned are necessary conditions; to obtain sufficient conditions, we consider small values of $\overline{\mathbf{u}}$. In fact, the stability conditions must be satisfied for small deviations from the fundamental state.

Suppose now that the necessary conditions are satisfied; thus the most important terms in (2.3.42) are

$$P_2[\overline{\mathbf{u}}], \quad P_{21}[a_h\mathbf{u}_h, \overline{\mathbf{u}}], \quad P_4[a_h\mathbf{u}_h],$$

namely, for $\|\overline{\mathbf{u}}\| = O(\|a_{(h)}, \mathbf{u}_{(h)}\|^2)$ these terms are of the same order of magnitude, whereas all other terms have integrands of order $O(\|a_{(h)}, \mathbf{u}_{(h)}\|^n)$, $n \geq 5$.

We now first consider the minimum problem,

$$\operatorname*{Min}_{\substack{T_{11}[\mathbf{u}_k, \overline{\mathbf{u}}]=0 \\ a_k \text{ const.} \\ k\in(1,\ldots,m)}} \left(P_2[\overline{\mathbf{u}}] + P_{21}[a_h\mathbf{u}_h, \overline{\mathbf{u}}]\right). \tag{2.3.44}$$

This is a meaningful minimum problem because P_2 is a positive-definite functional and P_{21} is a functional linear in $\overline{\mathbf{u}}$. Let $\overline{\mathbf{u}} = \mathbf{v}$ be the solution to this problem; thus for a variation of this field $\mathbf{v} + \varepsilon\eta$ we have

$$P_2[\mathbf{v}] + \varepsilon P_{11}[\mathbf{v}, \eta] + \varepsilon^2 P_2[\eta] + P_{21}[a_h\mathbf{u}_h, \mathbf{v}] + \varepsilon P_{21}[a_h\mathbf{u}_h, \eta] \geq P_2[\mathbf{v}] + P_{21}[a_h\mathbf{u}_h, \mathbf{v}]$$

or

$$\varepsilon\left(P_{11}[\mathbf{v}, \eta] + P_{21}[a_h\mathbf{u}_h, \eta]\right) + \varepsilon^2 P_2[\eta] \geq 0. \tag{2.3.45}$$

Because this inequality must hold for all sufficiently small values of ε, it follows that

$$P_{11}[\mathbf{v}, \eta] + P_{21}[a_h\mathbf{u}_h, \eta] = 0, \tag{2.3.46}$$

which is an equation for $\mathbf{v}$.

To answer the question of whether the solution to (2.3.44) is unique, we consider a second solution $\mathbf{v}^*$. Subtracting the equation obtained from (2.3.46) by replacing $\mathbf{v}$ with $\mathbf{v}^*$, from (2.3.46) we find

$$P_{11}[\mathbf{v} - \mathbf{v}^*, \eta] = 0. \tag{2.3.47}$$

By arguments similar to those following (2.3.37), we may also replace η by ζ, where ζ is an arbitrary displacement field.† Further, we have

$$T_{11}[a_h\mathbf{u}_h, \mathbf{v}-\mathbf{v}^*] = 0, \tag{2.3.48}$$

i.e., $\mathbf{v}-\mathbf{v}^*$ is orthogonal with respect to the linear combination of buckling modes, and due to (2.3.47),

$$P_2[\mathbf{v}-\mathbf{v}^*] = 0. \tag{2.3.49}$$

However, this result contradicts our assumption (2.3.43) that $\lambda_{m+1} > 0$. This completes our proof that $\mathbf{v}$ is unique.

Equation (2.3.46) is quadratic in the buckling modes but linear in $\mathbf{v}$, so the solution can be written as

$$\mathbf{v}(\mathbf{x}) = a_h a_k \mathbf{v}_{hk}(\mathbf{x}), \quad \mathbf{v}_{hk} = \mathbf{v}_{kh}. \tag{2.3.50}$$

Substitution of this expression into (2.3.46) yields

$$P_{11}[\mathbf{v}_{hk}, \zeta] + \frac{1}{2}P_{111}[\mathbf{u}_h, \mathbf{u}_k, \zeta] = 0, \tag{2.3.51}$$

where we have used the relation

$$P_{21}[a_h\mathbf{u}_h; \zeta] = \frac{1}{2}a_h a_k P_{111}[\mathbf{u}_h, \mathbf{u}_k, \zeta].$$

For a known field $\mathbf{v} = a_h a_k \mathbf{v}_{hk}$, we may now determine the minimum value of the energy functional (2.3.42). Only retaining the most important terms, we find

$$\underset{a_k=\text{const.}}{\text{Min}} P_{11}[\mathbf{u}] = P_4[a_h\mathbf{u}_h] + P_2[a_h a_k \mathbf{v}_{hk}] + P_{21}[a_h\mathbf{u}_h, a_k a_\ell \mathbf{v}_{k\ell}] + \cdots. \tag{2.3.52}$$

Using (2.3.46) with $\zeta = a_h a_k \mathbf{v}_{hk}$ and making use of the relation

$$P_{11}[a_h a_k \mathbf{v}_{hk}, a_\ell a_m \mathbf{v}_{\ell m}] = 2P_2[a_h a_k \mathbf{v}_{hk}],$$

we find

$$2P_2[a_h a_k \mathbf{v}_{hk}] + P_{21}[a_h\mathbf{u}_h, a_k a_\ell \mathbf{v}_{k\ell}] = 0 \tag{2.3.53}$$

so that (2.3.52) may be rewritten to yield

$$\begin{aligned} \underset{(a_k=\text{const.})}{\text{Min}} P[\mathbf{u}] &= P_4[a_h\mathbf{u}_h] - P_2[a_h a_k \mathbf{v}_{hk}] + \cdots \\ &= P_4[a_h\mathbf{u}_h] + \frac{1}{2}P_{21}[a_h\mathbf{u}_h, a_k a_\ell \mathbf{v}_{k\ell}] + \cdots. \end{aligned} \tag{2.3.54}$$

It follows that the (necessary) condition $P_4[a_h\mathbf{u}_h] \geq 0$ is not a sufficient condition, because a positive-definite term $P_2[a_h a_k \mathbf{v}_{hk}]$ is subtracted. Introducing the notation

$$P_4[a_h\mathbf{u}_h] - P_2[a_h a_k \mathbf{v}_{hk}] \equiv A_{ijk\ell} a_i a_j a_k a_\ell, \tag{2.3.55}$$

† The equation with ζ is preferable because then we do not have the side condition of orthogonality.

our results may be summarized as follows:

- $A_{ijk\ell}a_i a_j a_k a_\ell \geq 0$ is a *necessary condition* for stability, and
- $A_{ijk\ell}a_i a_j a_k a_\ell$ as positive-definite (i.e., zero only for a displacement field $\mathbf{u} = \mathbf{0}$) is a *sufficient condition* for stability.

We shall now show that the results obtained are independent of the particular choice of the auxiliary functional T_2. Let T_2^* be a positive-definite functional ($T_2^* \neq T_2$), so then we may construct a minimizing displacement field

$$\overline{\mathbf{u}}^*_{\text{Min}} = a_h a_k \mathbf{v}^*_{hk}(\mathbf{x}). \tag{2.3.56}$$

This field satisfies the same equations as the field $\mathbf{v} = a_h a_k \mathbf{v}_{hk}(\mathbf{x})$ but is subjected to a different orthogonality condition. The difference of these fields must satisfy the homogeneous variational equation

$$P_{11}\left[\mathbf{v}_{hk} - \mathbf{v}^*_{hk}, \zeta\right] = 0, \tag{2.3.57}$$

which means that $\mathbf{v}_{hk} - \mathbf{v}^*_{hk}$ is a linear combination of buckling modes, say

$$\mathbf{v}_{hk} - \mathbf{v}^*_{hk} = c_{hk\ell}\mathbf{u}_\ell, \tag{2.3.58}$$

which does not satisfy one of the orthogonality conditions implied by T_2 and T_2^*. The minimum of the potential energy functional is now given by

$$\underset{(a_k=\text{const.})}{\text{Min}} \; P(\mathbf{u}) = P_4\left[a_h\mathbf{u}_h\right] - P_2\left[a_h a_k \mathbf{v}^*_{hk}\right] + \cdots, \tag{2.3.59}$$

where $P_2\left[a_h a_k \mathbf{v}^*_{hk}\right]$ may be written as

$$\begin{aligned} P_2\left[a_h a_k\left(\mathbf{v}_{hk} - c_{hkl}\mathbf{u}_l\right)\right] &= P_2\left[a_h a_k \mathbf{v}_{hk}\right] \\ &\quad - P_{11}\left[a_h a_k \mathbf{v}_{hk}, a_p a_q c_{pqr}\mathbf{u}_r\right] + P_2\left[a_h a_k c_{hk\ell}\mathbf{u}_\ell\right]. \end{aligned} \tag{2.3.60}$$

The second term on the right-hand side of (2.3.60) vanishes because $P_{11}\left[a_h\mathbf{u}_h, \zeta\right] = 0$ for all fields ζ, and hence also for $\zeta = \mathbf{v}$, and the last term vanishes because the argument is a linear combination of buckling modes. Hence it follows that the minimum value of $P[\mathbf{u}]$ is independent of the particular choice of the auxiliary functional.

2.4 Equilibrium states for loads in the neighborhood of the buckling load

In the foregoing analysis, we employed the fundamental state as a reference. Because we want to investigate equilibrium states for loads in the neighborhood of the buckling load, we must investigate the behavior of the structure for small but finite deflections. The fundamental state then depends on the load, so it cannot be used as a reference state. We need a fixed reference state, and it is suitable to choose the undeformed (stress-free) state as the reference state.

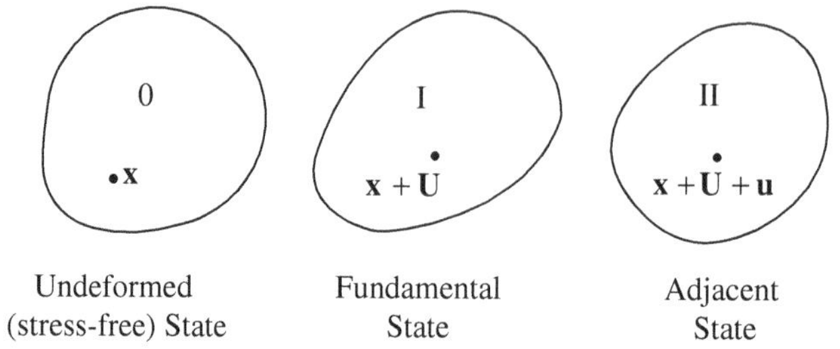

Figure 2.4.1

The strain tensor in state II is now given by

$$\begin{aligned}
&\frac{1}{2}(U_i+u_i)_{,j}+\frac{1}{2}(U_j+u_j)_{,i}+\frac{1}{2}(U_h+u_h)_{,i}(U_h+u_h)_{,j}\\
&\quad=\frac{1}{2}(U_{i,j}+U_{j,i})+\frac{1}{2}U_{h,i},U_{h,j}\\
&\quad+\frac{1}{2}(u_{i,j}+u_{j,i})+\frac{1}{2}u_{h,i}u_{h,j}+\frac{1}{2}U_{h,i},u_{h,j}+\frac{1}{2}U_{h,j},u_{h,i}\equiv\Gamma_{ij}+\gamma_{ij}
\end{aligned}\tag{2.4.1}$$

where

$$\Gamma_{ij}\equiv\frac{1}{2}(U_{i,j}+U_{j,i})+\frac{1}{2}U_{h,i}U_{h,j}\tag{2.4.2}$$

is the strain tensor going from the undeformed state to the fundamental state, and

$$\gamma_{ij}\equiv\frac{1}{2}(u_{i.j}+u_{j,i})+\frac{1}{2}u_{h,i}u_{h,j}+\frac{1}{2}U_{h,i}u_{h,j}+\frac{1}{2}U_{h,j}u_{h,i}\tag{2.4.3}$$

describes the strains going from the fundamental state to the adjacent state. Notice that this tensor also depends on **U**.

The elastic energy density in state II is given by

$$\frac{1}{2}E_{ijkl}(\Gamma_{ij}+\gamma_{ij})(\Gamma_{k\ell}+\gamma_{k\ell}).\tag{2.4.4}$$

The increment of elastic energy density going from state I to state II is given by

$$\begin{aligned}
\frac{1}{2}E_{ijk\ell}(\Gamma_{ij}+\gamma_{ij})(\Gamma_{k\ell}+\gamma_{k\ell})-\frac{1}{2}E_{ijk\ell}\Gamma_{ij}\Gamma_{k\ell}&=E_{ijk\ell}\Gamma_{ij}\gamma_{k\ell}+\frac{1}{2}E_{ijk\ell}\gamma_{ij}\gamma_{k\ell}\\
&=s_{k\ell}\gamma_{k\ell}+\frac{1}{2}E_{ijk\ell}\gamma_{ij}\gamma_{k\ell}.
\end{aligned}\tag{2.4.5}$$

This expression is of the same form as our original energy functional but the strain tensor is different.

We consider loads that can be written as a unit load multiplied by a load factor λ. This means that the displacement field in the fundamental state also depends on λ, i.e., $\mathbf{U}=\mathbf{U}(\mathbf{x};\lambda)$. The increment of the potential energy going from state I to state II is now given by

$$P_{\mathrm{II}}-P_{\mathrm{I}}=P[\mathbf{u}(\mathbf{x});\lambda]=P_2[\mathbf{u}(\mathbf{x});\lambda]+P_3[\mathbf{u}(\mathbf{x});\lambda]+\cdots.\tag{2.4.6}$$

Notice that this expression starts with quadratic terms because the fundamental state is an equilibrium configuration, and also that this expression depends on **U**, i.e., it refers to the fundamental state.

For sufficiently small loads, the potential energy will be positive-definite and the equilibrium configuration will be stable and unique (Kirchhoff's uniqueness theorem). Now let $\lambda \geq 0$ be monotonically increasing; then for a certain value of λ, say $\lambda = \lambda_1$, the potential energy will become semi-definite-positive. (The case $\lambda < 0$ is a different stability problem that may be treated similarly.)

We shall now expand the potential energy with respect to the load parameter λ in the vicinity of $\lambda = \lambda_1$ (assuming that such an expansion is possible), e.g.,

$$P_2\left[\mathbf{u}(\mathbf{x});\lambda\right] = P_2\left[\mathbf{u}(\mathbf{x});\lambda_1\right] + (\lambda - \lambda_1)\,P_2'\left[\mathbf{u}(\mathbf{x});\lambda_1\right] + \cdots \tag{2.4.7}$$

where $()' \equiv \partial()/\partial\lambda$.

As mentioned previously, the potential energy depends on $\mathbf{U}(\mathbf{x},\lambda)$. In the following, we shall assume that the fundamental state is known and the question is whether there exist equilibrium configurations in the vicinity of the fundamental state. Therefore stationary values of $P\left[\mathbf{u}(\mathbf{x});\lambda\right]$ are considered for nonzero displacement increments from the fundamental state.

A necessary condition for equilibrium is that

$$P_{11}\left[\mathbf{u}(\mathbf{x}),\zeta(\mathbf{x});\lambda\right] + P_{21}\left[\mathbf{u}(\mathbf{x}),\zeta(\mathbf{x});\lambda\right] + \cdots = 0. \tag{2.4.8}$$

Notice that this equation is satisfied for $\mathbf{u}(\mathbf{x}) = \mathbf{0}$, the fundamental state, as it should be. At the critical load, the infinitesimal displacement field is a linear combination of buckling modes, i.e., $\mathbf{u} = a_h\mathbf{u}_h$. Assuming that for small but finite displacements the field can also (approximately) be represented by this expression, we write

$$\mathbf{u}(\mathbf{x}) = a_h\mathbf{u}_h(\mathbf{x}) + \bar{\mathbf{u}}(\mathbf{x}), \quad T_{11}\left[\mathbf{u}_k,\bar{\mathbf{u}}\right] = 0, \tag{2.4.9}$$

which is always possible.

The energy functional now becomes

$$\begin{aligned} P\left[\mathbf{u}(\mathbf{x});\lambda\right] = {} & P_2\left[a_h\mathbf{u}_h;\lambda_1\right] + P_{11}\left[a_h\mathbf{u}_h,\bar{\mathbf{u}};\lambda_1\right] \\ & + P_2\left[\bar{\mathbf{u}};\lambda_1\right] + (\lambda-\lambda_1)\,P_2'\left[a_h\mathbf{u}_h;\lambda_1\right] + (\lambda-\lambda_1)\,P_{11}'\left[a_h\mathbf{u}_h,\bar{\mathbf{u}};\lambda_1\right] \\ & + (\lambda-\lambda_1)\,P_2'\left[\bar{\mathbf{u}};\lambda_1\right] + \cdots + P_3\left[a_h\mathbf{u}_h;\lambda_1\right] + P_{21}\left[a_h\mathbf{u}_h,\bar{\mathbf{u}};\lambda_1\right] + \cdots \\ & + (\lambda-\lambda_1)\,P_3'\left[a_h\bar{\mathbf{u}}_h;\lambda_1\right] + (\lambda-\lambda_1)\,P_{21}'\left[a_h\mathbf{u}_h,\bar{\mathbf{u}};\lambda_1\right] + \cdots \\ & + P_4\left[a_h\mathbf{u}_h;\lambda_1\right] + \cdots. \end{aligned} \tag{2.4.10}$$

By known arguments, the first two terms on the right-hand side of (2.4.10) vanish. We shall now first consider the terms depending on $\bar{\mathbf{u}}$. First of all, we have the positive-definite term $P_2\left[\bar{\mathbf{u}};\lambda_1\right]$, thus the term $(\lambda-\lambda_1)\,P_{11}'\left[a_h\mathbf{u}_h,\bar{\mathbf{u}};\lambda_1\right]$ is important because it contains $\bar{\mathbf{u}}$ linearly, and is also linear in $a_h\mathbf{u}_h$ and $\lambda - \lambda_1$. The term $(\lambda-\lambda_1)\,P_2'\left[\bar{\mathbf{u}};\lambda_1\right]$ may be neglected because it is quadratic in $\bar{\mathbf{u}}$ and is multiplied by $\lambda - \lambda_1$. The next important term depending on $\bar{\mathbf{u}}$ is $P_{21}\left[a_h\mathbf{u}_h,\bar{\mathbf{u}};\lambda_1\right]$, which is linear in $\bar{\mathbf{u}}$ and quadratic in $a_h\mathbf{u}_h$. Other terms depending on $\bar{\mathbf{u}}$ may be neglected because they are either of higher order in $\bar{\mathbf{u}}$ or they contain higher-order terms in $a_h\mathbf{u}_h$ or $(\lambda - \lambda_1)$. We now determine the minimum with respect to $\bar{\mathbf{u}}$ of the three terms mentioned,

$$\underset{\text{w.r.t. }\bar{\mathbf{u}}}{\text{Min}}\; P_2\left[\bar{\mathbf{u}};\lambda_1\right] + (\lambda-\lambda_1)\,P_{11}'\left[a_h\mathbf{u}_h,\bar{\mathbf{u}};\lambda_1\right] + P_{21}\left[a_h\mathbf{u}_h,\bar{\mathbf{u}};\lambda_1\right]. \tag{2.4.11}$$

The solution to this minimum problem is unique. Suppose that $\overline{\mathbf{u}}^*$ is a second solution. Then the following equation must holds,

$$P_{11}\left[\overline{\mathbf{u}}-\overline{\mathbf{u}}^*, \eta\right]=0 \quad \text{for } \forall\eta|\ T_{11}\left[\overline{\mathbf{u}}, \eta\right]=T_{11}\left[\overline{\mathbf{u}}^*, \eta\right]=0. \tag{2.4.12}$$

The solutions of this equation are linear combinations of the buckling modes, but $\overline{\mathbf{u}}$ and $\overline{\mathbf{u}}^*$ are orthogonal with respect to these buckling modes, so we must have $\overline{\mathbf{u}}-\overline{\mathbf{u}}^*=\mathbf{0}$. Guided by the structure of (2.4.11), we try a solution of the form

$$\overline{\mathbf{u}}_{\text{Min}}(\mathbf{x})=(\lambda-\lambda_1)\, a_h \mathbf{v}_h'(\mathbf{x})+a_h a_k \mathbf{v}_{hk}(\mathbf{x}). \tag{2.4.13}$$

Because $\overline{\mathbf{u}}_{\text{Min}}(\mathbf{x})$ is the solution to (2.4.11), we have for a variation of this field, say $\overline{\mathbf{u}}_{\text{Min}}+\varepsilon\eta$,

$$\begin{aligned} &P_2\left[\overline{\mathbf{u}}_{\text{Min}}+\varepsilon\eta;\lambda_1\right]+(\lambda-\lambda_1)\,P_{11}'\left[a_h\overline{\mathbf{u}}_h, \overline{\mathbf{u}}_{\text{Min}}+\varepsilon\eta;\lambda_1\right]+P_{21}\left[a_h\mathbf{u}_h, \overline{\mathbf{u}}_{\text{Min}}+\varepsilon\eta;\lambda_1\right]\\ &\quad\geq P_2\left[\overline{\mathbf{u}}_{\text{Min}};\lambda_1\right]+(\lambda-\lambda_1)\,P_{11}'\left[a_h\overline{\mathbf{u}}_h, \overline{\mathbf{u}}_{\text{Min}};\lambda_1\right], \quad \text{for } \forall\varepsilon. \end{aligned} \tag{2.4.14}$$

This inequality must be satisfied for all sufficiently small values of $\varepsilon\in\mathbb{R}$, which implies

$$P_{11}\left[\overline{\mathbf{u}}_{\text{Min}}, \eta;\lambda_1\right]+(\lambda-\lambda_1)\,P_{11}'\left[a_h\overline{\mathbf{u}}_h, \eta;\lambda_1\right]+P_{21}\left[a_h\overline{\mathbf{u}}_h, \eta;\lambda_1\right]=0. \tag{2.4.15}$$

Substitution of (2.4.13) into (2.4.15) yields

$$\begin{aligned} &(\lambda-\lambda_1)\,P_{11}\left[a_h\mathbf{v}_h', \eta;\lambda_1\right]+P_{11}\left[a_h a_k\mathbf{v}_{hk}, \eta;\lambda_1\right]\\ &\quad+(\lambda-\lambda_1)\,P_{11}'\left[a_h\overline{\mathbf{u}}_h, \eta\right]+P_{21}\left[a_h\overline{\mathbf{u}}_h, \eta;\lambda_1\right]=0. \end{aligned} \tag{2.4.16}$$

Because this equation must hold for all $(\lambda-\lambda_1)$, we must have

$$P_{11}\left[\mathbf{v}_h', \eta;\lambda_1\right]+P_{11}'\left[\mathbf{u}_h, \eta;\lambda_1\right]=0 \tag{2.4.17}$$

and

$$P_{11}\left[\mathbf{v}_{hk}, \eta;\lambda_1\right]+\frac{1}{2}P_{111}'\left[\mathbf{u}_h, \mathbf{u}_k;\lambda_1\right]=0, \tag{2.4.18}$$

where for (2.4.18) we have used the relation

$$P_{21}\left[a_h\mathbf{u}_h, \eta;\lambda_1\right]=\frac{1}{2}a_h a_k P_{111}\left[\mathbf{u}_h, \mathbf{u}_k, \eta;\lambda_1\right].$$

From these linear equations, $\mathbf{v}_h'$ and $\mathbf{v}_{hk}$ can be determined.

In the following, we shall need the following property:

$$\text{Min}(F_2[\mathbf{u}]+F_1[\mathbf{u}])=-F_2[\mathbf{u}^*], \tag{2.4.19}$$

where $F_2[\mathbf{u}]$ is a positive-definite functional and $\mathbf{u}^*$ is the minimizing vector field. To show this, we first notice that the minimizing displacement field satisfies the equation

$$F_{11}\left[\mathbf{u}^*, \zeta\right]+F_1\left[\zeta\right]=0 \quad \text{for } \forall\zeta. \tag{2.4.20}$$

Now choose $\zeta=\mathbf{u}^*$, then

$$F_1\left[\mathbf{u}^*\right]=-F_{11}\left[\mathbf{u}^*, \mathbf{u}^*\right].$$

With

$$F_2[\mathbf{u}] = F_2\left[\frac{1}{2}\mathbf{u} + \frac{1}{2}\mathbf{u}\right] = \frac{1}{2}F_2[\mathbf{u}] + \frac{1}{4}F_{11}[\mathbf{u}, \mathbf{u}]$$

or

$$F_{11}[\mathbf{u}, \mathbf{u}] = 2F_2[\mathbf{u}]$$

we find

$$F_1[\mathbf{u}^*] = -2F_2[\mathbf{u}^*],$$

so that

$$F_2[\mathbf{u}^*] + F_1[\mathbf{u}^*] = -F_2[\mathbf{u}^*].$$

Using this property, we find that the minimum of (2.4.11) is

$$\begin{aligned}
&\underset{\text{w.r.t.}\,\bar{\mathbf{u}}}{\text{Min}}\ P_2[\bar{\mathbf{u}};\lambda_1] + (\lambda-\lambda_1)P'_{11}[a_h\mathbf{u}_h, \bar{\mathbf{u}};\lambda_1] + P_{21}[a_h\mathbf{u}_h, \bar{\mathbf{u}};\lambda_1] \\
&= -P_2[(\lambda-\lambda_1)a_h\mathbf{v}'_h(\mathbf{x}) + a_h a_k \mathbf{v}_{hk}(\mathbf{x})] \\
&= -(\lambda-\lambda_1)^2 P_2[a_h\mathbf{v}'_h(\mathbf{x});\lambda_1] - (\lambda-\lambda_1)P_{11}[a_h\mathbf{v}'_h(\mathbf{x}), a_h a_k\mathbf{v}_{hk}(\mathbf{x});\lambda_1] \\
&-P_2[a_h a_k \mathbf{v}_{hk}(\mathbf{x});\lambda_1].
\end{aligned} \tag{2.4.21}$$

Let us now first consider the case that $P_3[a_h\mathbf{u}_h;\lambda_1] \not\equiv 0$. In this case, the term $P_4[a_h\mathbf{u}_h;\lambda_1]$ is small compared to the cubic term, and may thus be neglected in (2.4.10). The minimum of the remaining terms with $\bar{\mathbf{u}}$ is given in (2.4.21), and here the term $P_2[a_h a_k v_{hk}(\mathbf{x});\lambda_1]$ may be neglected because it is of fourth degree in the amplitudes of the buckling modes. Further, the term $(\lambda-\lambda_1)P_{11}[a_h\mathbf{v}'_h(\mathbf{x}),$ $a_h a_k \mathbf{v}_h(\mathbf{x});\lambda_1]$ may be neglected because it is of third degree in the amplitudes of the buckling modes and multiplied by $(\lambda-\lambda_1)$. Finally, we may also neglect the first term on the right-hand side of (2.4.21) because it is of order $O[(\lambda-\lambda_1)^2 a^2]$, where a represents the order of magnitude of the amplitudes of the buckling mode. The only remaining term in (2.4.10) that still is to be discussed is $(\lambda-\lambda_1)P'_2[a_h\mathbf{u}_h;\lambda_1]$. This term is of order $O[(\lambda-\lambda_1)a^2]$ and may thus be important. Our final result is now given by

$$\begin{aligned}
\underset{(a_h=\text{const})}{\text{Min}}\ P(\mathbf{u};\lambda) &= (\lambda-\lambda_1)P'_2[a_h\mathbf{u}_h;\lambda_1] \\
&+ P_3[a_h\mathbf{u}_h;\lambda_1] + O(a^4, (\lambda-\lambda_1)a^3, (\lambda-\lambda_1)^2a^2).
\end{aligned} \tag{2.4.22}$$

In other words, to a first approximation we might have neglected the terms containing $\bar{\mathbf{u}}$ to obtain the same result.

Let us now consider the case $P_3[a_h\mathbf{u}_h;\lambda_1] \equiv 0$. Now we must take into account the term $P_4[a_h\mathbf{u}_h;\lambda_1]$, and hence also the last term in (2.4.21). Further, the term $(\lambda-\lambda_1)P'_2[a_h\mathbf{u}_h;\lambda_1]$ is important, whereas the first two terms on the right-hand side of (2.4.21) may be neglected compared to this term (under the assumption that $P'_2[a_h\mathbf{u}_h;\lambda_1] \not\equiv 0$).

In this case, our result is

$$\begin{aligned}
\underset{(a_h=\text{const.})}{\text{Min}}\ P(\mathbf{u};\lambda) &= (\lambda-\lambda_1)P'_2[a_h\mathbf{u}_h;\lambda_1] + P_4[a_h\mathbf{u}_h;\lambda_1] - P_2[a_h a_k\mathbf{v}_{hk};\lambda_1] \\
&+ O[a^5, (\lambda-\lambda_1)a^3, (\lambda-\lambda_1)^2a^2].
\end{aligned} \tag{2.4.23}$$

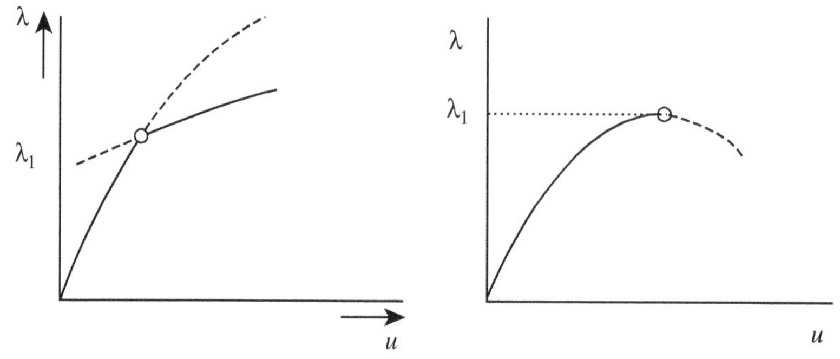

Equilibrium possible for $\lambda > \lambda_1$

No equilibrium possible for $\lambda > \lambda_1$, no linear terms in $\lambda - \lambda_1$

Figure 2.4.2

Notice that the last two terms in this expression are the terms that decide on the stability of neutral equilibrium (cf. 2.3.55).

In our discussion, we have restricted ourselves to cases where $P_2' \left[a_h \mathbf{u}_h; \lambda_1\right] \not\equiv 0$, i.e., we have assumed that the derivative of $P_2 \left[a_h \mathbf{u}_h; \lambda\right]$ with respect to the load parameter λ exists for $\lambda - \lambda_1$ (the fundamental state). This means that we have assumed that there are equilibrium states in the vicinity of the fundamental state. This is not always the case (see Figure 2.4.2).

In the following, we shall see that the present theory only enables us to treat branch points. Finally, we notice that when $P_2' \left[a_h \mathbf{u}_h; \lambda_1\right] \not\equiv 0$, this term must be negative because for $\lambda - \lambda_1 < 0$, the equilibrium in the fundamental state is stable, i.e.,

$$P_2 \left[a_h \mathbf{u}_h; \lambda\right] = P_2 \left[a_h \mathbf{u}_h; \lambda_1\right] + (\lambda - \lambda_1)\, P_2' \left[a_h \mathbf{u}_h; \lambda_1\right] + \cdots \geq 0 \quad \text{for } \lambda - \lambda_1 < 0.$$

The first term on the right-hand side is zero, and hence

$$P_2' \left[a_h \mathbf{u}_h, \lambda_1\right] < 0. \tag{2.4.24}$$

Let us now consider the simple case that $m = 1$, i.e., there is only one buckling mode, say $\mathbf{u}(\mathbf{x}) = a\mathbf{u}_1(\mathbf{x})$. Let us further introduce the notation

$$P_3 \left[\mathbf{u}_1; \lambda_1\right] = A_3, \quad P_2' \left[\mathbf{u}_1; \lambda_1\right] = A_2'. \tag{2.4.25}$$

First, consider the case $A_3 \not\equiv 0$, $A_2' < 0$. Then we obtain from (2.4.22)

$$\underset{a_k = \text{const.}}{\text{Min}}\; P\left[\mathbf{u}(\mathbf{x}), \lambda\right] = (\lambda - \lambda_1)\, A_2' a^2 + A_3 a^3. \tag{2.4.26}$$

The condition for equilibrium is obtained by differentiating this expression with respect to a and equating this result to zero,

$$2(\lambda - \lambda_1)\, A_2' a + 3A_3 a^2 = 0. \tag{2.4.27}$$

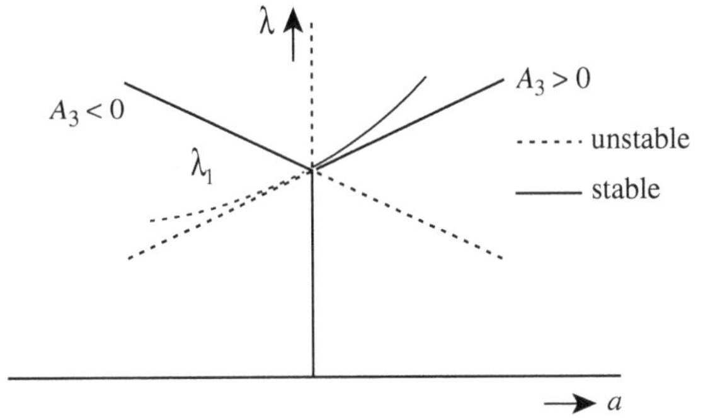

Figure 2.4.3

The solutions are

$$a_1 = 0 \qquad \text{(the fundamental state)}$$
$$a_2 = -\frac{2}{3}(\lambda - \lambda_1)\frac{A_2'}{A_3} \qquad \text{(branched equilibrium state).} \tag{2.4.28}$$

The stability condition is that the second derivative of (2.4.26) is non-negative,

$$2(\lambda - \lambda_1)A_2' + 6A_3 a \geq 0. \tag{2.4.29}$$

It follows that the fundamental state is stable for $\lambda < \lambda_1$ and unstable for $\lambda > \lambda_1$ and $\lambda = \lambda_1$ (because $A_3 \neq 0$). Substitution of a_2 into (2.4.29) yields

$$-2(\lambda - \lambda_1)A_2' \geq 0, \tag{2.4.30}$$

which implies that the branched solution is stable for $\lambda > \lambda_1$ and unstable for $\lambda < \lambda_1$ and $\lambda = \lambda_1$ (because $A_3 \neq 0$). These results are plotted in Figure 2.4.3.

The curved line is the behavior of the exact solution. An example of a structure with such a behavior is the two-bar structure shown in Figure 2.4.4 (to be discussed later).

For λ sufficiently close to λ_1, our approximate solution will be a good approximation to the actual solution. However, it is not possible to assess the accuracy of the approximation.

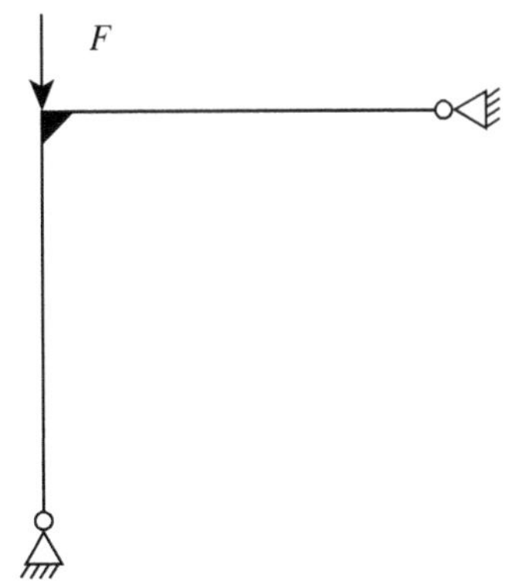

Figure 2.4.4

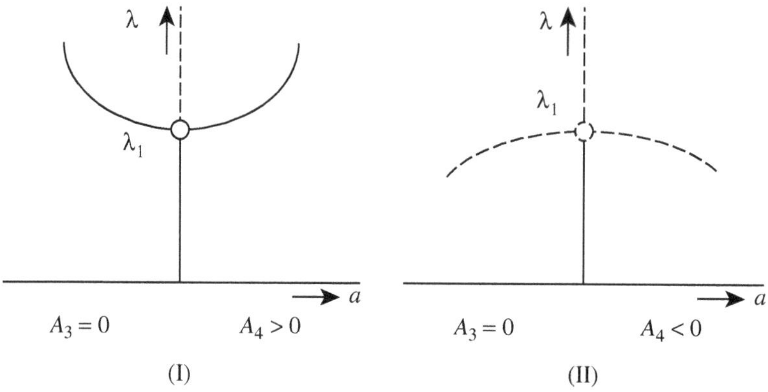

Figure 2.4.5

Let us now consider the case $A_3 = 0$. From (2.4.23) we obtain

$$\underset{a_k=\text{const.}}{\text{Min}}\ P[\mathbf{u}, \lambda] = (\lambda - \lambda_1) A_2' a^2 + A_4 a^4, \tag{2.4.31}$$

where

$$A_4 \equiv P_4[\mathbf{u}_h, \lambda_1] - P_2[\mathbf{v}_{hk}; \lambda_1].$$

The branch point is stable when $A_4 > 0$ and unstable when $A_4 < 0$.

The equilibrium condition reads

$$2(\lambda - \lambda_1) A_2' a + 4A_4 a^3 = 0, \tag{2.4.32}$$

from which

$$\begin{aligned} a_1 &= 0 \quad \text{(fundamental state)} \\ a_2^2 &= -\frac{1}{2}(\lambda - \lambda_1)\frac{A_2'}{A_4}. \end{aligned} \tag{2.4.33}$$

Because for real values of a_2 the left-hand side is positive, the right-hand side must be positive, i.e.,

$$\lambda \gtrless \lambda_1 \quad \text{for } A_4 \gtrless 0. \tag{2.4.34}$$

The stability condition reads

$$2(\lambda - \lambda_1) A_2' + 12A_4 a^2 \geq 0, \tag{2.4.35}$$

so the fundamental state is stable for ($\lambda < \lambda_1$) and unstable for $\lambda > \lambda_1$.

Substitution of a_2^2 into this condition yields

$$-4(\lambda - \lambda_1) A_2' \geq 0, \tag{2.4.36}$$

from which follows that the branched equilibrium state is stable for $\lambda > \lambda_1$ and $A_4 > 0$, and unstable for $\lambda < \lambda_1$ and $A_4 < 0$. These results are plotted in Figure 2.4.5 and Figure 2.4.6.

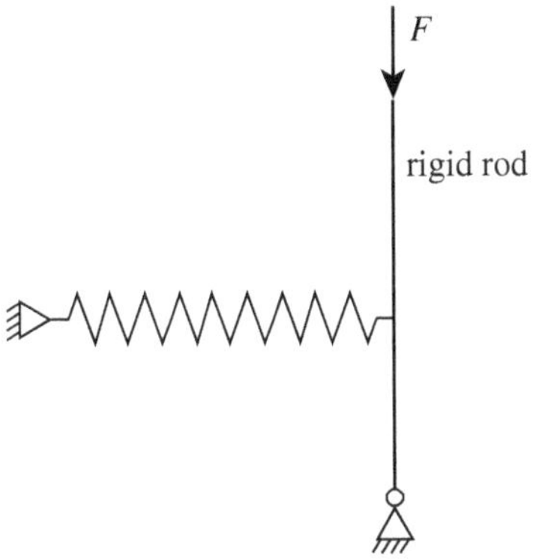

Figure 2.4.6

Examples of structures with the behavior of (I) are the Euler column and plates loaded in their plane. An example of a structure with the behavior of (II) is given in Figure 2.4.6.

From Figure 2.4.3 and Figure 2.4.5, we see that branched equilibrium states for loads below the critical load are unstable. Later, we shall prove that this is always the case.† The converse that branched equilibrium states would always be stable for loads exceeding the critical load is not true (except for the single mode case).

Let us now proceed with the discussion of $\mathrm{Min}_{a_h=\mathrm{const}} P[\mathbf{u}, \lambda]$, which is given by (2.4.32) when $P_3[a_h\mathbf{u}_h; \lambda_1] \not\equiv 0$ and by (2.4.23) when this term is identically equal to zero. The term $P_2'[a_h\mathbf{u}_h; \lambda_1]$ is quadratic in the buckling modes and may be transformed to a form containing only quadratic terms, which will be normalized so that the coefficients are all equal to -1 (remember $P_2'[a_h\mathbf{u}; \lambda_1] \leq 0$), so that we may write

$$P_2'\left[a_h\mathbf{u}_h; \lambda_1\right] = -a_i a_i. \tag{2.4.37}$$

Further, we introduce the notation

$$\begin{aligned} P_3\left[a_h\mathbf{u}_h; \lambda_1\right] &\equiv A_{ijk} a_i a_j a_k, \\ P_4\left[a_h\mathbf{u}_h; \lambda_1\right] - P_2\left[a_h a_k \mathbf{v}_{hk}; \lambda_1\right] &\equiv A_{ijkl} a_i a_j a_k a_l. \end{aligned} \tag{2.4.38}$$

We may normalize the load factor λ so that at the critical load, $\lambda = \lambda_1 = 1$. Denoting $\mathrm{Min}\ P[\mathbf{u}; \lambda]$ by $F[a_i; \lambda]$, our discussion of stability is reduced to the discussion of the quadratic form

$$F(a_i; \lambda) = (1 - \lambda) a_i a_i + \begin{cases} A_{ijk} a_i a_j a_k \\ A_{ijk\ell} a_i a_j a_k a_\ell \end{cases} \tag{2.4.39}$$

where the term on top is important when $A_{ijk} a_i a_j a_k 0$, else the quartic term must be used.

† W. T. Koiter, Some properties of (completely) symmetric multilinear forms.... Rep. Lab. For *Appl. Mech. Delft*, 587 (1975).

Let δa_i be a variation of a_i,

$$F(a_i + \delta a_i; \lambda) = (1-\lambda) a_i a_i + 2(1-\lambda) a_i \delta a_i + (1-\lambda) \delta a_i \delta a_i$$
$$+ \begin{cases} A_{ijk}(a_i a_j a_k + 3a_j a_k \delta a_i + 3a_k \delta a_i \delta a_j + \delta a_i \delta a_j \delta a_k) \\ A_{ijkl}(a_i a_j a_k a_l + 4a_j a_k a_l \delta a_i + 6a_k a_l \delta a_i \delta a_j + 4a_l \delta a_i \delta a_j \delta a_k + \delta a_i \delta a_j \delta a_k \delta a_l) \end{cases}$$
$$= F(a_i; \lambda) + \delta F(a_i; \lambda) + \delta^2 F(a_i; \lambda) + \cdots \qquad (2.4.40)$$

so that

$$\delta F(a_i; \lambda) = \left[2(1-\lambda) a_i + \begin{cases} 3A_{ijk} a_j a_k \\ 4A_{ijk\ell} a_j a_k a_\ell \end{cases} \right] \delta a_i, \qquad (2.4.41)$$

$$\delta^2 F(a_i; \lambda) = \left[1 - \lambda + \begin{cases} 3A_{ijk} a_k \\ 6A_{ijk\ell} a_k a_\ell \end{cases} \right] \delta a_i \delta a_j . \qquad (2.4.42)$$

The equilibrium equations are obtained from $\delta F = 0$, for all variations δa_i, so that

$$2(1-\lambda) a_i + \begin{cases} 3A_{ijk} a_j a_k \\ 4A_{ijk\ell} a_j a_k a_\ell \end{cases} = 0. \qquad (2.4.43)$$

The condition for stability is that $\delta^2 F \geq 0$,

$$(1-\lambda) \delta a_i \delta a_j + \begin{cases} 3A_{ijk} a_k \delta a_i \delta a_j \\ 6A_{ijk\ell} a_k a_\ell \delta a_i \delta a_j \end{cases} \geq 0. \qquad (2.4.44)$$

Let $\delta a_h = b_h$ be a unit vector, i.e., $\|\mathbf{b}\| = b_i b_i = 1$, and let $a_h = a e_h$, where $\mathbf{e}$ is a unit vector and a is the amplitude of $\mathbf{a}$. The equilibrium equations now read

$$2(1-\lambda) e_i + \begin{cases} 3aA_{ijk} e_j e_k \\ 4a^2 A_{ijk\ell} e_j e_k e_\ell \end{cases} = 0, \qquad (2.4.45)$$

and the stability condition is now given by

$$1 - \lambda + \begin{cases} 3aA_{ijk} e_i b_i b_j \\ 6a^2 A_{ijk\ell} e_k e_\ell b_i b_j \end{cases} \geq 0 \qquad (\forall\, \mathbf{b} \mid b_i b_i = 1). \qquad (2.4.46)$$

We are now in a position to prove the statement that the branched equilibrium states are unstable for loads below the critical load, i.e., for $\lambda < 1$.

To show this, we first notice that (2.4.46) must hold for all unit vectors $\mathbf{b}$, and hence also for $\mathbf{b} = \mathbf{e}$, so that

$$1 - \lambda + \begin{cases} 3aA_{ijk} e_i e_j e_j \\ 6a^2 A_{ijk\ell} e_i e_j e_k e_\ell \end{cases} \geq 0. \qquad (2.4.47)$$

Multiplying (2.4.45) by e_i, we obtain

$$2(1-\lambda) + \begin{cases} 3aA_{ijk} e_i e_j e_k \\ 4a^2 A_{ijk\ell} e_i e_j e_k e_\ell \end{cases} = 0, \qquad (2.4.48)$$

from which

$$3aA_{ijk} e_i e_j e_k = -2(1-\lambda), \qquad (2.4.49)$$

or when the cubic term is missing,

$$a^2 A_{ijk\ell} e_i e_j e_k e_\ell = -\frac{1}{2}(1-\lambda). \tag{2.4.50}$$

Substitution of these expressions into the corresponding form of the stability condition (2.4.47) yields

$$-(1-\lambda) \geq 0 \quad \text{resp.} \quad -2(1-\lambda) \geq 0,$$

and hence $\lambda \geq 1$. It follows that for $\lambda < 1$ the equilibrium state is unstable.

Let us now consider stationary values of the multilinear forms

$$A_{ijk} t_i t_j t_k \quad \text{resp.} \quad A_{ijk\ell} t_i t_j t_k t_\ell$$

on the unit sphere $\|\mathbf{t}\| = 1$. Introducing a Lagrangean multiplier μ, we look for stationary values of

$$A_{ijk} t_i t_j t_j + \mu (t_i t_i - 1) \quad \text{resp.} \quad A_{ijk\ell} t_i t_j t_k t_\ell + \mu (t_i t_i - 1), \tag{2.4.51}$$

i.e., the first variation of these forms must vanish. This yields

$$3A_{ijk} t_j t_k - 2\mu t_i = 0 \quad \text{resp.} \quad 4A_{ijk\ell} t_j t_k t_\ell - 2\mu t_i = 0. \tag{2.4.52}$$

These equations are of the same form as our equilibrium equations – in other words, the equilibrium equations yield stationary directions on the unit sphere $\|\mathbf{e}\| = 1$. Among these stationary directions, there are minimizing directions. Let $\mathbf{t} = \mathbf{t}^*$ be a minimizing direction, and let

$$A_{ijk} t_i^* t_j^* t_k^* = A_3(\mathbf{t}^*) = A_3^* \leq 0 \quad \text{and} \quad A_{ijk\ell} t_i^* t_j^* t_k^* t_\ell^* = A_4(\mathbf{t}^*) = A_4^*. \tag{2.4.53}$$

For $A_3^* \neq 0$, i.e., $A_3^* < 0$, the equilibrium state is unstable. For $A_3^* \equiv 0$ and $A_4^* > 0$, the equilibrium state is stable. To show this, we need the following property[†]:

$$\min_{\|\mathbf{e}\|=\|\mathbf{b}\|=1} A_{ijkl} e_i e_j b_k b_l = \min_{\|\mathbf{t}\|=1} A_{ijk\ell} t_i t_j t_k t_\ell. \tag{2.4.54}$$

Using this property, we obtain from the stability condition (2.4.46)

$$1 - \lambda + 6a^2 A_4^* \geq 0. \tag{2.4.55}$$

From the equilibrium equation (2.4.45) for $\mathbf{e} = \mathbf{t}^*$ and multiplied by $\mathbf{t}_i^*$, we obtain

$$2(1-\lambda) + 4a^2 A_4^* = 0, \tag{2.4.56}$$

so we may rewrite (2.4.56) to yield

$$-2(1-\lambda) \geq 0. \tag{2.4.57}$$

Hence the branched equilibrium state is stable for ($\lambda > 1$). This property has been proved for a minimizing direction $\mathbf{t}^*$. For other solutions of the equilibrium equations, the question of stability cannot be established in general.

† W. T. Koiter, Some properties of (completely) symmetric multi-linear forms.... Rep. Lab. For *Appl. Mech. Delft*, 587 (1975).

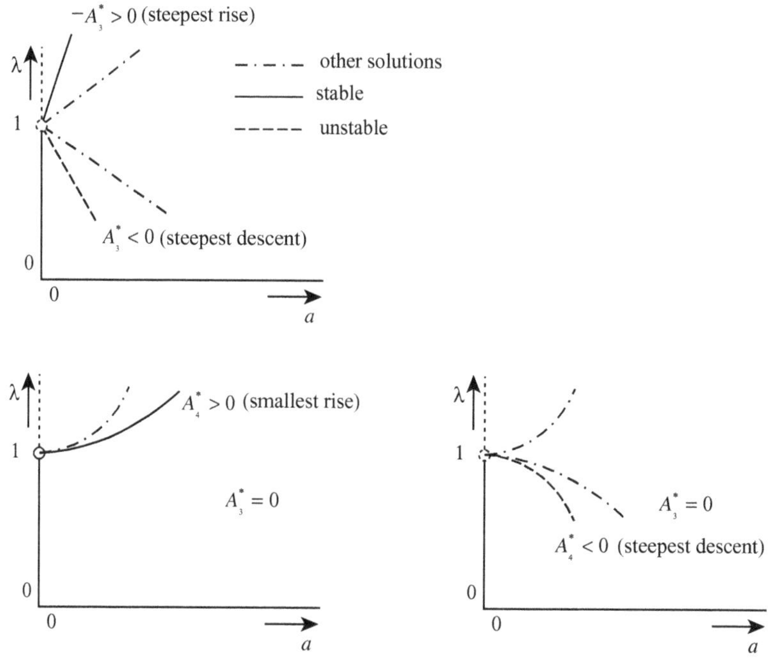

Figure 2.4.7

From the equilibrium equations (2.4.45) with $\mathbf{e} = \mathbf{t}^*$ and multiplied by $\mathbf{t}_i^*$, we find

$$\begin{aligned} a &= -\frac{2}{3}\frac{1-\lambda}{A_3^*} \qquad (A_3^* < 0) \\ a^2 &= -\frac{1}{2}\frac{1-\lambda}{A_4^*} \qquad (A_3^* = 0)\,. \end{aligned} \tag{2.4.58}$$

Our results are shown in Figure 2.4.7.

When $A_3^* < 0$, the dashed line corresponding to the minimizing solution is a line of steepest descent because the amplitude a for fixed λ is monotonically increasing for increasing values of $A_3 \in [A_3^*, 0) \subset \mathbb{R}^-$. Similarly, the dashed curve for $A_4^* < 0$ is a curve of steepest descent. The solid curve for the minimizing solution when $A_4^* > 0$ is a curve of smallest rise because for fixed λ, the amplitude a is monotonically decreasing with increasing values of $A_4 \in [A_4^*, \infty) \subset \mathbb{R}^+$. We emphasize here once again that our positive verdict about stability only holds for curves of smallest rise, and that for other rising solutions stability must be examined in each particular case. Descending solutions are unstable.

The behavior of the structure for minimizing solutions may be presented more clearly. To this end, we consider the generalized displacement corresponding to the external load. This generalized displacement is such that the product of the load and the displacement is equal to the increase of the elastic energy in the structure. For example, consider an elastic bar (Figure 2.4.3),

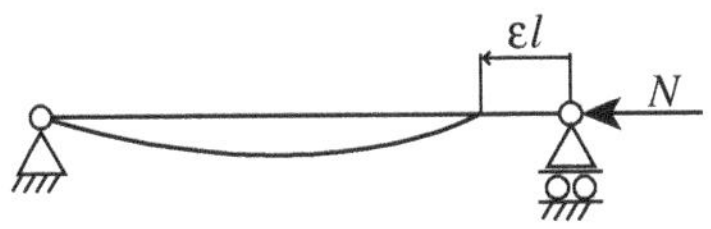

Figure 2.4.8

where $\varepsilon\ell$ is the generalized displacement. The behavior of this structure is plotted in Figure 2.4.9.

Here the curve I corresponds to the fundamental state, which is stable up to the critical load N_1 and unstable for loads exceeding N_1. The branched solution II is stable.

In the fundamental state, for $N = N_B$ the elastic energy is given by the area of ODB, and the energy of the load is given by $N_B(\varepsilon\ell)_B$, the area of the square $ODBN_B$. The potential energy in B is thus given by the negative of the area ΔOBN_B.

In the branched state for $N = N_B$ the elastic energy is given by the area of $OACE$, and the energy of the load is given by $N_B(\varepsilon\ell)_C$, the area of the square $OECN_B$. The potential energy is then given by the negative of the area $OACN_B$. Hence the increment of the potential energy going from the fundamental state B to the branched state C is given by

$$P\,[\mathbf{u}^{\mathrm{eq}}; N] = -\text{ area } ABC, \tag{2.4.59}$$

where $\mathbf{u}^{\mathrm{eq}}$ indicates that $\mathbf{u}$ is the displacement from the equilibrium state B to the equilibrium state C. The first variation of $P\,[\mathbf{u}^{\mathrm{eq}}; \lambda]$ is given by

$$\delta P\,[\mathbf{u}^{\mathrm{eq}}; \lambda] = -\,\delta\,(\text{area } ABC) = -\,(\Delta\varepsilon l)\,\delta N \tag{2.4.60}$$

or

$$\Delta\varepsilon\ell = -\frac{\partial P\,[\mathbf{u}^{\mathrm{eq}}; N]}{\partial N}. \tag{2.4.61}$$

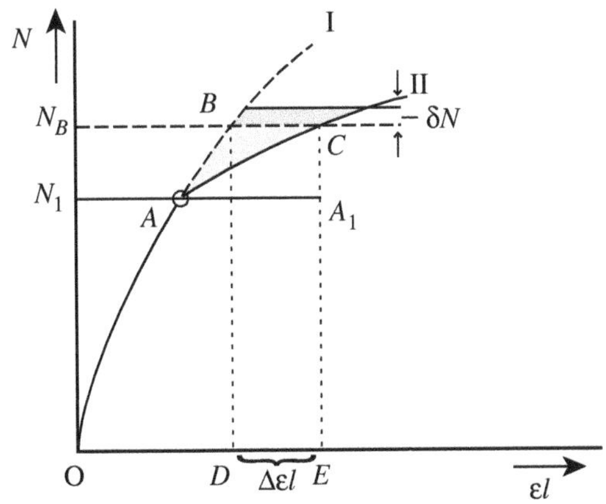

Figure 2.4.9

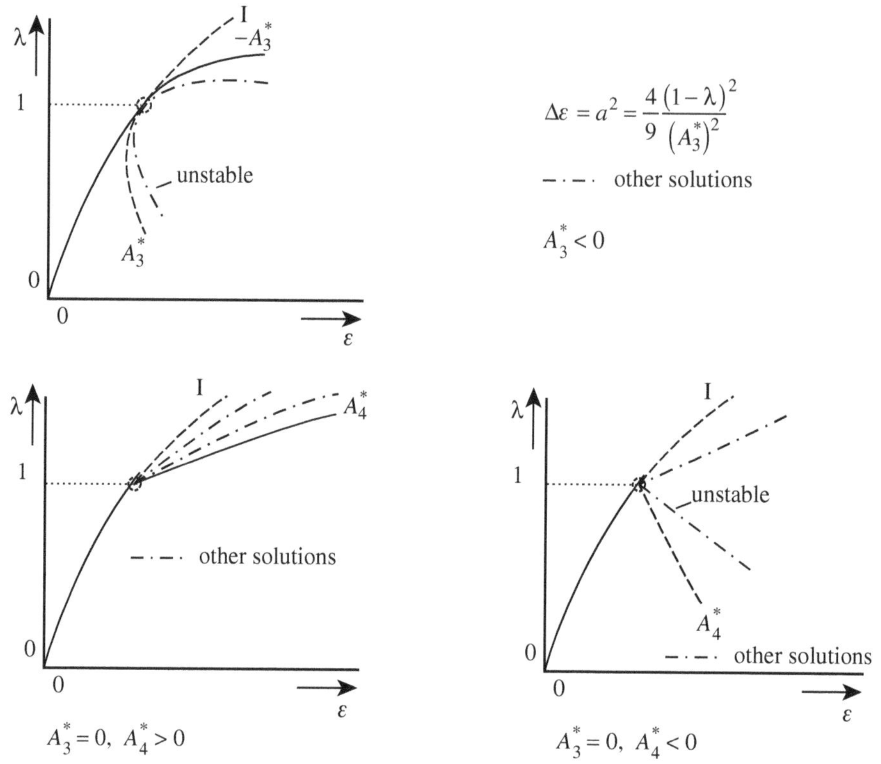

Figure 2.4.10

In words: The increment of the generalized displacement is equal to the negative of the derivative of the potential energy with respect to the external load, evaluated in the branched equilibrium state.

In general, denoting the generalized displacement by ε, we may now write

$$\Delta\varepsilon = -\frac{d}{d\lambda}F\left(a_i^{\text{eq}};\lambda\right) = -\frac{\partial F}{\partial\lambda}\left(a_i^{\text{eq}};\lambda\right) = a_i a_i = a^2. \qquad (2.4.62)$$

It follows that a^2 is a natural measure for the increment of the generalized displacement $\Delta\varepsilon$, and this justifies our earlier notation $\boldsymbol{a} = a\boldsymbol{e}$.

We can now make the graphs shown in Figure 2.4.10.

When $A_3^* = 0$ and $A_4^* \neq 0$, we have from (2.4.58)

$$\Delta\varepsilon = a^2 = -\frac{1}{2}\frac{1-\lambda}{A_4^*}. \qquad (2.4.63)$$

Let us make some final remarks on the results obtained so far:

i. Our results are only valid in the immediate vicinity of the branch point. It is not possible to assess the accuracy of the results for finite deflections from the fundamental state.
ii. A structure will (unless prohibited) follow the path of steepest descent or weakest ascent. This suggests that for a structure with geometrical imperfections,

the imperfections coinciding with minimizing directions are the most dangerous imperfections. It is therefore meaningful to consider these imperfections in particular.

iii. The fact that the slope of the branched equilibrium path is always smaller than that of the fundamental path implies that the stiffness of the structure after buckling is always decreased.

2.5 The influence of imperfections

In the foregoing discussions, it was assumed that the geometry of the structure and the distribution of the loads were known exactly. Further, it was assumed that the material was elastic and followed the generalized Hooke's Law. However, in practice the quantities are not known exactly, e.g., it is impossible to manufacture a perfectly straight column of constant cross section, or a perfect cylindrical shell. The deviations from the "idealized structure" in the absence of loads are called INITIAL IMPERFECTIONS. Some structures are extremely sensitive to initial imperfection, e.g., cylindrical shells under axial compressive loads.

The primary effect of initial imperfection is that the fundamental state of the idealized perfect structure, described by the displacement vector **U** from the undeformed state, does not represent a configuration of equilibrium of the actual imperfect structure. The bifurcation point of the perfect structure can also not be a bifurcation point of the structure with imperfections.

In our previous discussions, we have seen that for structures under a dead–weight load in an equilibrium configuration, the expansion of the potential energy functional starts with terms quadratic in the displacements. However, in the presence of imperfections the fundamental state is not an equilibrium configuration, and hence the expansion of the potential energy functional will start with a term linear in the displacements. Furthermore, if we restrict ourselves to small initial imperfections, this term will also be linear in these imperfections. This implies that our energy functional $F(a_i;\lambda)$ is replaced by the functional $F^*(a_i;\lambda)$, defined by

$$F^*(a_i;\lambda) = (1-\lambda)\,a_i a_i + \begin{cases} A_{ijk} a_i a_j a_k \\ A_{ijk\ell} a_i a_j a_k a_\ell \end{cases} + \mu B_i a_i, \tag{2.5.1}$$

where μ is a measure for the magnitude of the imperfections and **B** is determined by the shape of the imperfections. It should be noted that our approach is valid under the restrictions

$$|a_i| \ll 1, \quad |1-\lambda| \ll 1, \quad |\mu| \ll 1. \tag{2.5.2}$$

The equilibrium equations now follow from $\delta F^*(a_i;\lambda) = 0$, which yields

$$2(1-\lambda)\,a_i + \begin{cases} 3A_{ijk} a_j a_k \\ 4A_{ijk\ell} a_j a_k a_\ell \end{cases} + \mu B_i = 0. \tag{2.5.3}$$

The stability condition is not affected by the linear term $\mu B_i a_i$, and is still given by (2.4.44). Writing $a_i = ae_i$, $|\mathbf{e}| = 1$, the equilibrium equations become

$$2(1-\lambda)\,ae_i + \begin{cases} 3a^2 A_{ijk} e_j e_k \\ 4a^3 A_{ijk\ell} e_j e_k e_\ell \end{cases} + \mu B_i = 0, \tag{2.5.4}$$

and the stability condition becomes

$$1 - \lambda + \begin{cases} 3aA_{ijk} e_k b_i b_j \\ 6a^2 A_{ijk\ell} e_k e_\ell b_i b_j \end{cases} \geq 0, \tag{2.5.5}$$

where we have put $\delta a_i = b_i$, $\|\mathbf{b}\| = 1$.

The equations (2.5.4) are m equations for e_i, $i \in (1, m) \subset \mathbb{N}$, and the amplitude a., i.e., $m - 1$ unknowns because $\|\mathbf{e}\| = 1$. In general, this problem is extremely complicated, but for the most dangerous imperfections, the ones with directions corresponding to the minimizing directions t^*, the problem is simplified enormously.

We shall now treat this problem in more detail. First, choose the imperfections in the directions of stationary values on the unit ball, i.e., $B_i = t_i$, and choose $e_i = t_i$. Let us now first consider the case $P_3\,[a_h \mathbf{u}_h; \lambda]\,0$. The equilibrium equation now reads

$$2(1-\lambda)\,at_i + \mu t_i + 3a^2 A_{ijk} t_j t_k = 0. \tag{2.5.6}$$

Multiplying this equation by t_i, summing over i, and using the notation $A_{ijk} t_i t_j t_k = A_3(\mathbf{t})$, we obtain

$$2(1-\lambda)\,a + \mu + 3a^2 A_3(t) = 0. \tag{2.5.7}$$

This equation is to be solved under the side condition $a \in \mathbb{R}^+$. The stability condition reads

$$1 - \lambda + 3aA_{ijk} b_i b_j t_k \geq 0 \quad \text{for} \quad \forall\, \mathbf{b} |\, \|\mathbf{b}\| = 1. \tag{2.5.8}$$

Now consider the particular case that the stationary directions are minimizing directions, i.e., $\mathbf{t} = \mathbf{t}^*$. Using the property

$$\underset{\|\mathbf{a}\|=\|\mathbf{b}\|=\|\mathbf{c}\|=1.}{\text{Min}} A_{ijk} a_i b_j c_k = A_{ijk} t_i^* t_j^* t_k^* = A_3(\mathbf{t}^*) = A_3^* < 0, \tag{2.5.9†}$$

we obtain from (2.5.8)

$$1 - \lambda + 3aA_3^* \geq 0, \quad \Rightarrow \lambda < 1. \tag{2.5.10}$$

The stability limit follows from

$$1 - \lambda^* + 3a^* A_3^* = 0, \quad \Rightarrow \lambda^* < 1. \tag{2.5.11}$$

† W. T. Koiter, Some properties of (completely) symmetric multi-linear forms... Rep. Lab. For *Appl. Mech. Delft,* 587 (1975).

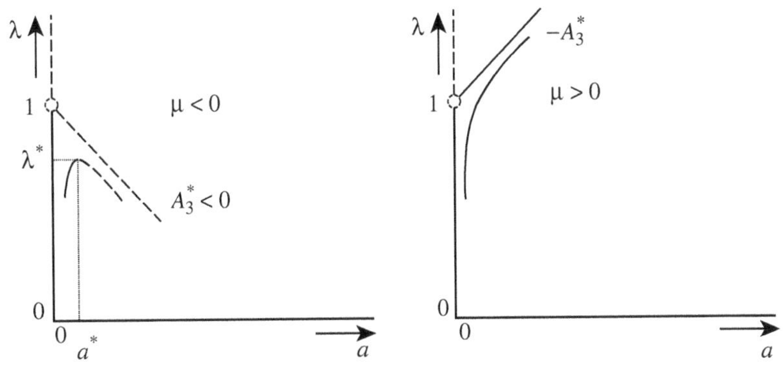

Figure 2.5.1

For $\mathbf{t} = \mathbf{t}^*$, we obtain from (2.5.7)

$$a^* \left[2(1 - \lambda^*) + 3a^* A_3^*\right] + \mu = 0. \tag{2.5.12}$$

Because the term between the square brackets is positive, this equation can only be satisfied for $\mu < 0$. From (2.5.11), we find for the amplitude at the stability limit

$$a^* = -\frac{1 - \lambda^*}{3A_3^*}, \tag{2.5.13}$$

and substituting this expression into (2.5.12), we find for the load factor at the stability limit

$$(1 - \lambda^*)^2 - 3\mu A_3^*. \tag{2.5.14}$$

The general expression for the amplitude of the buckling modes in the presence of imperfections in the minimizing directions follows from (2.5.7) with $\mathbf{t} = \mathbf{t}^*$,

$$a_{1,2} = \frac{1}{3A_3^*}\left[-(1 - \lambda) \pm \sqrt{(1 - \lambda)^2 - 3\mu A_3^*}\right]. \tag{2.5.15}$$

These amplitudes are real for $(1 - \lambda)^2 > 3\mu A_3^*$, and they are both positive when $\mu < 0$. When $\mu > 0$, only the negative sign can be used because $a \in \mathbb{R}^+$. These results are shown in Figure 2.5.1.

One easily shows that for $\mu < 0$, the path is stable for $a < a^*$ and unstable for $a > a^*$, and that for $\mu > 0$ the path is stable. Notice that due to the presence of imperfections, the branch point is not reached. The stability limit is now a limit point and not a branch point, as was the case for the perfect structure.

In the foregoing discussion, we have made it plausible that imperfections in the minimizing directions are the most harmful ones. To show this, we return to (2.5.4). Multiplying this equation by e_i and summing over i, we find

$$2(1 - \lambda)a + 3a^2 A_3(\mathbf{e}) + \mu B_i e_i = 0. \tag{2.5.16}$$

The stability condition (2.5.5) reads

$$1 - \lambda + 3aA_{ijk}b_i b_j e_k \geq 0 \quad \text{for} \quad \forall\, \mathbf{b} |\, \|\mathbf{b}\| = 1. \tag{2.5.17}$$

Let us now assume that $\mathbf{e}$ is known, and let the minimum of the trilinear form in (2.5.17) for fixed $\mathbf{e}$ be $C_3(\mathbf{e})$. We then have the following inequalities;

$$A_3^* = A_3(\mathbf{t}^*) \leq C_3(\mathbf{e}) \leq A_3(\mathbf{e}). \tag{2.5.18}$$

The stability limit now follows from

$$1 - \lambda + 3aC_3(\mathbf{e}) = 0; \tag{2.5.19}$$

hence,

$$a = -\frac{1-\lambda}{3C_3(\mathbf{e})}. \tag{2.5.20}$$

Because we are interested in the reduction of the critical load due to imperfections ($1 < \lambda$), it suffices to consider only negative values of $C_3(\mathbf{e})$. Substitution of (2.5.20) into (2.5.16) yields

$$(1-\lambda)^2 \left[2 - \frac{A_3(\mathbf{e})}{C_3(\mathbf{e})}\right] = 3\mu B_i e_i C_3(\mathbf{e}). \tag{2.5.21}$$

For the right-hand side of (2.5.21), we have the estimate

$$\left|3\mu B_i e_i C_3(\mathbf{e})\right| \leq \left|3\mu C_3(\mathbf{e})\right| \leq \left|3\mu A_3^*\right|. \tag{2.5.22}$$

The term between the square brackets in (2.5.21) is always larger than 1, for $A_3(\mathbf{e}) > 0$ it is larger than 2, and for $A_3(\mathbf{e}) < 0$ we have $A_3(\mathbf{e})/C_3(\mathbf{e}) \leq 1$ as then $A_3(\mathbf{e}) \in [C_3(\mathbf{e}), 0) \subset \mathbb{R}^-$. The critical load occurs for $\mu < 0$, so we may write

$$(1-\lambda)^2 \geq 3\mu A_3^* = (1-\lambda^*)^2, \tag{2.5.23}$$

which implies $\lambda^* < \lambda$, i.e., λ^* is a lower bound for the critical load.

Let us now consider the case $P_3[a_h\mathbf{u}_h;\lambda] \equiv 0$ and $A_{ijk\ell}a_i a_j a_k a_\ell$ indefinite. The equilibrium equation now reads

$$2(1-\lambda)ae_i + 4A_{ijk\ell}e_j e_k e_\ell a^3 + \mu B_i = 0 \tag{2.5.24}$$

and the stability condition is given by

$$1 - \lambda + ba^2 A_{ijk\ell} b_i b_j e_k e_\ell \geq 0 \quad \text{for} \quad \forall\, \mathbf{b} \,|\, \|\mathbf{b}\| = 1. \tag{2.5.25}$$

Because the fourth degree form is indefinite, we have

$$\operatorname*{Min}_{\|\mathbf{t}\|=1} A_{ijk\ell}t_i t_j t_k t_\ell = A_4(\mathbf{t}^*) = A_4^* < 0. \tag{2.5.26}$$

Now we consider imperfections in the direction of the minimizing direction, i.e., $B_i = t_i^*$. From (2.5.24) and with $\mathbf{e} = \mathbf{t}^*$, we now obtain

$$[2(1-\lambda)a + \mu]\, t_i^* + 4A_{ijk\ell}t_i^* t_k^* t_\ell^* a^3 = 0. \tag{2.5.27}$$

Multiplying by t_i^* and summing over i, we obtain for the amplitude

$$2(1-\lambda)a + 4A_4^* a^3 + \mu = 0. \tag{2.5.28}$$

The stability condition (2.5.25) must hold for all $\mathbf{b}$ with $\|\mathbf{b}\| = 1$, and hence also for $\mathbf{b} = \mathbf{t}^*$, so

$$1 - \lambda + 6A_4^* a^2 \geq 0. \tag{2.5.29}$$

The stability limit follows from

$$1 - \lambda^* + 6A_4^* a^{*2} = 0, \tag{2.5.30}$$

from which follows

$$a_{1,2}^* = \pm\sqrt{-\frac{1-\lambda^*}{6A_4^*}} \quad (\lambda^* < 1,\ \text{because } A_4^* < 0)\,. \tag{2.5.31}$$

Substitution into (2.5.28) yields

$$(1 - \lambda^*)^{3/2} = \mp\frac{3}{4}\mu\sqrt{-6A_4^*}. \tag{2.5.32}$$

As $\lambda^* < 1$, the left-hand side of (2.5.32) is positive and hence we must choose the negative sign in (2.5.31) when $\mu < 0$. This implies that (2.5.32) can be rewritten to yield

$$(1 - \lambda^*)^{3/2} = \frac{3}{4}\,|\mu|\,\sqrt{-6A_4^*}. \tag{2.5.33}$$

Equation (2.5.33) is a formula for the reduction of the stability limit due to imperfections in the direction of the minimizing direction, corresponding to the line of steepest descent.

We shall now show that this reduction is the largest possible reduction. Consider imperfections in an arbitrary direction. Multiplying the equilibrium equation (2.5.24) by e_i and summing over i, we obtain

$$2\,(1 - \lambda)\,a + 4A_4\,(\mathbf{e})\,a^3 + \mu B_i e_i = 0. \tag{2.5.34}$$

Introducing $C_4\,(\mathbf{e})$ defined by

$$\underset{\substack{\text{w.r.t. } \mathbf{b},\ \|\mathbf{b}\|=1 \\ \mathbf{e} \text{ fixed}}}{\text{Min}} \quad A_{ijk\ell} b_i b_j e_k e_\ell = C_4\,(\mathbf{e})\,, \tag{2.5.35}$$

we obtain from the stability condition (2.5.25)

$$1 - \lambda + 6a^2 C_4\,(\mathbf{e}) \geq 0. \tag{2.5.36}$$

Further, we note the inequalities

$$A_4^* \leq C_4\,(\mathbf{e}) \leq A_4\,(\mathbf{e})\,. \tag{2.5.37}$$

Because we are interested in the reduction of the critical load due to imperfections, it suffices to restrict ourselves to values of λ for which $\lambda < 1$. From (2.5.36), it then follows that $C_4\,(\mathbf{e}) < 0$. With the stability limit, we have

$$1 - \lambda + 6a^2 C_4\,(\mathbf{e}) = 0, \tag{2.5.38}$$

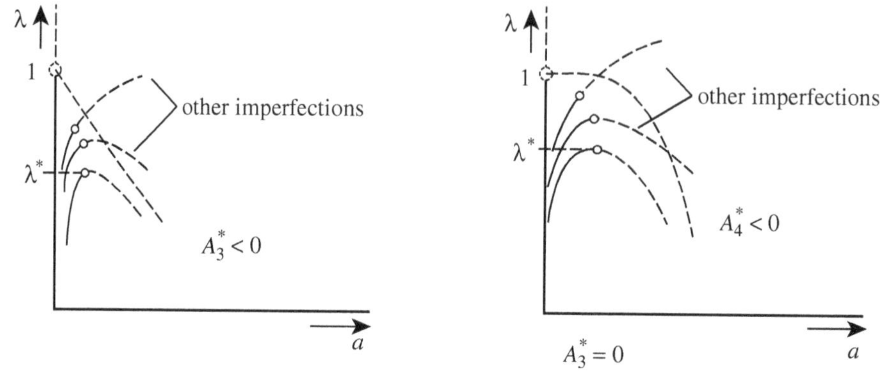

Figure 2.5.2

so that

$$a_{1,2} = \pm\sqrt{-\frac{1-\lambda}{6C_4(\mathbf{e})}}\ . \tag{2.5.39}$$

Substitution into (2.5.34) yields

$$2(1-\lambda)^{3/2}\left[1-\frac{A_4(\mathbf{e})}{3C_4(\mathbf{e})}\right] = \pm\mu B_i e_i\sqrt{-6C_4(\mathbf{e})}, \tag{2.5.40}$$

which determines the reduction of the critical load at the stability limit in the presence of imperfections in arbitrary directions. Let us start the discussion of this expression with the right-hand member. Because $\mathbf{B}$ and $\mathbf{e}$ are unit vectors, we have $B_i e_i \leq 1$. Because $C_4(\mathbf{e}) \in \left[A_4^*, 0\right) \subset \mathbb{R}^-$ we have, by virtue of (2.5.37), $\sqrt{-6C_4(\mathbf{e})} \leq \sqrt{-6A_4^*}$, so that $B_i e_i\sqrt{-6C_4(\mathbf{e})} \leq \sqrt{-6A_4^*}$. On the left-hand side of (2.5.40), the term between the square brackets is positive and greater than 1 when $A_4(\mathbf{e})$ is positive, and greater than or equal to 2/3 for $A_4(\mathbf{e}) < 0$, so that we can write

$$(1-\lambda)^{3/2} \geq \mp\frac{3}{4}\mu\sqrt{-6A_4^*}. \tag{2.5.41}$$

Because the left-hand member is positive, we must choose the negative sign in (2.5.39) when $\mu < 0$ and the positive sign when $\mu > 0$. Hence, we may write

$$(1-\lambda)^{3/2} \geq \frac{3}{4}|\mu|\sqrt{-6A_4^*}. \tag{2.5.42}$$

Comparing this expression with (2.5.33), we find that λ^* yields the largest possible reduction of the critical load, and is thus a lower bound for the critical load.

Summarizing our results, we conclude that imperfections in the direction of the minimizing directions always cause the largest reduction of the critical load.† The results obtained so far are shown in Figure 2.5.2.

Notice that for imperfections in the direction of the minimizing direction, the critical load is obtained as a limit point, whereas for other imperfections the critical load $\lambda > \lambda^*$ may be obtained as either a bifurcation point or a limit point.

† This theorem was first proved by D. Ho (*Int. J. Sol. Struct.* **10**, 1315–1330, 1974) for cubic systems.

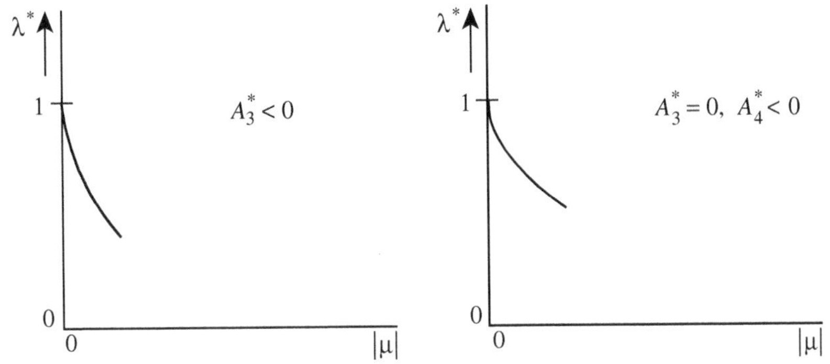

Figure 2.5.3

Let us finally consider the influence of the magnitude of the imperfections on the reduction of the critical load for imperfections in the direction of the minimizing direction. When $P_3\,[a_h\mathbf{u}_h;\lambda] \not\equiv 0$, the stability limit is given by (2.5.14), from which

$$\frac{d\lambda^*}{d\,|\mu|} = \frac{3A_3^*}{1-\lambda^*} \quad (<0)\,, \tag{2.5.43}$$

i.e., for $\lambda^* = 1$ the tangent to the $\lambda^* - |\mu|$ curve is vertical. Similarly, for the case that the cubic terms are identically equal to zero, we obtain from (2.5.33)

$$\frac{d\lambda^*}{d\,|\mu|} = -\frac{\sqrt{-6A_3^*}}{2\sqrt{1-\lambda^*}}, \tag{2.5.44}$$

which implies that also in this case the tangent is vertical for $\lambda^* = 1$.

This explains why very small imperfections may lead to a considerable reduction of the critical load.

2.6 On the determination of the energy functional for an elastic body

In the foregoing section we have given the general theory while assuming that the elastic energy functional is known. In this section, we shall determine the energy functional for a new class of problems.

Our line of thought is depicted in Figure 2.6.1:

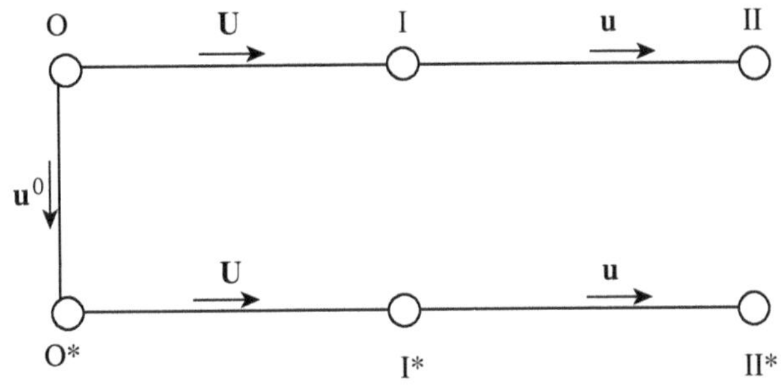

Figure 2.6.1

Let O be the undeformed (stress-free) state for the idealized (perfect) structure. The position of a material point in this configuration is given by the Cartesian components of the position vector $\mathbf{x}$. The position of this point in the fundamental state I (equilibrium configuration) is given by $\mathbf{x}+\mathbf{U}$, and the position in an adjacent state II (not necessarily an equilibrium configuration) by $\mathbf{x}+\mathbf{U}+\mathbf{u}$.

Let O^* be the undeformed (stress-free) state for the actual (imperfect) structure, obtained by superposition of a displacement field $\mathbf{u}^0$ on the state O, without introducing stresses. The position of a material point is now given by the Cartesian components of the position vector $\mathbf{x}+\mathbf{u}^0$, where $\mathbf{u}^0=\mathbf{u}^0(\mathbf{x})$. In the state I* (not an equilibrium configuration), the position of the material point is $\mathbf{x}+\mathbf{u}^0+\mathbf{U}$. In the adjacent state II*, the position is $\mathbf{x}+\mathbf{u}^0+\mathbf{U}+\mathbf{u}$, and we shall try to determine $\mathbf{u}$ so that the state II* is an equilibrium state.

In our discussion, we shall need the metric tensors in the various states,

$$g^0_{ij}=\delta_{ij} \tag{2.6.1}$$

$$g^{0*}_{ij}=\delta_{ij}+u^0_{i,j}+u^0_{j,i}+u^0_{h,i}u^0_{h,j} \tag{2.6.2}$$

$$g^{\mathrm{I}}_{ij}=\delta_{ij}+U_{i,j}+U_{j,i}+U_{h,i}U_{h,j}\equiv\delta_{ij}+2\Gamma_{ij} \tag{2.6.3}$$

$$\begin{aligned}g^{\mathrm{I}*}_{ij}&=\delta_{ij}+U_{i,j}+U_{j,i}+U_{h,i}U_{h,j}+u^0_{i,j}+u^0_{j,i}+u^0_{h,i}u^0_{h,j}\\&\quad+U_{h,i}u^0_{h,j}+U_{h,j}u^0_{h,i}\equiv\delta_{ij}+u^0_{i,j}+u^0_{j,i}+u^0_{h,i}u^0_{h,j}+2\Gamma^*_{ij}\end{aligned} \tag{2.6.4}$$

$$\begin{aligned}g^{\mathrm{II}}_{ij}&=\delta_{ij}+U_{i,j}+U_{j,i}+U_{h,i}U_{h,j}+u_{i,j}+u_{j,i}+u_{h,i}u_{h,j}\\&\quad+U_{h,i}u_{h,j}+U_{h,j}u_{h,i}\equiv\delta_{ij}+2\Gamma_{ij}+2\gamma_{ij}+U_{h,i}u_{h,j}+U_{h,j}u_{h,i}\end{aligned} \tag{2.6.5}$$

$$\begin{aligned}g^{\mathrm{II}*}_{ij}&=\delta_{ij}+U_{i,j}+U_{j,i}+U_{h,i}U_{h,j}+u_{i,j}+u_{j,i}+u_{h,i}u_{h,j}\\&\quad+u^0_{i,j}+u^0_{j,i}+u^0_{h,i}u^0_{h,j}+U_{h,i}u_{h,j}+U_{h,j}u_{h,i}\\&\quad+U_{h,i}u^0_{h,j}+U_{h,j}u^0_{h,i}+u_{h,i}u^0_{h,j}+u_{h,j}u_{h,i}\\&\equiv\delta_{ij}+2\Gamma^*_{ij}+2\gamma^*_{ij}+u^0_{i,j}+u^0_{j,i}+u^0_{h,i}u^0_{h,j}.\end{aligned} \tag{2.6.6}$$

For convenience, we note the relations

$$\begin{aligned}\Gamma_{ij}&=\frac{1}{2}(U_{i,j}+U_{j,i}+U_{h,i}U_{h,j})\\\Gamma^*_{ij}&=\Gamma_{ij}+\frac{1}{2}(U_{h,i}u^0_{h,j}+U_{h,j}u^0_{h,i})\\\gamma_{ij}&=\frac{1}{2}(u_{i,j}+u_{j,i}+u_{h,i}u_{h,j})\\\gamma^*_{ij}&=\gamma_{ij}+\frac{1}{2}\left(u_{h,i}u^0_{h,j}+u_{h,j}u^0_{h,i}\right)+\frac{1}{2}(U_{h,i}u_{h,j}+U_{h,j}u_{h,i}).\end{aligned} \tag{2.6.7}$$

Further, we introduce the following notation for the linearized strain tensors and the linearized rotation tensors:

$$\begin{aligned} \theta_{ij} &= \frac{1}{2}(u_{i,j} + u_{j,i}) \quad \Theta_{ij} = \frac{1}{2}(U_{i,j} + U_{j,i}) \\ \omega_{ij} &= \frac{1}{2}(u_{j,i} - u_{i,j}) \quad \Omega_{ij} = \frac{1}{2}(U_{j,i} - U_{i,j}). \end{aligned} \tag{2.6.8}$$

We shall now derive an approximate expression for the energy functional based on the following assumptions:

1. The elastic energy density follows the generalized Hooke's law.
2. The imperfections are (infinitesimally) small.
3. The displacements **u** are small (but finite).
4. The loads are dead-weight loads.
5. The fundamental state is (approximately) linear, i.e., Γ, Θ, and Ω are of the same order of magnitude.

Let us now first consider the idealized (perfect) structure.

The increment of the elastic energy density going from the fundamental state I to the adjacent state II is

$$\begin{aligned} W_{\mathrm{II}} - W_{\mathrm{I}} &= \frac{1}{2}E_{ijkl}\left[\Gamma_{ij} + \gamma_{ij} + \frac{1}{2}(U_{h,i}u_{h,j} + U_{h,j}u_{h,i})\right] \\ &\quad \times \left[\Gamma_{k\ell} + \gamma_{k\ell} + \frac{1}{2}(U_{m,k}u_{m,\ell} + U_{m,\ell}u_{m,k})\right] - \frac{1}{2}E_{ijk\ell}\Gamma_{ij}\Gamma_{k\ell} \\ &= E_{ijk\ell}\Gamma_{k\ell}\left[\gamma_{ij} + \frac{1}{2}(U_{h,i}u_{h,j} + U_{h,j}u_{h,i})\right] \\ &\quad + \frac{1}{2}E_{ijk\ell}\left[\gamma_{ij} + \frac{1}{2}(U_{h,i}u_{h,j} + U_{h,j}u_{h,i})\right]\left[\gamma_{k\ell} + \frac{1}{2}(U_{m,k}u_{m,\ell} + U_{m,\ell}u_{m,k})\right]. \end{aligned} \tag{2.6.9}$$

Because we are dealing with dead-weight loads, the terms linear in **u** are balanced by the energy of the external loads, so the potential energy is given by

$$\begin{aligned} P[\mathbf{u}] &= \int_V \left\{\frac{1}{2}E_{ijk\ell}\Gamma_{k\ell}u_{h,i}u_{h,j} + \frac{1}{2}E_{ijk\ell}\left[\gamma_{ij} + \frac{1}{2}(U_{h,i}u_{h,j} + U_{h,j}u_{h,i})\right]\right. \\ &\quad \left. \times \left[\gamma_{k\ell} + \frac{1}{2}(U_{m,k}u_{m,\ell} + U_{m,\ell}u_{m,k})\right]\right\} dV. \end{aligned} \tag{2.6.10}$$

This expression is fully exact for dead-weight loads and a material that follows the generalized Hooke's Law. To simplify this functional, we need estimates for the various terms. Using the relations

$$u_{h,i} = \theta_{ih} + \omega_{ih} \quad \text{and} \quad U_{h,i} = \Theta_{ih} + \Omega_{ih}, \tag{2.6.11}$$

we have the following estimates:

$$\gamma_{ij} = O(\theta, \omega^2) \quad \frac{1}{2}(U_{h,i}u_{h,j} + U_{h,j}u_{h,i}) = O(\Omega\omega, \Gamma\theta) = O(\Gamma\omega) \tag{2.6.12}$$

where

$$\theta = |\theta_{ij}|_{\max}, \quad \omega = |\omega_{ij}|_{\max}, \quad \text{and so on.} \tag{2.6.13}$$

Notice that for the estimate of the second term in (2.6.12), we have made use of the fact that the fundamental state is linear. We now have the following estimates:

$$\begin{aligned}
\gamma_{ij}\gamma_{k\ell} &= O(\theta^2, \theta\omega^2, \omega^4)\gamma_{ij}\frac{1}{2}(U_{h,i}u_{h,j} + U_{h,j}u_{h,i}) \\
&= O(\theta\Gamma\omega, \Gamma\omega^3)\frac{1}{4}(U_{h,i}u_{h,j} + U_{h,j}u_{h,i})(U_{m,k}u_{m,\ell} + U_{m,\ell}u_{m,k}) \quad (2.6.14)\\
&= O(\Gamma^2\omega^2)\frac{1}{2}\Gamma_{kl}u_{h,i}u_{h,j} = O(\Gamma\omega^2).
\end{aligned}$$

Because buckling in the elastic range only occurs in slender structures, we have $\omega \gg \theta$.

Now consider the linear terms in (2.6.9),

$$E_{ijk\ell}\Gamma_{k\ell}\,(u_{i,j} + U_{h,i}u_{h,j}) = E_{ijk\ell}\Gamma_{k\ell}\,(\delta_{ih} + U_{h,i})\,u_{h,j}\,.$$

Because in the energy functional these terms are balanced by the energy of the external loads, we find for the equilibrium equations

$$[E_{ijk\ell}\Gamma_{k\ell}(\delta_{hi} + U_{h,i})]_{,j} + \rho X_h = 0. \tag{2.6.15}$$

Because $U_{h,i} = O(\Gamma)$, we can say that

$$E_{ijk\ell}\Gamma_{k\ell} = S_{ij} \tag{2.6.16}$$

is proportional to the external loads. Here S_{ij} is the symmetric stress tensor of KIRCHHOFF-TREFFTZ.† It follows that we may write

$$S_{ij} = \lambda S_{ij}^{(1)}, \tag{2.6.17}$$

where $S_{ij}^{(1)}$ is the stress tensor for a unit load. Further, we shall normalize λ so that at the critical load, $\lambda = 1$.

The first term in (2.6.10) is now proportional to $\Gamma\omega^2$, so that of the other terms in (2.6.14), the term of $O(\theta^2)$ may be comparable. Competition of the terms of $O(\Gamma\omega^2)$ and $O(\theta^2)$ is only possible when $\theta = O(\omega\sqrt{\Gamma})$. It follows that the term $O(\theta\,\Gamma\omega)$ is of order $O(\Gamma^{3/2}\omega^2)$ and may thus be neglected compared to $O(\Gamma\omega^2)$. Terms of $O(\Gamma\omega^3)$ and $O(\Gamma^2\omega^2)$ in (2.6.14) may also be neglected. This means that the energy functional (2.6.10) can be simplified to yield

$$P[\mathbf{u}] = \int_V \left[\frac{1}{2}\lambda S_{ij}^{(1)} u_{h,i}u_{h,j} + \frac{1}{2}E_{ijk\ell}\gamma_{ij}\gamma_{k\ell}\right] dV \tag{2.6.18}$$

with a relative error of $O(\sqrt{\Gamma})$. Notice that this expression is linear in the load factor λ.

† Notice that this is not the same stress tensor as the one introduced in (2.2.7).

Let us now consider the actual (imperfect) structure. The increment in the elastic energy density going from state I* to state II* is

$$\begin{aligned}W^*_{\mathrm{II}^*} - W^*_{\mathrm{I}^*} &= \frac{1}{2}E^*_{ijk\ell}(\Gamma^*_{ij} + \gamma^*_{ij})(\Gamma^*_{k\ell} + \gamma^*_{k\ell}) - \frac{1}{2}E^*_{ijk\ell}\Gamma^*_{ij}\Gamma^*_{k\ell} \\ &= E^*_{ijk\ell}\Gamma^*_{k\ell}\gamma^*_{ij} + \frac{1}{2}E^*_{ijk\ell}\gamma^*_{ij}\gamma^*_{k\ell} = E^*_{ijk\ell}\left[\Gamma_{k\ell} + \frac{1}{2}\left(U_{m,k}u^0_{m,\ell} + U_{m,k}u^0_{m,\ell}\right)\right] \\ &\times\left[\gamma_{ij} + \frac{1}{2}\left(u_{h,i}u^0_{h,j} + u_{h,j}u^0_{h,i}\right) + \frac{1}{2}\left(U_{h,i}u_{h,j} + U_{h,j}u_{h,i}\right)\right] \\ &+\frac{1}{2}E^*_{ijk\ell}\left[\gamma_{ij} + \frac{1}{2}\left(u_{h,i}u^0_{h,j} + u_{h,j}u^0_{h,i}\right) + \frac{1}{2}\left(U_{h,i}u_{h,j} + U_{h,j}u_{h,i}\right)\right] \\ &\times\left[\gamma_{k\ell} + \frac{1}{2}\left(u_{m,k}u^0_{m,\ell} + u_{m,l}u^0_{m,k}\right) + \frac{1}{2}\left(U_{m,k}u_{m,\ell} + U_{m,\ell}u_{m,k}\right)\right],\end{aligned} \tag{2.6.19}$$

where $E^*_{ijk\ell}$ is the tensor of elastic moduli in O^*,

$$E^*_{ijk\ell} = E_{ijk\ell} + O(\gamma^0 E), \tag{2.6.20}$$

so we may replace $E^*_{ijk\ell}$ by $E_{ijk\ell}$. For our discussion of the terms with the imperfections, we shall need the following expressions,

$$\theta^0 = \frac{1}{2}\left|u^0_{i,j} + u^0_{j,i}\right|_{\max} \quad \text{and} \quad \omega^0 = \frac{1}{2}\left|u^0_{j,i} - u^0_{i,j}\right|_{\max}. \tag{2.6.21}$$

We now have the following estimates,

$$\begin{aligned}\Gamma_{kl}(u_{h,i}u^0_{h,j} + u_{h,j}u^0_{h,i}) &= O(\Gamma\omega\omega^0, \Gamma\theta\omega^0, \Gamma\omega\theta^0, \Gamma\theta\theta^0) \\ \gamma_{ij}(u^0_{h,i}u_{h,j} + u^0_{h,j}u_{h,i}) &= O(\theta\omega^0\omega, \omega^3\omega^0) \\ \gamma_{ij}(U_{m,k}u^0_{m,\ell} + U_{m,\ell}u^0_{m,k}) &= O(\theta\Gamma\omega^0, \omega^2\Gamma\omega^0).\end{aligned} \tag{2.6.22}$$

Further, (2.6.19) contains terms that are quadratic in the imperfections, which can be neglected. At the critical load, we have $\omega \gg \theta$. We know that Γ is small but finite, and we can always choose deviations from the fundamental state as small as we need so that $\theta \ll \Gamma$. The leading term in (2.6.22) is of order $\Gamma\omega\omega^0$. The only other term that may be important is the term $O(\Gamma\omega\theta^0)$, and all other terms are small compared to these terms. Using (2.6.20) and (2.6.17) and the fact that we have dead-weight loads, we can simplify the energy functional corresponding to (2.6.19) to yield

$$P^*[\mathbf{u}] = P[\mathbf{u}] + \int_V \lambda S^{(1)}_{ij} u^0_{h,i}u_{h,j}\, dV, \tag{2.6.23}$$

where $P[\mathbf{u}]$ is the functional for the perfect structure defined in (2.6.18). Notice that the term with the imperfections depends linearly on the load factor λ.

Let us now consider the effect of eccentric loads. In this case, only the functional of the external loads must be modified. Let $\mathbf{x}$ denote the position of a material point, and let $\rho\mathbf{X}$ be the force per unit volume at this point. The potential energy of the dead-weight load is then $-\rho\mathbf{X}\cdot\mathbf{u}(\mathbf{x})$, where $\mathbf{u}$ denotes the displacement of the point

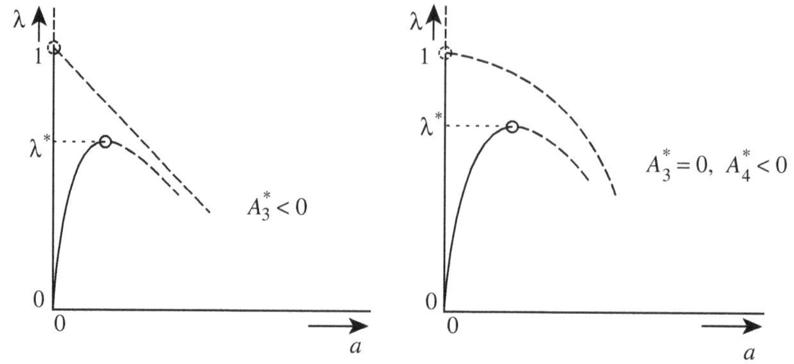

Figure 2.6.2

in question. When the load is applied eccentrically, say, in a point $\mathbf{y}$ close to $\mathbf{x}$, the potential energy becomes

$$-\rho\mathbf{X}\cdot\mathbf{u}(\mathbf{y}) = -\rho\mathbf{X}\cdot[\mathbf{u}(\mathbf{x}) + (y_i - x_i)\,\mathbf{u}_{,i}(\mathbf{x}) + \cdots], \tag{2.6.24}$$

where $y_i - x_i$ determines the eccentricity of the loads. To a first approximation, the correction of the functional of the external loads is linear in the eccentricity of the loads and linear in the loads. It follows that for both the case of imperfection and the case of eccentric loads, the additional term in the energy functional is of the form $\lambda Q[\mathbf{u}]$.

It is now convenient to slightly modify our function $F^*(a_i, \lambda)$,

$$F^*(a_i,\lambda) = (1-\lambda)\,a_i a_i + \begin{cases} A_{ijk}a_i a_j a_k \\ A_{ijk\ell}a_i a_j a_k a_\ell \end{cases} + \lambda\mu B_i a_i \tag{2.6.25}$$

so that it explicitly shows the linear dependence on the load factor in the last term. This expression is valid for $\lambda \geq 0$, e.g., the equilibrium equations, when $P_3(a_h\mathbf{u}_h, \lambda) \not\equiv 0$, now read

$$2(1-\lambda)\,a_i + 3A_{ijk}a_j a_k + \lambda\mu B_i = 0. \tag{2.6.26}$$

For $\lambda = 0$, the solution is $a_i = 0$, and for $\lambda \ll 1$, we have

$$a_i = -\frac{\lambda\mu B_i}{2(1-\lambda)}. \tag{2.6.27}$$

Our figures relating λ to a now are shown in Figure 2.6.2. The accuracy of these curves increases with decreasing values of λ.

In the case that we only have to deal with geometrical imperfections, we can slightly modify the functional (2.6.25). To this end, the imperfections are represented by

$$\mathbf{u}^0 = a_h^0\mathbf{u}_h + \overline{\mathbf{u}}^0 \tag{2.6.28}$$

and the displacement field is written as

$$\mathbf{u} = a_h\mathbf{u}_h + \overline{\mathbf{u}}. \tag{2.6.29}$$

We now return to (2.6.23), which fully written out, reads

$$P^*[\mathbf{u}] = \int_V \left[\lambda S_{ij}^{(1)}\left(\frac{1}{2}u_{h,i}u_{h,j} + \overset{0}{u}_{h,i}u_{h,j}\right) + \frac{1}{2}E_{ijk\ell}\gamma_{ij}\gamma_{k\ell}\right] dV. \quad (2.6.30)$$

The second variation is

$$P_2[\mathbf{u}, \lambda] = \int_V \left[\lambda S_{ij}^{(1)}\frac{1}{2}u_{h,i}u_{h,j} + \frac{1}{2}E_{ijk\ell}\theta_{ij}\theta_{k\ell}\right] dV. \quad (2.6.31)$$

Expanding (2.6.31) with respect to λ about $\lambda = 1$, and using the fact that $P_2[\mathbf{u}, 1] = 0$, we obtain

$$P_2[\mathbf{u}, \lambda] = (1-\lambda)\int_V S_{ij}^{(1)}\frac{1}{2}u_{h,i}u_{h,j}\, dV + \cdots \equiv (1-\lambda)P_2'[\mathbf{u}, 1] + \cdots \quad (2.6.32)$$

Substitution of (2.6.29) into (2.6.32) yields for the integral

$$\int_V \frac{1}{2}S_{ij}^{(1)}\left[a_k a_l u_{kh,i}u_{\ell h,i} + a_k u_{kh,i}\overline{u}_{h,j} + a_l u_{\ell h,j}\overline{u}_{h,i} + \overline{u}_{h,i}\overline{u}_{h,j}\right] dV. \quad (2.6.33)$$

The orthogonality of $\overline{\mathbf{u}}$ with respect to the buckling modes is now defined by

$$\int_V \frac{1}{2}S_{ij}^{(1)}a_k u_{kh,i}\overline{u}_{h,j}\, dV = 0. \quad (2.6.34)$$

The last term in (2.6.33) may be neglected because it is small compared to the first term. The buckling modes are now normalized so that

$$\int_V \frac{1}{2}S_{ij}^{(1)}a_k a_l u_{kh,i}u_{\ell h,i}\, dV = a_k a_l \int_V S_{ij}^{(1)}u_{kh,i}u_{\ell h,j}\, dV = a_k a_l \delta_{k\ell} = a_k a_k. \quad (2.6.35)$$

Using (2.6.28) and (2.6.29), we find for the term with the imperfections in (2.6.30),

$$\int_V \lambda S_{ij}^{(1)}\left[\overset{0}{a}_k a_l u_{kh,i}u_{\ell h,j} + \overset{0}{a}_k u_{kh,i}\overline{u}_{h,j} + a_\ell u_{\ell h,j}\overline{\overset{0}{u}}_{h,i} + \overline{\overset{0}{u}}_{h,i}\overline{\overset{0}{u}}_{h,j}\right] dV. \quad (2.6.36)$$

The second term vanishes by virtue of (2.6.34), and the third term vanishes by virtue of the orthogonality condition,

$$\int_V S_{ij}^{(1)}a_\ell u_{\ell h}\overline{\overset{0}{u}}_{h,i}\, dV = 0. \quad (2.6.37)$$

The last term in (2.6.36) may be neglected because it is small compared to the first term, so we have

$$\lambda \overset{0}{a}_k a_l \int_V S_{ij}^{(1)}u_{kh,i}u_{\ell h,j}\, dV = 2\lambda \overset{0}{a}_k a_l \delta_{k\ell} = 2\lambda \overset{0}{a}_k a_k, \quad (2.6.38)$$

where we have made use of (2.6.35).

In (2.6.25), the influence of the imperfections is accounted for by the term $\lambda\mu B_i a_i$, where $\mu < 0$, and is equivalent to (2.6.38). Hence, we may modify (2.6.25) to yield

$$F^*(a_h, \lambda) = (1 - \lambda)\, a_i a_i + \begin{cases} A_{ijk} a_i a_j a_k \\ A_{ijk\ell} a_i a_j a_k a_\ell \end{cases} - 2\lambda a_i^0 a_i. \tag{2.6.39}$$

It is this form of $F^*(a_h, \lambda)$ that is most suitable for application.

3

Applications

3.1 The incompressible bar (the problem of the elastica)

We consider a simply supported incompressible bar loaded by compressive forces N at its ends, as shown in Figure 3.1.1. Its length is denoted by ℓ.

Let φ denote the angle between the horizontal and the tangent at a point of the deflected bar. The following relations then hold:

$$\begin{aligned} \sin\varphi &= \frac{dw}{dx} = w' \\ \cos\varphi \cdot \varphi' &= w'' \\ \varphi' &= \frac{w''}{\cos\varphi} = \frac{w''}{\sqrt{1-w'^2}}, \end{aligned} \tag{3.1.1}$$

where φ' denotes the curvature of the bar at the corresponding point. The displacement of the moveable end is now given by

$$-\Delta\ell = \ell - \int_0^\ell \cos\varphi\, dx = \int_0^\ell \left[1 - \sqrt{1-w'^2}\right] dx. \tag{3.1.2}$$

Here we have made use of the fact that the bar is incompressible, i.e., inextensional.

An explicit condition for incompressibility reads

$$[(1+u')^2 + w'^2](dx)^2 = (dx)^2 \tag{3.1.3}$$

or

$$u' = \sqrt{1-w'^2} - 1. \tag{3.1.4}$$

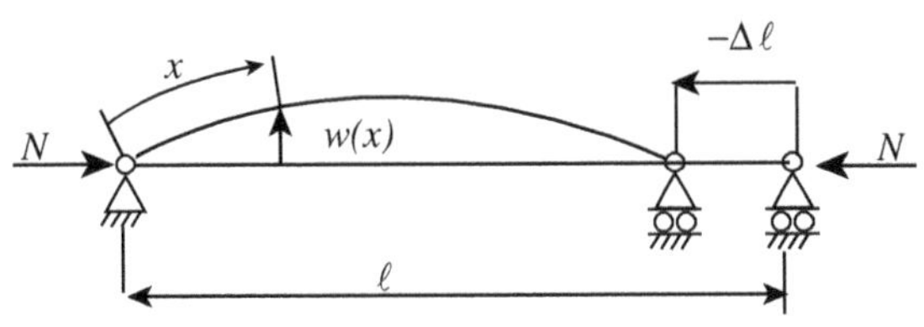

Figure 3.1.1

The potential energy of this system is now given by

$$P[w;N] = \int_0^{\ell} \frac{1}{2} EI\varphi'^2 \, dx + N\Delta\ell = \int_0^{\ell} \left[\frac{1}{2} EJ \frac{w''^2}{1 - w'^2} + N\left(\sqrt{1 - w'^2} - 1\right) \right] dx. \tag{3.1.5}$$

The expression (3.1.5) was already given in (2.2.25), and it was discussed there that the potential energy of the external load is a nonlinear function of the displacement due to the incompressibility condition imposed.

Using the expansion

$$\begin{aligned} \frac{1}{1 - w'^2} &= 1 + w'^2 + w'^4 + \cdots \\ \sqrt{1 - w'^2} &= 1 - \frac{1}{2} w'^2 - \frac{1}{8} w'^4 - \cdots, \end{aligned} \tag{3.1.6a}$$

we may rewrite (3.1.5) to yield

$$P[w;N] = \int_0^{\ell} \left[\frac{1}{2} EIw''^2 (1 + w'^2 + \cdots) - N\left(\frac{1}{2} w'^2 + \frac{1}{8} w'^4 + \cdots \right) \right] dx, \tag{3.1.6b}$$

and hence we have

$$P_2[w;N] = \int_0^{\ell} \left[\frac{1}{2} EIw''^2 - \frac{1}{2} Nw'^2 \right] dx \tag{3.1.7}$$

$$P_3[w;N] = 0 \tag{3.1.8}$$

$$P_4[w;N] = \int_0^{\ell} \left[\frac{1}{2} EIw''^2 w'^2 - \frac{1}{8} Nw'^4 \right] dx. \tag{3.1.9}$$

The variational equation for neutral equilibrium now follows from (3.1.7),

$$P_{11}[w, \zeta; N] = \int_0^{\ell} [EIw''\zeta'' - Nw'\zeta'] \, dx = 0. \tag{3.1.10}$$

Integration by parts yields

$$EIw''\zeta' \Big|_0^{\ell} - (EIw''' + Nw')\,\zeta \Big|_0^{\ell} + \int_0^{\ell} (EIw'''' + Nw'')\,\zeta dx = 0. \tag{3.1.11}$$

Because this equation must hold for all kinematically admissible displacement fields, we have

$$EIw'''' + Nw'' = 0 \tag{3.1.12}$$

$$EIw''(\ell) = EIw''(0) = 0, \tag{3.1.13}$$

while the kinematic conditions at the supports lead to the requirement

$$w(\ell) = w(0) = 0. \tag{3.1.14}$$

The general solution to (3.1.12) reads

$$w = c_1 + c_2 x + c_3 \cos \alpha x + c_4 \sin \alpha x, \tag{3.1.15}$$

where we have introduced the parameter

$$\alpha^2 = \frac{N}{EI}. \tag{3.1.16}$$

The boundary conditions (3.1.13) and (3.1.14) admit a nonzero value for c_4 only. This value exists if and only if

$$\sin \alpha \ell = 0, \tag{3.1.17}$$

so $\alpha\ell = k\pi$ for $k = 1, 2, \ldots$. The constant c_4 remains undetermined. The critical load now follows from the smallest root,

$$N_1 = \frac{\pi^2 EI}{\ell^2}. \tag{3.1.18}$$

We shall now investigate whether this equilibrium state is stable. The necessary condition for stability $P_3[w; N] = 0$ is satisfied. The displacement field for small deflections from the equilibrium configuration is now written as

$$w(x) = a w_1(x) + \overline{w}(x), \tag{3.1.19}$$

where $w_1(x)$ is the buckling mode and $\overline{w}(x)$ is the orthogonal remainder. We must now consider the minimum problem, as per (2.3.44),

$$\underset{T_{11}[\mathbf{w}, \overline{\mathbf{w}}]=0}{\text{Min}} \; P_2[\overline{w}] + P_{21}[w_1, \overline{w}] \tag{3.1.20}$$

but the last term vanishes because $P_3[w; N] \equiv 0$. Conseqently, the quantity that governs stability in the critical state is $P_4[w_1; N_1] = 0$ (see 2.3.54). Taking $w_1(x) = 1 \cdot \sin \pi x/\ell$, we find

$$\begin{aligned} P_4[w_1; N_1] &= \int_0^\ell \left[\frac{1}{2} EI \frac{\pi^6}{\ell^6} \cos^2 \frac{\pi x}{\ell} \sin^2 \frac{\pi x}{\ell} - \frac{1}{8} N_1 \frac{\pi^4}{\ell^4} \cos^4 \frac{\pi x}{\ell} \right] dx \\ &= \frac{1}{2} N_1 \frac{\pi^4}{\ell^4} \int_0^\ell \left[\frac{1}{4} \sin^2 \frac{2\pi x}{\ell} - \frac{1}{4} \cdot \frac{1}{4} \left(1 + \cos^2 \frac{2\pi x}{\ell} \right)^2 \right] dx = \frac{1}{64} \frac{\pi^4}{\ell^4} N_1, \end{aligned} \tag{3.1.21}$$

which is positive, and hence equilibrium at the buckling load is stable.

Let us now consider the behavior of the structure for loads slightly exceeding N_1. The increment of the second variation is

$$P_2[w_1; N] - P_2[w_1; N_1] = \frac{1}{2} \int_0^\ell (N_1 - N) w_1'^2 dx = \frac{\pi^4}{4\ell^2} (N_1 - N). \tag{3.1.22}$$

The function $F(a;N)$ is now given by

$$F(a;N) = \frac{\pi^4}{4\ell^2}(N_1 - N)\,a^2 + \frac{1}{64}\frac{\pi^4}{\ell^4}N_1 a^4 = \frac{\pi^2 N_1}{4\ell^2}\left[(1-\lambda)\,a^2 + \frac{\pi^2}{16\ell^2}a^4\right], \quad (3.1.23)$$

where $\lambda = N/N_1$. The equilibrium condition follows from

$$\frac{\partial F}{\partial a} = 0 = \frac{\pi^2 N_1}{2\ell^2}a\left[(1-\lambda) + \frac{\pi^2}{8\ell^2}a^2\right]. \quad (3.1.24)$$

Besides the fundamental solution $a = 0$, (3.1.23) has the solution

$$a = \frac{2\sqrt{2}}{\pi}\ell\sqrt{\lambda - 1}.^{\dagger} \quad (3.1.25)$$

Let us now consider the displacement $-\Delta\ell$,

$$-\Delta\ell = \frac{1}{2}\int_0^\ell \left[1 - \sqrt{1 - w'^2}\right]dx \approx \int_0^\ell \frac{1}{2}w'^2\,dx$$
$$= \frac{1}{2}a^2\int_0^\ell \frac{\pi^2}{\ell^2}\cos^2\frac{\pi x}{\ell}\,dx = \frac{\pi^2 a^2}{\ell^2} = 2\ell\,(\lambda - 1), \quad (3.1.26)$$

where we have made use of (3.1.25). The effective relative shortening is

$$-\frac{\Delta\ell}{\ell} = 2\,(\lambda - 1) = 2\left(\frac{N}{N_1} - 1\right). \quad (3.1.27)$$

Further, we have

$$\frac{\partial}{\partial N}\left(-\frac{\Delta\ell}{\ell}\right) = \frac{2}{N_1} = \frac{2\ell^2}{\pi^2 EI} = \frac{1}{EA}\frac{2\ell^2}{\pi^2 i^2}, \quad (3.1.28)$$

where $i = \sqrt{I/A}$ is the radius of gyration of the cross section. Our results are now plotted in Figure 3.1.2.

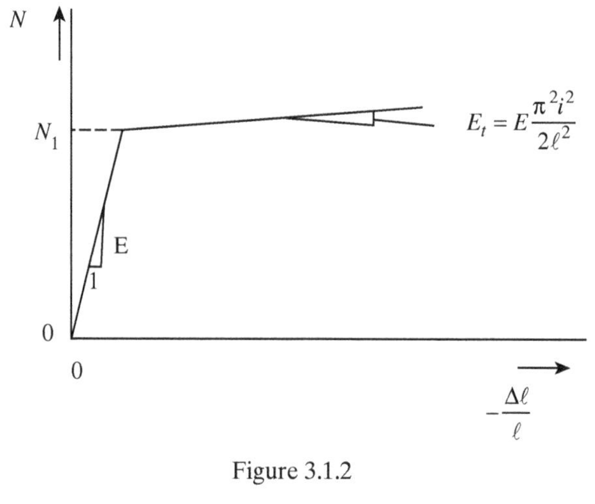

Figure 3.1.2

† $\lambda = 1.01$ already gives a deflection $a \approx 0.1\ell$.

It should be noted that these results are only valid for small deflections. Comparison with the exact solution shows that the results are valid up to deflections of the middle of the bar of about 20% of its length.†

3.2 Bar with variable cross section and variable load distribution

Consider a bar with variable cross section loaded in compression by a normal force $N(x)$ and a load distribution $P(x)$, clamped at one end (see Figure 3.2.1). The bar is considered to be incompressible.

Consider an infinitesimal part of the bar in the deflected configuration (see Figure 3.2.2).

Due to the deflection of the bar, the load $P(x)\,dx$ will move over a distance $\int_0^x dx\,[1-\cos\varphi(x)]$, and hence the contribution to the potential energy is given by

$$\int_0^\ell -P(x)\,dx \int_0^x [1-\cos\varphi(\xi)]\,d\xi,$$

so the total potential energy is given by

$$\begin{aligned} P[\varphi(x);\lambda] = &\int_0^\ell \frac{1}{2}B(x)\varphi'^2(x)\,dx + \lambda\int_0^\ell P(x)\left\{\int_0^x [\cos\varphi(\xi)-1]\,d\xi\right\}dx \\ &+\lambda N\int_0^\ell [\cos\varphi(x)-1]\,dx, \end{aligned} \tag{3.2.1}$$

where $B(x)$ denotes the bending stiffness. The normal force in the bar is given by

$$N(x) = N + \int_x^\ell P(\xi)\,d\xi, \tag{3.2.2}$$

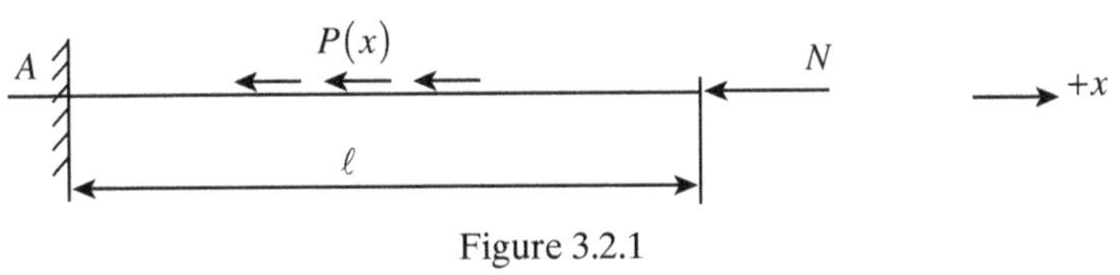

Figure 3.2.1

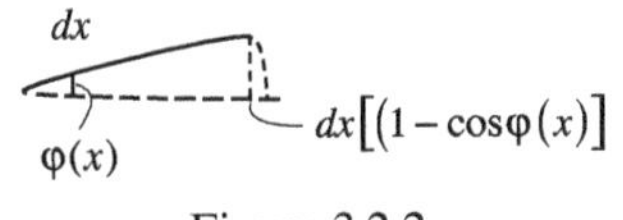

Figure 3.2.2

† Cf. W. T. Koiter, *Stability of Elastic Equilibrium*, Sect. 6 (Thesis, Delft).

and hence we have $N'(x) = -P(x)$. Using this relation to rewrite the second term in (3.2.1), we find

$$\int_0^\ell -N'(x) \int_0^x [\cos\varphi(\xi) - 1]\, d\xi\, dx$$

$$= -N(x) \int_0^x [\cos\varphi(\xi) - 1]\, d\xi \Big|_0^\ell + \int_0^\ell N(x)[\cos\varphi(x) - 1]\, dx$$

$$= N(\ell) \int_0^\ell [\cos\varphi(x) - 1]\, dx + \int_0^\ell N(x)[\cos\varphi(x) - 1]\, dx,$$

so that (3.2.1) can be rewritten to yield

$$P[\varphi(x);\lambda] = \int_0^\ell \left\{ \frac{1}{2} B(x)\varphi'^2(x) + \lambda N(x)[\cos\varphi(x) - 1] \right\} dx. \tag{3.2.3}$$

The second, third, and fourth variations are given by

$$P_2[\varphi(x);\lambda] = \int_0^\ell \left[\frac{1}{2} B(x)\varphi'^2(x) - \frac{1}{2}\lambda N(x)\varphi^2(x) \right] dx \tag{3.2.4}$$

$$P_3[\varphi(x);\lambda] \equiv 0 \tag{3.2.5}$$

$$P_4[\varphi(x);\lambda] = \int_0^\ell \frac{1}{24}\lambda N(x)\varphi^4(x)\, dx. \tag{3.2.6}$$

Because $P_3[\varphi(x);\lambda] \equiv 0$ at the critical load factor, and where $P_2[\varphi(x);\lambda_1]$ is semi-definite-positive, stability is governed by the sign of $P_4[\varphi(x);\lambda_1]$. The condition $P_{11}[\varphi(x), \psi(x);\lambda_1] = 0$ yields

$$P_{11}[\varphi(x), \psi(x);\lambda] = \int_0^\ell [B(x)\varphi'(x)\psi'(x) - \lambda N(x)\varphi(x)\psi(x)]\, dx = 0, \tag{3.2.7}$$

where $\psi(x)$ is a kinematically admissible rotation field.

Integrating by parts, we obtain

$$B(x)\varphi'(x)\psi(x)\big|_0^\ell - \int_0^\ell \{[B(x)\varphi'(x)]' + \lambda N(x)\varphi(x)\}\psi(x)\, dx = 0. \tag{3.2.8}$$

The first term vanishes when the ends of the bar are hinged or clamped. From the second term, we obtain

$$[B(x)\varphi'(x)]' + \lambda N(x)\varphi(x) = 0. \tag{3.2.9}$$

The boundary conditions are

$$\begin{cases} \varphi(0) = 0 & \text{(when } A \text{ is clamped)} \\ B(0)\varphi'(0) = 0 & \text{(when } A \text{ is hinged)} \end{cases}$$
$$B(\ell)\varphi'(\ell) = 0. \tag{3.2.10}$$

From the equation for neutral equilibrium (3.2.9) we obtain at the buckling load

$$\lambda_1 N(x)\,\varphi_1(x) = -\left[B(x)\,\varphi_1'(x]\right)', \tag{3.2.11}$$

so that

$$P_4[\varphi_1(x);\lambda_1] = \int_0^\ell -\frac{1}{24}\,[B(x)\,\varphi_1'(x)]'\,\varphi_1^3(x)\,dx$$
$$= -\frac{1}{24}B(x)\,\varphi_1'(x)\,\varphi_1^3(x)\Big|_0^\ell + \int_0^\ell \frac{1}{24}B(x)\,\varphi_1'(x)\cdot 3\varphi_1^2(x)\,\varphi_1'(x)\,dx.$$

The first term vanishes when the ends are either clamped or hinged, so

$$P_4[\varphi_1(x);\lambda_1] = \frac{1}{8}\int_0^\ell B(x)\varphi_1'^2(x)\,\varphi_1^2(x)\,dx > 0, \tag{3.2.12}$$

and hence the equilibrium at the buckling mode is stable. Notice that $N(x)$ may also be a locally tensile force. In the case that A is an elastic hinge, we must add an additional term to the potential energy given by $\frac{1}{2}C\varphi^2(0)$. The boundary condition for $x = 0$ then becomes

$$B(0)\,\varphi'(0) = C\varphi(0) \tag{3.2.13}$$

and in the expression for $P_4[\varphi_1(x);\lambda_1]$, from the stock term we now have a contribution at $x = 0$,

$$\frac{1}{24}B(0)\,\varphi_1'(0)\,\varphi_1^3(0) > 0, \tag{3.2.14}$$

so that for a elastically hinged bar the equilibrium at buckling is also stable.

3.3 The elastically supported beam†

Consider a uniform elastically supported hinged-hinged beam loaded in compression by a normal force N (see Figure 3.3.1).

† Cf., J. G. LEKKERKERKER, *On the stability of an elastically supported beam*... (*Proc. Kon. Ned. Ak. Wet.* B**65**, no. 2, 190–197 (1962).

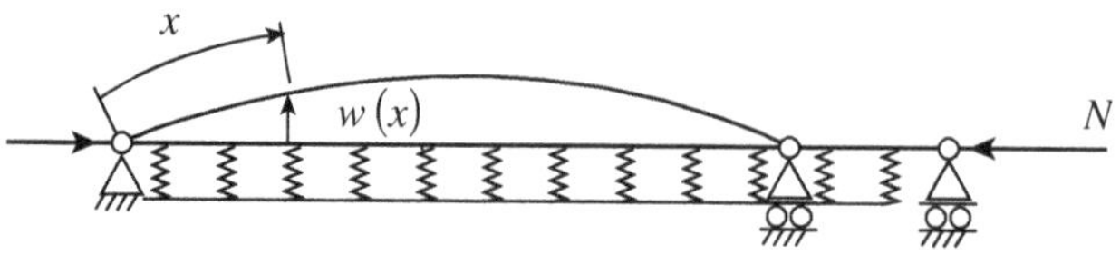

Figure 3.3.1

The reaction of the springs per unit length of the beam is cw, and the corresponding energy is $\frac{1}{2}cw^2(x)$. The potential energy of this system is now given by

$$P[w(x);N] = \int_0^\ell \left[\frac{1}{2}B\frac{w''^2}{1-w'^2} + \frac{1}{2}cw^2 + N\left(\sqrt{1-w'^2}-1\right)\right]dx, \tag{3.3.1}$$

and hence

$$P_2[w;N] = \int_0^\ell \left(\frac{1}{2}Bw''^2 + \frac{1}{2}cw^2 - \frac{1}{2}Nw'^2\right)dx \tag{3.3.2}$$

$$P_3[w;N] \equiv 0 \tag{3.3.3}$$

$$P_4[w;N] = \int_0^\ell \left(\frac{1}{2}Bw''^2w'^2 - \frac{1}{8}Nw'^4\right)dx. \tag{3.3.4}$$

The variational equation for neutral equilibrium follows from (3.3.2),

$$P_{11}[w,\zeta;N] = \int_0^\ell (Bw''^2\zeta'' + cw\zeta - Nw'\zeta')\,dx = 0. \tag{3.3.5}$$

From integration by parts, we obtain

$$\begin{aligned} &Bw''\zeta'\big|_0^\ell + \int_0^\ell (-Bw'''\zeta' + cw\zeta + Nw''\zeta)\ dx - Nw'\zeta\big|_0^\ell \\ &= Bw''\zeta'\big|_0^\ell - Bw'''\zeta\big|_0^\ell - Nw'\zeta\big|_0^\ell + \int_0^\ell (Bw'''' + Nw'' + cw)\,\zeta dx = 0, \end{aligned} \tag{3.3.6}$$

from which follows

$$Bw'''' + Nw'' + cw = 0, \tag{3.3.7}$$

with the dynamic boundary conditions

$$Bw''(\ell) = Bw''(0) = 0. \tag{3.3.8}$$

Further, we have the kinematic boundary conditions

$$w(0) = w(\ell) = 0. \tag{3.3.9}$$

The deflection $w(x)$ is now written as a Fourier sine series,

$$w(x) = \sum_{k=1}^{\infty} a_k \sin \frac{k\pi x}{\ell}. \tag{3.3.10}$$

Notice that each term in this series satisfies the kinematic boundary conditions. To determine the stability limit, we might substitute this series into (3.3.7); however, then we need the fourth derivative of this series, and it is unknown beforehand whether the series obtained is convergent. It is therefore more suitable to return to the second variation as this only requires the second derivative of this series, which is expected to be convergent in view of the fact that the dynamic boundary conditions also contain a second derivative.

Substitution of (3.3.10) into (3.3.2) yields

$$P_2[w;N] = \int_0^{\ell} \left[\frac{1}{2} B \left(\sum_{k=1}^{\infty} -\frac{k^2\pi^2}{\ell^2} a_k \sin \frac{k\pi x}{\ell} \right)^2 + \frac{1}{2} c \left(\sum_{k=1}^{\infty} a_k \sin \frac{k\pi x}{\ell} \right)^2 - \frac{1}{2} N \left(\sum_{k=1}^{\infty} \frac{k\pi}{\ell} a_k \cos \frac{k\pi x}{\ell} \right)^2 \right] dx = \frac{\ell}{4} \sum_{k=1}^{\infty} \left[B \frac{k^4\pi^4}{\ell^4} + c - N \frac{k^2\pi^2}{\ell^2} \right] a_k^2. \tag{3.3.11}$$

This expression is positive when all its coefficients are positive. The coefficient of a_k becomes zero for

$$N_1 = \frac{k^2\pi^2}{\ell^2} B + \frac{\ell^2}{k^2\pi^2} C. \tag{3.3.12}$$

This is the stability limit. To obtain the lowest (critical) value of N, we must minimize (3.3.12) with respect to k,

$$\frac{dN_1}{dk} = 2\frac{k\pi}{\ell^2} B - \frac{2\ell^2}{k^3\pi^2} C = 0, \tag{3.3.13}$$

from which

$$\frac{k\pi}{\ell} = \left(\frac{C}{B} \right)^{1/4}. \tag{3.3.14}$$

However, here we have assumed that $k \in \mathbb{R}^+$, but $k \in \mathbb{N}^+$, so k cannot generally take on the value in (3.3.14). For $k \in \mathbb{R}^+$, we have from (3.3.14) and (3.3.12)

$$N_{\text{Min}} = 2\sqrt{BC}. \tag{3.3.15}$$

It is now convenient to plot $N/2\sqrt{BC}$ versus $(C/B)^{1/4}\,\ell/\pi$ for various values of $k \in \mathbb{N}^+$ (see Figure 3.3.2).

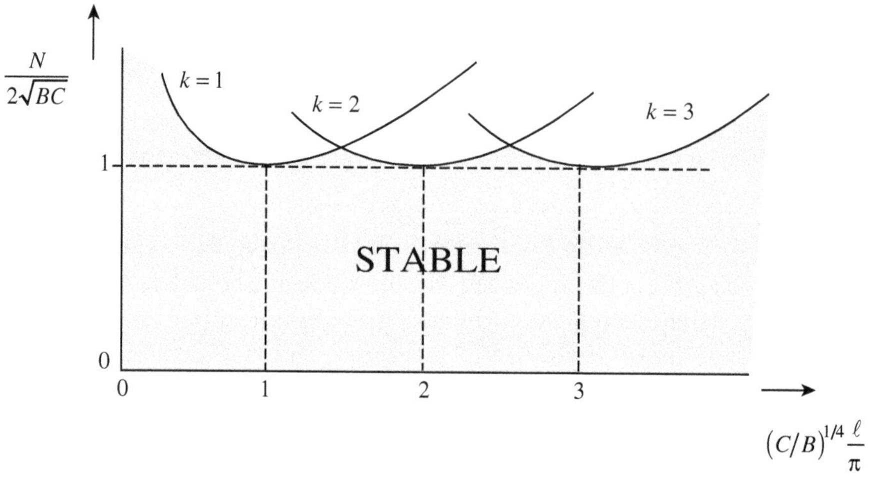

Figure 3.3.2

Suppose now that we have $\ell/\pi\,(C/B)^{1/4} = 3.5$. We then investigate what the value of N is for $k = 3$ and $k = 4$. From (3.3.12), we obtain for $k = 3$

$$N = \frac{9}{(3.5)^2}\sqrt{BC} + \frac{1}{9}(3.5)^2\sqrt{BC} = 2.096\sqrt{BC},$$

whereas the actual minimizing value is $N = 2\sqrt{BC}$. Similarly, we obtain for $k = 4$, $N = 2.072\sqrt{BC}$.

In general, for a sufficiently high wave number k we may assume that $k \in \mathbb{R}^+$, and then N_{Min} is given by (3.3.15). As indicated in Figure 3.3.2, equilibrium is stable in the shaded region but not necessarily at the stability limit, i.e., on the curves bounding the shaded region. Because $P_3\,[w; N] \equiv 0$, stability is governed by the sign of $P_4\,[w_1; N_1]$.

With N_1 given by (3.3.12) and

$$w_1(x) = a_k \sin\frac{k\pi x}{\ell}, \qquad k \in \mathbb{N}^+, \tag{3.3.16}$$

we obtain from (3.3.4)

$$\begin{aligned} P_4\,[w_1; N_1] &= \int_0^\ell \left[\frac{1}{2}B\left(-\frac{k^2\pi^2}{\ell^2}a_k \sin\frac{k\pi x}{\ell}\right)^2 \left(\frac{k\pi}{\ell}a_k \cos\frac{k\pi x}{\ell}\right)^2 - \frac{1}{8}N_1\left(a_k\frac{k\pi}{\ell}\cos\frac{k\pi x}{\ell}\right)^2\right]dx \\ &= \int_0^\ell \left(\frac{1}{8}B\frac{k^6\pi^6}{\ell^6}a_k^4 \sin^2\frac{2k\pi x}{\ell} - \frac{1}{8}N_1\frac{k^4\pi^4}{\ell^4}a_k^4\cos^4\frac{k\pi x}{\ell}\right)dx \qquad (3.3.17) \\ &= \frac{1}{16}B\frac{k^6\pi^6}{\ell^5}a_k^4 - \frac{3}{64}N_1\frac{k^6\pi^6}{\ell^3}a_k^4 = \frac{1}{64}\frac{k^2\pi^2}{\ell}a_k^4\left(\frac{k^4\pi^4}{\ell^4}B - 3C\right). \end{aligned}$$

The sign of this expression depends on C, and the critical value is

$$C = \frac{k^4\pi^4}{3\ell^4}B. \tag{3.3.18}$$

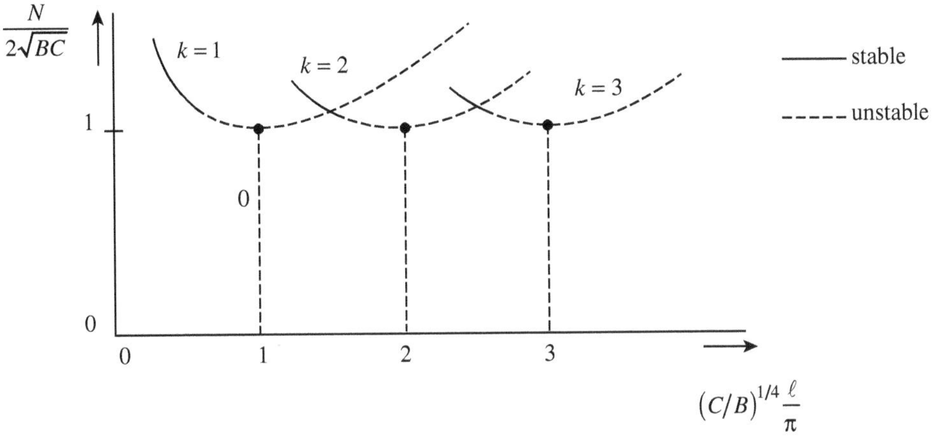

Figure 3.3.3

For the minimizing value of $k\pi/\ell$ given in (3.3.14), we obtain

$$P_4\left[w_1; N_{\text{Min}}\right] = -\frac{1}{32}\ell a_k^4\sqrt{\frac{C}{B}}C < 0, \tag{3.3.19}$$

which means that at the minimizing values of $k\pi/\ell$, the equilibrium is unstable.

We now plot our results in Figure 3.3.3.

For values of N exceeding N_1, the increment of the second variation is given by

$$\begin{aligned} P_2\left[w_1; N\right] &= \int_0^\ell \left[-\frac{1}{2}(N - N_1)\, a_k^2\frac{k^2\pi^2}{\ell^2}\cos^2\frac{k\pi x}{\ell}\right] dx \\ &= \frac{1}{4}(N_1 - N)\, a_k^2\frac{k^2\pi^2}{\ell} \\ &= \frac{1}{4}(1-\lambda)\, a_k^2 N_1\frac{k^2\pi^2}{\ell}, \end{aligned} \tag{3.3.20}$$

and for $N_1 = N_{\min}$ and $k\pi/\ell$ given by (3.3.14), we obtain

$$P_2\left[w_1; N\right] = \frac{1}{2}(1-\lambda)\, a_k^2 c\ell. \tag{3.3.21}$$

The function $F(a_k; \lambda)$ is now given by

$$F(a_k; \lambda) = \frac{1}{2}c\ell\left[(1-\lambda)\, a_k^2 - \frac{1}{16}\sqrt{\frac{C}{B}}a_k^4\right] = \frac{1}{2}c\ell\left[(1-\lambda)\, a_k^2 - \frac{1}{16}\frac{k^2\pi^2}{\ell^2}a_k^4\right], \tag{3.3.22}$$

Let us now consider the influence of imperfections on the behavior of this structure. In (2.6.39), we saw that for imperfections of the shape of the buckling mode, we must add to (3.3.22) a term $-2\lambda\bar{a}_k^0 a_k$, hence

$$F^*(a_k; \lambda) = \frac{1}{2}c\ell\left[(1-\lambda)\, a_k^2 - \frac{1}{16}\frac{k^2\pi^2}{\ell^2}a_k^4 - 2\lambda a_k^0 a_k\right]. \tag{3.3.23}$$

The equilibrium condition is

$$\frac{\partial F^*}{\partial a_k}(a_k;\lambda) = 0 = \frac{1}{2}c\ell\left[2(1-\lambda)a_k - \frac{1}{4}\frac{k^2\pi^2}{\ell^2}a_k^3 - 2\lambda a_k^0\right] \tag{3.3.24}$$

and the stability condition is

$$\frac{\partial^2 F^*}{\partial a_k^2}(a_k;\lambda) = \frac{1}{2}c\ell\left[2(1-\lambda) - \frac{3}{4}\frac{k^2\pi^2}{\ell^2}a_k^2\right] \geq 0. \tag{3.3.25}$$

The stability limit is reached for

$$a_k^2 = \frac{8}{3}(1-\lambda^*)\frac{\ell^2}{k^2\pi^2}. \tag{3.3.26}$$

Substitution into the equilibrium condition yields

$$\frac{4}{3}(1-\lambda^*)^{3/2}\sqrt{6}\frac{\ell}{k\pi} - \frac{4}{9}\sqrt{6}(1-\lambda^*)^{3/2}\frac{\ell}{k\pi} - 2\lambda^* a_k^0 = 0$$

or

$$(1-\lambda^*)^{3/2} = \frac{3}{8}\sqrt{6}\frac{k\pi}{\ell}\lambda^* a_k^0. \tag{3.3.27}$$

This result shows that for small imperfections the critical load is reduced appreciably, e.g., $ka_k^0/\ell \approx 0.01$ yields $\lambda^* \approx 0.91$, i.e., a reduction of about 10% in the critical load. Summarizing, we may say that the elastic foundation increases the buckling load but it also increases the imperfection sensitivity.

Finally, let us calculate the tangent modulus, which determines the stiffness of the structure after buckling. The extra shortening after buckling is

$$-\Delta\ell = \int_0^\ell \left[\sqrt{1-w'^2} - 1\right]dx \approx \int_0^\ell \frac{1}{2}w'^2\,dx$$

$$= \frac{1}{2}\int_0^\ell a_k^2\frac{k^2\pi^2}{\ell^2}\cos^2\frac{k\pi x}{\ell}\,dx = \frac{1}{4}\frac{k^2\pi^2}{\ell}a_k^2 = \frac{1}{4}\sqrt{\frac{C}{B}}a_k^2\ell. \tag{3.3.28}$$

With (3.3.26), we find

$$-\Delta\ell = \frac{1}{4}\sqrt{\frac{C}{B}}\frac{8}{3}(1-\lambda^*)\frac{\ell^2}{k^2\pi^2}\ell$$

or

$$-\frac{\Delta\ell}{\ell} = \frac{2}{3}\sqrt{\frac{C}{B}}(1-\lambda^*)\sqrt{\frac{B}{C}} = \frac{2}{3}(1-\lambda^*), \tag{3.3.29}$$

so that

$$\frac{\partial(-\Delta\ell/\ell)}{\partial N} = -\frac{2}{3N_{\min}} = \frac{1}{E_t A}. \tag{3.3.30}$$

Hence,

$$E_t = -\frac{3}{2}\frac{N_{\min}}{A} = -3\frac{\sqrt{BC}}{A}. \tag{3.3.31}$$

This result is shown in Figure 3.3.4.

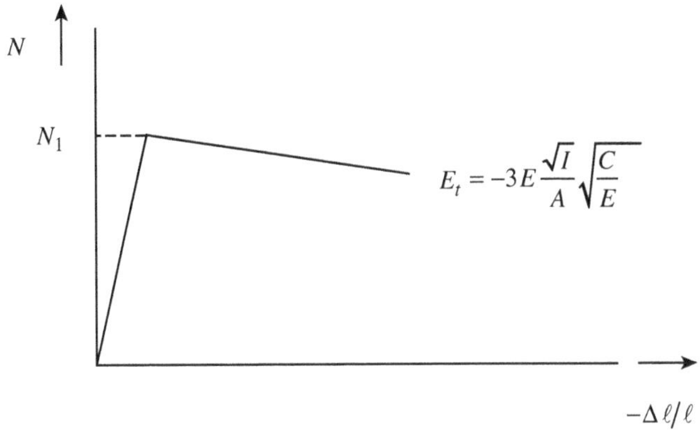

Figure 3.3.4

3.4. Simple two-bar frame

We consider a structure consisting of two incompressible bars of length ℓ, rigidly attached to each other on one end and hinged on the other end (see Figure 3.4.1), loaded by a force N.

Let $()'$ denote the derivative with respect to x and $()^{\cdot}$ the derivative with respect to y. At point A, the following kinematic relations then hold:

$$w'(\ell) = v^{\cdot}(\ell) = -\sin\theta \tag{3.4.1}$$

$$w(\ell) = \int_0^{\ell} \left(\sqrt{1 - v^{\cdot 2}} - 1\right) dy \tag{3.4.2}$$

$$v(\ell) = \int_0^{\ell} \left(1 - \sqrt{1 - w'^2}\right) dx. \tag{3.4.3}$$

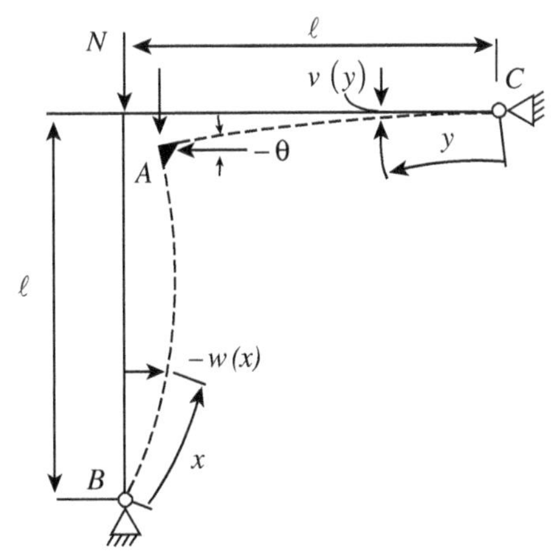

Figure 3.4.1

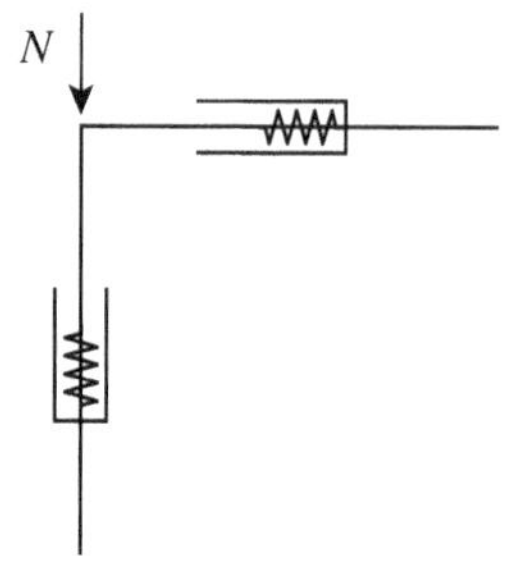

Figure 3.4.2

The potential energy is now given by

$$P\left[v\left(y\right), w\left(x\right); N\right] = \int_0^\ell \left[\frac{1}{2}B\frac{w''^2}{1-w'^2} + N\left(\sqrt{1-w'^2}-1\right)\right]dx + \int_0^\ell \frac{1}{2}B\frac{v^{\cdot\cdot 2}}{1-v^{\cdot 2}}\,dy, \tag{3.4.4}$$

where B is the bending stiffness of the bars. However, it is inconvenient to formulate the problem this way due to the nonlinear kinematic side conditions (3.4.2) and (3.4.3). It is more convenient to replace the rigid connection in A by the elastic element of Figure 3.4.2, where both springs have a stiffness c. The corresponding additional energy (penalty function) is given by

$$\frac{1}{2}c\left[w\left(\ell\right) - \int_0^\ell \left(1-v^{\cdot 2}-1\right)dy\right]^2 + \frac{1}{2}c\left[w\left(\ell\right) - \int_0^\ell \left(1-\sqrt{1-w'^2}\right)dx\right]^2. \tag{3.4.5}$$

Using the appropriate expansions, we find the following expressions for the second and third variation:

$$P_2\left[v, w; N, c\right] = \int_0^\ell \left(\frac{1}{2}Bw''^2 - \frac{1}{2}Nw'^2\right)dx + \int_0^\ell \frac{1}{2}Bv^{\cdot\cdot 2}\,dy + cw^2\left(\ell\right) + \frac{1}{2}cv^2\left(\ell\right) \tag{3.4.6}$$

$$P_3\left[v, w; N, c\right] = \frac{1}{2}cw\left(\ell\right)\int_0^\ell v^{\cdot 2}dy - \frac{1}{2}cv\left(\ell\right)\int_0^\ell w'^2\,dy. \tag{3.4.7}$$

In this case, we have $P_3 \not\equiv 0$.

The condition for neutral equilibrium follows from (3.4.6),

$$P_{11}\left[v, \eta, w, \zeta; N, c\right] = \int_0^\ell \left(\frac{1}{2}Bw''\zeta'' - \frac{1}{2}Nw'\zeta'\right)dx$$

$$+ \int_0^\ell \frac{1}{2}Bv^{\cdot\cdot}\eta^{\cdot\cdot}\,dy + cw\left(\ell\right)\zeta(\ell) + cv\left(\ell\right)\eta(\ell) = 0. \tag{3.4.8}$$

From integration by parts, we obtain

$$0 = Bw''\zeta'\,\Big|_0^\ell - Bw'''\zeta\Big|_0^\ell - Nw'\zeta\Big|_0^\ell + \int_0^\ell (Bw'''' + Nw'')\,\zeta dx$$
$$+ Bv^{\cdot\cdot}\eta^{\cdot}\,\Big|_0^\ell - Bv^{\cdot\cdot\cdot}\eta\Big|_0^\ell + \int_0^\ell Bv^{\cdot\cdot\cdot\cdot}\eta\, dy + cw\,(\ell)\,\zeta(\ell) + cv\,(\ell)\,\eta(\ell)\,, \tag{3.4.9}$$

from which the following differential equations are obtained:

$$Bw'''' + Nw'' = 0, \quad Bv^{\cdot\cdot\cdot\cdot} = 0. \tag{3.4.10}$$

For the evaluation of the stock terms, we must bear in mind that η and ζ are kinematically admissible functions, and that they must satisfy the kinematic condition (3.4.1), i.e.,

$$\zeta'\,(\ell) = \eta^{\cdot}\,(\ell)\,. \tag{3.4.11}$$

We then obtain

$$Bw''\,(\ell) + Bv^{\cdot\cdot}\,(\ell) = 0 \tag{3.4.12}$$

$$\begin{aligned} Bw''\,(0) = Bv^{\cdot\cdot}\,(0) &= 0 \\ -\,Bw'''\,(\ell) - Nw'\,(\ell) + cw\,(\ell) &= 0 \\ -\,Bv^{\cdot\cdot\cdot}\,(\ell) + cv\,(\ell) &= 0. \end{aligned} \tag{3.4.13}$$

In addition to these dynamic boundary conditions, we have the kinematic boundary conditions

$$\begin{aligned} w\,(0) = v\,(0) &= 0 \\ w'(\ell) = v^{\cdot}\,(\ell) &= -\sin\theta. \end{aligned} \tag{3.4.14}$$

These boundary conditions and the differential equations are also homogeneous and linear, so they always have the trivial solution $v = w = 0$. We now introduce the parameter

$$N/B = \alpha^2. \tag{3.4.15}$$

Choosing solutions for u and v so that the first two conditions of (3.4.13) and the first two conditions of (3.4.14) are satisfied, we obtain

$$\begin{aligned} w(x) &= A_1 \sin\alpha x + A_2 x \\ v(x) &= A_3 y + A_4 y^3. \end{aligned} \tag{3.4.16}$$

From the remaining conditions, we obtain the following set of equations:

$$\begin{aligned} &-\alpha^2 A_1 \sin\alpha\ell + 6A_4\ell = 0 \\ &B\alpha^3 A_1 \cos\alpha\ell - N\,(\alpha A_1 \cos\alpha\ell + A_2) + c\,(A_1 \sin\alpha\ell + A_2\ell) = 0 \\ &-6BA_4 + c\left(A_3\ell + A_4\ell^3\right) = 0 \\ &\alpha A_1 \cos\alpha\ell + A_2 = A_3 + 3A_4\ell^2 = -\sin\theta. \end{aligned} \tag{3.4.17}$$

The second of these conditions can be simplified because the first two terms cancel, so that

$$A_1 \sin \alpha\ell + A_2\ell\left(1 - \frac{N}{\ell c}\right) = 0. \tag{3.4.18}$$

For $c \to \infty$, we obtain from this equation and the fourth equation of (3.4.17),

$$A_1 = \frac{\ell \sin\theta}{\sin\alpha\ell - \alpha\ell\cos\alpha\ell}, \quad A_2\ell = \frac{-\ell\sin\alpha\ell\sin\theta}{\sin\alpha\ell - \alpha\ell\cos\alpha\ell}. \tag{3.4.19}$$

From the third equation of (3.4.17) and the fourth equation of (3.4.17), for $c \to \infty$ we obtain

$$A_3 = \frac{1}{2}\sin\theta, \quad A_4\ell^2 = -\frac{1}{2}\sin\theta. \tag{3.4.20}$$

From the first of the equations of (3.4.17), we now obtain the bifurcation condition

$$\frac{\alpha^2\ell^2\sin\alpha\ell}{\sin\alpha\ell - \alpha\ell\cos\alpha\ell} + 3 = 0 \tag{3.4.21}$$

or

$$\tan\alpha\ell = \frac{\alpha\ell}{1 + \frac{1}{3}(\alpha\ell)^2}. \tag{3.4.22}$$

This bifurcation equation is valid for $c \to \infty$, i.e., for our actual structure.

We now determine the value of P_3 at the critical load,

$$P_3[v_1, w_1; N_1, c \to \infty] = \lim_{c\to\infty}\left[\frac{1}{2}cw_1(\ell)\int_0^\ell v_1'^2 dy - \frac{1}{2}cv_1(\ell)\int_0^\ell w_1'^2\, dx\right]. \tag{3.4.23}$$

From (3.4.16) and (3.4.18), we obtain

$$\lim_{c\to\infty} cw_1(\ell) = A_2 N = \frac{3N\sin\theta}{(\alpha\ell)^2} = \frac{3B\sin\theta}{(\ell)^2}, \tag{3.4.24}$$

where we have used (3.4.19), (3.4.22), and (3.4.15).

From (3.4.16) and the third of equation of (3.4.17), we obtain

$$\lim_{c\to\infty} cv_1(\ell) = 6BA_4 = -\frac{3B\sin\theta}{(\ell)^2}. \tag{3.4.25}$$

We now obtain

$$\begin{aligned} P_3[v_1, w_1; N_1, c \to \infty] &= \frac{3}{2}\frac{B\sin\theta}{\ell^2}\left[\int_0^\ell (A_3 + 3A_4y^2)^2\, dy + \int_0^\ell (\alpha_1 A_1\cos\alpha_1 x + A_2)^2\, dx\right] \\ &= \frac{3}{2}\frac{B\sin\theta}{\ell^2}\left[A_3\ell + 2A_3A_4\ell^3 + \frac{9}{5}A_4^2\ell^5\right. \\ &\quad \left. + \frac{1}{2}\alpha A_1^2\left(\ell + \frac{1}{2\alpha}\sin 2\alpha_1\ell\right) + 2A_1A_2\sin\alpha_1\ell + A_2^2\ell\right]. \end{aligned} \tag{3.4.26}$$

Using (3.4.20) and

$$A_1 = \frac{3\ell \sin\theta}{-(\alpha\ell)^2 \sin\alpha\ell}, \quad A_2\ell = \frac{3\ell\sin\theta}{(\alpha\ell)^2}, \tag{3.4.27}$$

which follow from (3.4.19) and (34.22), after some algebra we find

$$P_3\left[v_1, w_1; N_1, c \to \infty\right] = \frac{3N_1 \sin^3\theta}{(\alpha_1\ell)^2}\left[\frac{7}{20} + \frac{9}{2(\alpha_1\ell)^2}\right]. \tag{3.4.28}$$

The smallest positive root of (3.4.22) is

$$\alpha_1\ell = 1.1861\pi, \tag{3.4.29}$$

and using this value we finally find

$$P_3\left[v_1, w_1; N_1, c \to \infty\right] = 0.14565 N_1 \ell \sin^3\theta. \tag{3.4.30}$$

This means that we have the behavior shown in Figure 3.4.3.

The equilibrium path is stable for positive values of $w_{\max}$ and unstable for negative values of $w_{\max}$.

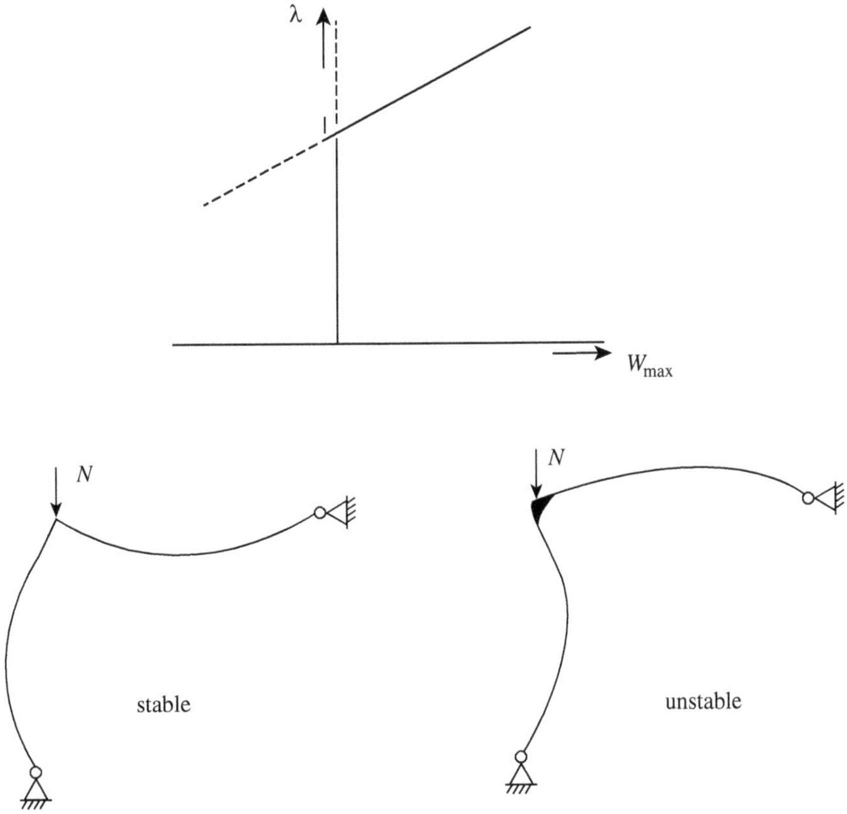

Figure 3.4.3

For a further discussion of this problem, we refer the reader to W. T. KOITER, Postbuckling analysis of a simple two-bar frame, *Recent Progress in Applied Mech., The Folke Odquist Volume* (337–354).

3.5 Simple two-bar frame loaded symmetrically

We consider again the simple two-bar frame, but now loaded symmetrically by two compressive forces N (see Figure 3.5.1).

The stiff joint is again replaced by an elastic element, and the potential energy is now given by

$$\begin{aligned} P[v(y), w(x); N] = {} & \int_0^\ell \left[\frac{1}{2}B\frac{w''^2}{1-w'^2} + N\left(\sqrt{1-w'^2}-1\right)\right] dx \\ & + \int_0^\ell \left[\frac{1}{2}B\frac{\ddot{v}^2}{1-\dot{v}^2} + N\left(\sqrt{1-\dot{v}^2}-1\right)\right] dy \\ & + \frac{1}{2}c\left[w(\ell) - \int_0^\ell \left[\sqrt{1-\dot{v}^2}-1\right] dy\right]^2 \\ & + \frac{1}{2}c\left[v(\ell) - \int_0^\ell \left[1-\sqrt{1-w'^2}\right] dx\right]^2 . \end{aligned} \tag{3.5.1}$$

The second variation is now given by

$$\begin{aligned} P_2[v, w; N, c] = {} & \int_0^\ell \left(\frac{1}{2}Bw''^2 - \frac{1}{2}Nw'^2\right) dx \\ & + \int_0^\ell \left(\frac{1}{2}B\ddot{v}^2 - \frac{1}{2}N\dot{v}^2\right) dy + \frac{1}{2}cw^2(\ell) + \frac{1}{2}cv^2(\ell) . \end{aligned} \tag{3.5.2}$$

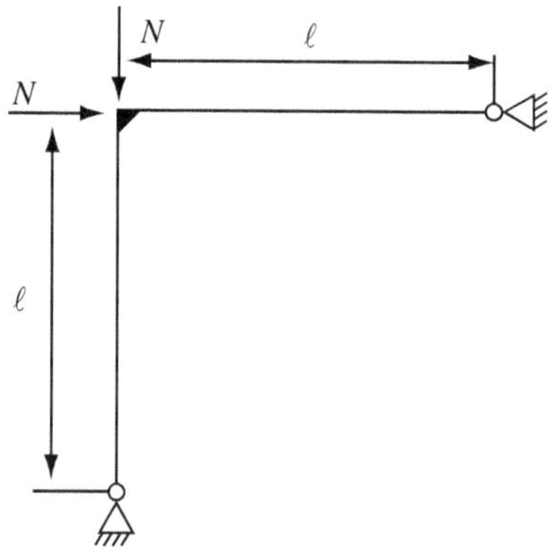

Figure 3.5.1

The third variation is still given by (3.4.7) but vanishes due to the symmetry of the structure and its loading. We will also need the fourth variation,

$$P_4[v,w;N,c] = \int_0^\ell \left(\frac{1}{2}Bw''^2 w'^2 - \frac{1}{8}Nw'^4\right)dx + \int_0^\ell \left(\frac{1}{2}Bv^{\cdot\cdot 2} v^{\cdot 2} - \frac{1}{8}v^{\cdot 4}\right) \\ + \frac{1}{2}c\left[\int_0^\ell \frac{1}{2}v^{\cdot 2}\,dy\right]^2 + \frac{1}{2}c\left[\int_0^\ell \frac{1}{2}w'^2\,dx\right]^2. \tag{3.5.3}$$

From (3.5.2), similar to the previous example we now obtain

$$Bw'''' + Nw'' = 0, \quad Bv^{\cdot\cdot\cdot\cdot} + Nv^{\cdot\cdot} = 0. \tag{3.5.4}$$

The kinematic condition (3.4.12) is now automatically satisfied because the bending moments vanish at $x = y = \ell$.

The dynamic boundary conditions are

$$\begin{aligned} Bw''(0) = Bv^{\cdot\cdot}(0) &= 0 \\ -Bw'''(\ell) - Nw'(\ell) + cw(\ell) &= 0 \\ -Bv^{\cdot\cdot\cdot}(\ell) - Nv^{\cdot}(\ell) + cv(\ell) &= 0. \end{aligned} \tag{3.5.5}$$

Further, we have the kinematic boundary conditions

$$\begin{aligned} w(0) = v(0) &= 0 \\ w'(\ell) = v^{\cdot}(\ell) &= -\sin\theta. \end{aligned} \tag{3.5.6}$$

The differential equations and the boundary conditions are satisfied by

$$w_1 = a\sin\frac{\pi x}{\ell}, \quad v_1 = a\sin\frac{\pi y}{\ell}, \quad \frac{N_1}{B} = \frac{\pi^2}{\ell^2}, \tag{3.5.7}$$

and we have

$$cw_1(\ell) = cv_1(\ell) = 0 \tag{3.5.8}$$

$$a = \frac{\ell}{\pi}\sin\theta. \tag{3.5.9}$$

To investigate the stability of this neutral equilibrium state, for small displacements from this configuration we write

$$w(x) = a\sin\frac{\pi x}{\ell} + \overline{w}(x), \quad v(x) = a\sin\frac{\pi y}{\ell} + \overline{v}(y), \tag{3.5.10}$$

where $\overline{w}$ and $\overline{v}$ are orthogonal to w_1 and v_1, respectively.

We now consider (see 2.4.11)

$$\underset{\text{w.r.t. } \overline{v}, \overline{w}}{\text{Min}} \{P_2[\overline{v}, \overline{w}; N_1, c\to\infty] + (N-N_1)P'_{11}[v_1, w_1, \overline{v}, \overline{w}; N_1, c\to\infty] \\ + P_{21}[v_1, w_1, \overline{v}, \overline{w}; N_1, c\to\infty]\}. \tag{3.5.11}$$

Now

$$P'_{11}\left[v_1, w_1, \overline{v}, \overline{w}^{\cdot}; N_1, c \to \infty\right] = -\int_0^\ell w_1'\overline{w}'\,dx - \int_0^\ell v_1^{\cdot}\overline{v}^{\cdot}\,dy = 0 \tag{3.5.12}$$

due to the orthogonality conditions. From (3.5.2) we obtain

$$\begin{aligned} P_2\left[\overline{v}, \overline{w}^{\cdot}; N_1, c\right] = &\int_0^\ell \left(\frac{1}{2}B\overline{w}''^2 - \frac{1}{2}N\overline{w}'^2\right)dx + \int_0^\ell \left(\frac{1}{2}B\overline{v}^{\cdot\cdot 2} - \frac{1}{2}N\overline{v}^{\cdot 2}\right)dy \\ &+ \frac{1}{2}c\overline{w}^2(\ell) + \frac{1}{2}c\overline{v}^2(\ell). \end{aligned} \tag{3.5.13}$$

From (3.4.7), we obtain

$$\begin{aligned} P_3\left[v_1 + \overline{v}, w_1 + \overline{w}; N_1, c\right] &= \frac{1}{2}c\left[w_1(\ell) + \overline{w}(\ell)\right]\int_0^\ell \left(v_1^{\cdot 2} + 2v_1^{\cdot}\overline{v}^{\cdot} + \overline{v}^2\right)dy \\ &\quad - \frac{1}{2}c\left[v_1(\ell) + \overline{v}(\ell)\right]\int_0^\ell \left(w_1'^2 + 2w_1'\overline{w}' + \overline{w}^2\right)dx \\ &= \frac{1}{2}c\overline{w}_1(\ell)\int_0^\ell v_1^{\cdot 2}\,dy - \frac{1}{2}c\overline{v}_1(\ell)\int_0^\ell w_1'^2\,dx \\ &\quad + cw_1(\ell)\int_0^\ell v_1^{\cdot}\overline{v}^{\cdot}\,dy - cv_1(\ell)\int_0^\ell w_1'\overline{w}'\,dx + \cdots \\ &= \frac{1}{2}c\overline{w}(\ell)\int_0^\ell v_1^{\cdot 2}\,dy - \frac{1}{2}c\overline{v}(\ell)\int_0^\ell w_1'^2\,dx + \cdots \\ &= P_{21}\left[v_1, w_1, \overline{v}, \overline{w}^{\cdot}; N_1, c\right] + \cdots. \end{aligned} \tag{3.5.14}$$

Let $\overline{v}, \overline{w}$ be the minimizing solution. The minimizing solution now must satisfy the condition of 3.4.18

$$P_{11}\left[\overline{v}, \overline{w}, \eta, \zeta; N_1, c\right] + P_{21}\left[v_1, w_1, \eta, \zeta; N_1, c\right] = 0, \tag{3.5.15}$$

i.e.,

$$\begin{aligned} &\int_0^\ell \left(B\overline{w}''\zeta'' - N_1\overline{w}'\zeta'\right)dx + \int_0^\ell \left(B\overline{v}^{\cdot\cdot}\eta^{\cdot\cdot} - N_1\overline{v}^{\cdot}\eta^{\cdot}\right)dy \\ &+ c\overline{w}(\ell)\zeta(\ell) + c\overline{v}(\ell)\eta(\ell) \\ &+ \frac{1}{2}\left[c\zeta(\ell)\int_0^\ell v_1^{\cdot 2}\,dy - c\eta(\ell)\int_0^\ell w_1'^2\,dx\right] = 0. \end{aligned} \tag{3.5.16}$$

From integration by parts, we obtain

$$\begin{aligned} &B\overline{w}''\zeta' \Big|_0^\ell - B\overline{w}'''\zeta\Big|_0^\ell - N_1\overline{w}'\zeta\Big|_0^\ell + \int_0^\ell (B\overline{w}'''' + N\overline{w}'')\,\zeta dx \\ &+ B\overline{v}^{\cdot\cdot}\eta^{\cdot} \Big|_0^\ell - B\overline{v}^{\cdot\cdot\cdot}\zeta\Big|_0^\ell - N_1\overline{v}^{\cdot}\eta\Big|_0^\ell + \int_0^\ell (B\overline{v}^{\cdot\cdot\cdot\cdot} + N\overline{v}^{\cdot\cdot})\,\eta dy \\ &+ c\overline{w}(\ell)\,\zeta(\ell) + c\overline{v}(\ell)\,\eta(\ell) \\ &+ \left(\frac{1}{2}c\int_0^\ell v_1^{\cdot 2}\,dy\right)\zeta(\ell) - \left(\frac{1}{2}c\int_0^\ell w_1'^2 dx\right)\eta(\ell) = 0, \end{aligned} \tag{3.5.17}$$

which yields the differential equations

$$B\overline{w}'''' + N_1\overline{w}'' = 0, \quad B\overline{v}^{\cdot\cdot\cdot\cdot} + N_1\overline{v}^{\cdot\cdot} = 0. \tag{3.5.18}$$

For the evaluation of the stock terms, we must again note that η and ζ are kinematically admissible functions, and that they must satisfy the condition (3.5.6) at $x = y = \ell$, i.e.,

$$\eta^{\cdot}(\ell) = \zeta'(\ell). \tag{3.5.19}$$

Taking into account this condition, we find

$$B\overline{w}''(\ell) + B\overline{v}^{\cdot\cdot}(\ell) = 0 \tag{3.5.20}$$

$$\begin{aligned} B\overline{w}''(0) = B\overline{v}^{\cdot\cdot}(0) = 0 \\ -B\overline{w}'''(\ell) - N_1\overline{w}'(\ell) + c\overline{w}(\ell) + \frac{1}{2}c\int_0^\ell v_1^{\cdot 2}\,dy = 0 \\ -B\overline{v}^{\cdot\cdot\cdot}(\ell) - N_1\overline{v}^{\cdot}(\ell) + c\overline{v}(\ell) - \frac{1}{2}c\int_0^\ell w_1'^2 dx = 0. \end{aligned} \tag{3.5.21}$$

In addition to these dynamic boundary conditions, we have the kinematic boundary conditions

$$\begin{aligned} \overline{w}(0) = \overline{v}(0) = 0 \\ \overline{w}'(\ell) = \overline{v}^{\cdot}(\ell). \end{aligned} \tag{3.5.22}$$

With the exception of the third and fourth equations of (3.5.21), these differential equations and boundary conditions are the same as those for the determination of v_1 and w_1. However, notice that due to the fact that the third and fourth equations of (3.5.21) are inhomogeneous, a solution $\overline{v} = \overline{w} = 0$ is not possible. The inhomogeneous terms are $\pm\frac{1}{4}a^2\frac{\pi^2 c}{\ell}$. To satisfy these inhomogeneous conditions, we try a solution of the form

$$\begin{aligned} \overline{v}(y) = A_2^* y + A_2^{**}\sin\frac{\pi y}{\ell} \\ \overline{w}(x) = A_1^* x + A_1^{**}\sin\frac{\pi x}{\ell}. \end{aligned} \tag{3.5.23}$$

These solutions must be orthogonal to the solution v_1, w_1. The orthogonality condition corresponding to (3.5.12) is

$$\int_0^\ell v_1 \overline{v}\, dy + \int_0^\ell w_1 \overline{w}\, dx = 0, \tag{3.5.24}$$

which yields

$$(A_1^* + A_2^*) \frac{\ell^2}{\pi^2} + \frac{1}{2} (A_1^{**} + A_2^{**}) \ell = 0. \tag{3.5.25}$$

The boundary condition (3.5.20), the first two conditions of (3.5.21), and the first two conditions of (3.5.22) are automatically satisfied by (3.5.23). From the third equation of (3.5.21), we obtain

$$\left(-B\frac{\pi^2}{\ell^2} + N_1 \frac{\pi}{\ell}\right) A_1^{**} + (-N_1 + c\ell) A_1^* = -\frac{1}{4} a^2 \frac{\pi^2 c}{\ell},$$

where the first term vanishes due to (3.5.7), so that

$$A_1^* = \frac{-a^2 \pi^2}{4\ell^2 (1 - N_1/c\ell)}. \tag{3.5.26}$$

Similarly, we obtain from the fourth equation of (3.5.21),

$$A_2^* = \frac{a^2 \pi^2}{4\ell^2 (1 - N_1/c\ell)}. \tag{3.5.27}$$

Because $A_1^* + A_2^* = 0$, we obtain from the orthogonality condition (3.5.25),

$$A_1^{**} + A_2^{**} = 0. \tag{3.5.28}$$

From the third condition of (3.5.22), we obtain

$$A_1^* - A_2^* = (A_1^{**} - A_2^{**}) \frac{\pi}{\ell}, \tag{3.5.29}$$

from which

$$A_1^{**} = -A_2^{**} = -\frac{a^2 \pi}{4\ell (1 - N_1/c\ell)}. \tag{3.5.30}$$

According to (2.3.54), stability is now governed by the sign of

$$P_4 [v_1, w_1; N_1, c \to \infty] + \frac{1}{2} P_{21} [v_1, w_1, \overline{v}, \overline{w}; N_1, c \to \infty]. \tag{3.5.31}$$

From (3.5.3), we obtain

$$P_4 [v_1, w_1; N_1, c] = -\frac{1}{16} B a^4 \frac{\pi^6}{\ell^6} + \frac{1}{16} \frac{\pi^4}{\ell^2} a^4 c, \tag{3.5.32}$$

and from (3.5.14),

$$P_{21} [v_1, w_1, \overline{v}, \overline{w}; N_1, c] = -\frac{\pi^4 a^4 c}{16\ell^3 (1 - N_1/c\ell)}, \tag{3.5.33}$$

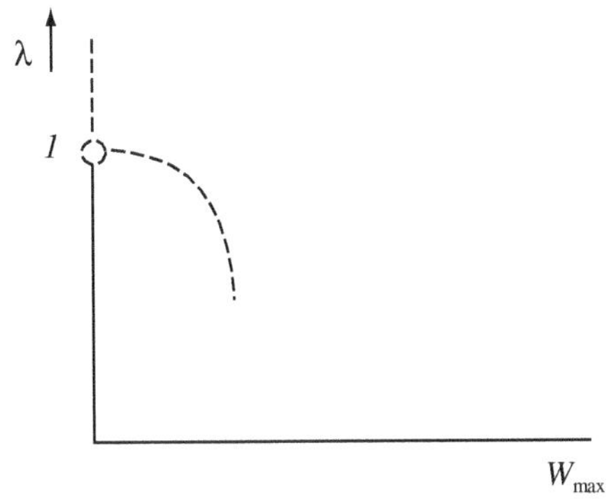

Figure 3.5.2

so that

$$\lim_{c\to\infty} P_4\left[v_1, w_1; N_1, c\right] + \frac{1}{2}P_{21}\left[v_1, w_1, \overline{v}.\overline{w}; N_1, c\right] = -\frac{\pi^6}{16}\frac{Ba^4}{\ell^6} < 0. \quad (3.5.34)$$

The equilibrium state is thus unstable. The behavior is shown in Figure 3.5.2.

At first sight, this result may seem surprising as both bars behave like the Euler bar. However, the result may be explained as shown in Figure 3.5.3.

Replace the forces N by two statically equivalent forces F ($F > N$). Let δ be the horizontal and vertical components of the corner point where the loads are applied; then

$$F \approx N + \frac{\delta}{\ell}N. \quad (3.5.35)$$

Let the deflection of the horizontal bar be

$$v(y) = a\sin\frac{\pi y}{\ell};$$

then δ is given by

$$\delta \approx \int_0^\ell \frac{1}{2}v^{\cdot 2}\,dy = \frac{1}{2}\frac{\pi^2}{\ell^2}a^2\int_0^\ell \cos^2\frac{\pi x}{\ell}\,dx = \frac{\pi^2 a^2}{4\ell^2}, \quad (3.5.36)$$

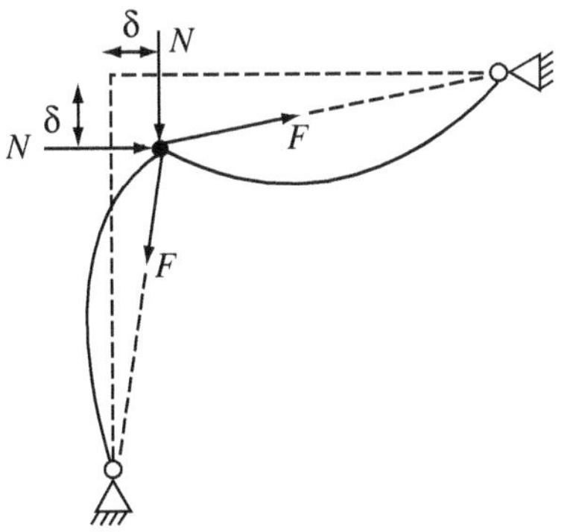

Figure 3.5.3

so that

$$F = N\left(1 + \frac{\pi^2 a^2}{4\ell^2}\right). \tag{3.5.37}$$

For the post-buckling behavior of the Euler bar, we had

$$F = F_1\left(1 + \frac{\pi^2 a^2}{8\ell^2}\right), \tag{3.5.38}$$

where F_1 is the critical load. From (3.5.37) and (3.5.38), we find

$$N = F_1\left(1 - \frac{\pi^2 a^2}{8\ell^2}\right). \tag{3.5.39}$$

This means that the equilibrium is already unstable for values of N smaller than the critical load F_1.

3.6 Bending and torsion of thin-walled cross sections under compression

We consider a beam with a thin-walled cross section, loaded by forces that cause bending and torsion, as shown in Figure 3.6.1. Let (y_0, z_0) be the center of shear.

The displacements at a point (x, y, z) are given by

$$\begin{aligned}
u &= -yv_0'(x) - zw_0'(x) + \alpha'(x)\psi_0(y,z) \\
v &= v_0(x) - (z - z_0)\alpha(x) + \frac{1}{2}\upsilon\left(y^2 - z^2\right)v_0''(x) + \upsilon yzw_0''(x) - \upsilon\Psi_0(y,z)\alpha''(x) \\
w &= w_0(x) + (y - y_0)\alpha(x) + \upsilon yzv_0''(x) + \frac{1}{2}\upsilon\left(z^2 - y^2\right)w_0''(x) - \upsilon\Phi_0(y,z)\alpha''(x).
\end{aligned} \tag{3.6.1}$$

Here $\alpha(x)$ is the specific angle of torsion, $\psi_0(y,z)$ and $\varphi(y,z)$ are the real and the imaginary parts of an analytic function $F(y + iz) = \psi_0(y,z) + i\varphi(y,z)$, and $\Phi_0(y,z)$ and $\Psi_0(y,z)$ are defined by

$$\frac{\partial\Psi_0}{\partial y} = \frac{\partial\Phi_0}{\partial z} = \psi_0(y,z), \qquad -\frac{\partial\Psi_0}{\partial z} = \frac{\partial\Phi_0}{\partial y} = \varphi(y,z). \tag{3.6.2}$$

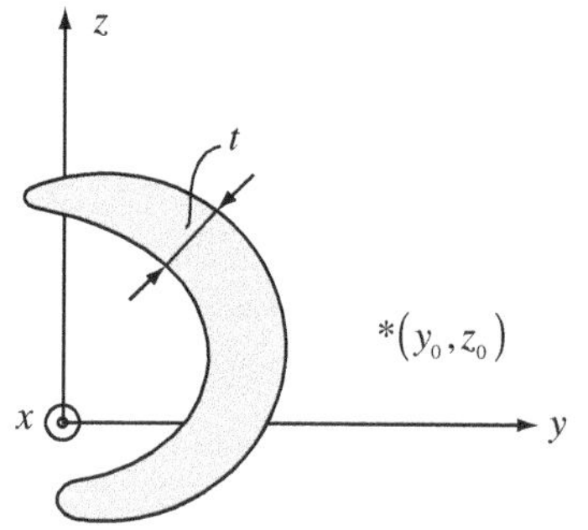

Figure 3.6.1

The approximations (3.6.1) are valid for $t^2/b^2 \ll 1$ and $b^2/L^2 \ll 1$, where t is the wall thickness of the cross section, L is a characteristic wavelength, and b is a characteristic dimension of the cross section. The center of shear is determined by

$$\int_A z\psi_0\,(y,z)\;dA = \int_A y\psi_0\,(y,z)\;dA = 0. \tag{3.6.3}$$

Further, the resultant of the normal stresses in the x-direction must vanish, which yields

$$\int_A \psi_0\,(y,z)\;dA = 0. \tag{3.6.4}$$

The expressions (3.6.1) further guarantee the vanishing of the stresses σ_y, σ_z, and σ_{yz}.

In our calculations, we shall make use of the estimates

$$\begin{aligned} &\alpha'(x) = O(\alpha/L), \alpha''(x) = O\left(\alpha/L^2\right), \quad \text{and so on} \\ &|y-y_0|\,,|z-z_0| = O\,(b)\,, \frac{\partial()}{\partial y}, \frac{\partial()}{\partial z} = O\,(()\,/b) \\ &|\psi_0|\,,|\varphi| = O\left(b^2\right), |\Phi_0|\,,|\psi_0| = O(b^3), \end{aligned} \tag{3.6.5}$$

where b is the width of the cross section. To simplify the calculations, we shall use the principal axes of inertia, i.e.,

$$\int_A yz\,dA = 0. \tag{3.6.6}$$

We now obtain the following expression for the potential energy:

$$\begin{aligned} P\,[\mathbf{u}] &= \int_V \frac{1}{2}\left(\sigma_x\frac{\partial u}{\partial x} + \tau_{xy}\gamma_{xy} + \tau_{xz}\gamma_{xz}\right) dV \\ &= \frac{1}{2}\int \Bigg\{E\left(-yv_0'' - zw_0'' + \psi_0\alpha''\right)^2 + \mathcal{G}\left[\alpha'\frac{\partial\psi_0}{\partial y} - (z-z_0)\,\alpha' + \cdots\right]^2 \\ &\quad + \mathcal{G}\left[\alpha'\frac{\partial\psi_0}{\partial z} + (y-y_0)\,\alpha' + \cdots\right]^2\Bigg\}\,dV, \end{aligned} \tag{3.6.7}$$

where the dots represent terms containing third-order derivatives with respect to x.

With the torsional stiffness S_t defined by

$$S_t = \int_A \mathcal{G}\left\{\left[\frac{\partial\psi_0}{\partial y} - (z-z_0)\right]^2 + \left[\frac{\partial\psi_0}{\partial z} + (y-y_0)\right]^2\right\} dA \tag{3.6.8}$$

and a warping constant Γ defined by

$$\Gamma = \int_A \psi_0^2\,(y,z)\;dA \tag{3.6.9}$$

and using the relations (3.6.3), (3.6.4), and (3.6.6) we may rewrite the potential energy expression to yield

$$P[\mathbf{u}] = \int_0^\ell \left(\frac{1}{2}EI_z v_0''^2 + \frac{1}{2}EI_y w_0''^2 + \frac{1}{2}E\Gamma\alpha''^2\right) dx + \int_0^\ell \frac{1}{2}S_t\alpha'^2\, dx. \tag{3.6.10}$$

Expression (3.6.10) is valid for $b^2/L^2 \ll 1$. In addition to this energy, we have the energy of the prestresses,

$$\int_V \frac{1}{2}S_{ij}u_{h,i}u_{h,j}\, dV = -\frac{1}{2}\sigma\int_V \left(u_{,x}^2 + v_{,x}^2 + w_{,x}^2\right) dV. \tag{3.6.11}$$

The term $\frac{1}{2}\int \sigma u_{,x}^2\, dV$ may be neglected compared to the term $\frac{1}{2}\int Eu_{,x}^2\, dV$ in (3.6.7). Hence, the additional energy is given by

$$\begin{aligned}
&-\frac{1}{2}\sigma\int_V \left\{[v_0' - (z - z_0)\alpha' + \cdots]^2 + [w_0' + (y - y_0)\alpha' + \cdots]^2\right\} dV \\
&= -\int_0^\ell \left(\frac{1}{2}Nv_0'^2 + \frac{1}{2}Nw_0'^2\right) dx - \frac{1}{2}\sigma\int_0^\ell\int_A \left[(y - y_0)^2 + (z - z_0)^2\right] dA\alpha'^2 dx \\
&\quad - \int_0^\ell (Nz_0\alpha' v_0' - Ny_0\alpha' w_0')\, dx
\end{aligned} \tag{3.6.12}$$

with

$$\int_A \left[(y - y_0)^2 + (z - z_0)^2\right] dA = I_{z_0} + I_{y_0} = \left[i_{z_0}^2 + i_{y_0}^2\right] A = i_0^2 A. \tag{3.6.13}$$

Finally, we obtain for the total potential energy

$$\begin{aligned}
P[\mathbf{u}, N] = \int_0^\ell \Big(&\frac{1}{2}EI_z v_0''^2 + \frac{1}{2}EI_y v_0''^2 + \frac{1}{2}E\Gamma\alpha''^2 - \frac{1}{2}Nv_0'^2 - \frac{1}{2}Nw_0'^2 - \frac{1}{2}Ni_0^2\alpha'^2 \\
&- Nz_0\alpha' v_0' + Ny_0\alpha' w_0' + \frac{1}{2}S_t\alpha'^2\Big) dx = P_2[\mathbf{u}; N].
\end{aligned} \tag{3.6.14}$$

Expression (3.6.14) shows that interaction between bending and torsion only occurs in the terms with $Nz_0\alpha' v_0'$ and $Ny_0\alpha' w_0'$. This condition implies that when the shear center coincides with the center of gravity, there is no interaction. This is the case when there are two axes of symmetry. When there is only one axis of symmetry, one of the coupling terms vanishes, i.e., we then have one axis of buckling mode without interaction and one with interaction.

We shall now restrict our discussion further to the case in which the values of I_y, I_z, Γ, S_t, and N are constant. The equations of neutral equilibrium follow from

$$\begin{aligned}
P_{11}[v_0, w_0, \alpha, \eta, \zeta, \chi; N] = \int_0^\ell &[EI_z v_0''\eta'' + EI_z w_0''{}_0\zeta'' + E\Gamma\alpha''\chi'' \\
&- Nv_0'\eta' - Nw_0'\zeta' - Ni_0^2\alpha'\chi' - Nz_0(\alpha'\eta' + \chi' v_0') \\
&+ Ny_0(\alpha'\zeta' + \chi' w_0') + S_t\alpha'\chi']\, dx = 0.
\end{aligned} \tag{3.6.15}$$

By integration by parts, we obtain

$$
\begin{aligned}
& EI_z\left(v_0''\eta'|_0^\ell - v_0'''\eta|_0^\ell + \int_0^\ell v_0''''\eta\,dx\right) + EI_y\left(w_0''\zeta'|_0^\ell - w_0'''\zeta|_0^\ell + \int_0^\ell w_0''''\zeta\,dx\right)\\
& + E\Gamma\left(\alpha''\chi'|_0^\ell - \alpha'''\chi|_0^\ell + \int_0^\ell \alpha''''\chi\,dx\right)\\
& - Nv_0'\eta'|_0^\ell + \int_0^\ell Nv_0''\eta\,dx - Nw_0'\zeta|_0^\ell + \int_0^\ell Nw_0''\zeta\,dx\\
& - Ni_0^2\alpha'\chi|_0^\ell + \int_0^\ell Ni_0^2\alpha''\chi\,dx - Nz_0\left(\alpha'\eta + \chi v_0'\right)|_0^\ell\\
& + Nz_0\int_0^\ell \left(\alpha''\eta + v_0''\chi\right)dx + Ny_0\left(\alpha'\zeta + w_0'\chi\right)|_0^\ell\\
& - Ny_0\int_0^\ell \left(\alpha''\zeta + w_0''\chi\right)dx + S_t\alpha'\chi|_0^\ell - \int_0^\ell S_t\alpha''\chi\,dx = 0.
\end{aligned}
\tag{3.6.16}
$$

We now consider a beam that is supported at $x = 0$ and $x = \ell$ so that

$$
v_0(0) = w_0(0) = \alpha(0) = v_0(\ell) = w_0(\ell) = \alpha(\ell) = 0. \tag{3.6.17}
$$

For a clamped-clamped beam, we have

$$
\begin{aligned}
\eta' = \zeta' = \chi' = 0 \quad & \text{at} \quad x = 0 \text{ and } x = \ell\\
\eta = \zeta = \chi = 0 \quad & \text{at} \quad x = 0 \text{ and } x = \ell,
\end{aligned}
\tag{3.6.18}
$$

and for a simply supported beam,

$$
\eta = \zeta = \chi = 0 \quad \text{at} \quad x = 0 \text{ and } x = \ell. \tag{3.6.19}
$$

The differential equations are

$$
\begin{aligned}
& v_0'''' + Nv_0'' + Nz_0\alpha'' = 0\\
& w_0'''' + Nw_0'' - Ny_0\alpha'' = 0\\
& \alpha'''' + Ni_0^2\alpha'' + Nz_0v_0'' - Ny_0w_0'' - S_t\alpha'' = 0.
\end{aligned}
\tag{3.6.20}
$$

For the clamped-clamped beam, the kinematic boundary conditions are

$$
u(0) = u(\ell) = 0, \quad v_0' = w_0' = \alpha' = 0 \quad \text{at } x = 0 \text{ and } x = \ell. \tag{3.6.21}
$$

For the simply supported beam, we have the dynamic boundary conditions

$$
EI_zv_0'' = EI_yw_0'' = E\Gamma\alpha'' = 0 \quad \text{at } x = 0 \text{ and } x = \ell, \tag{3.6.22}
$$

i.e., the vanishing of the bending moments M_z and M_y and of the bimoment.†

† The bimoment is not a real moment; its dimension is $[FL^2]$.

We shall now consider the case of a simply supported beam in more detail. Because the differential equations and the boundary conditions contain only even derivatives, we try a solution of the form

$$v_0 = B \sin \frac{\pi x}{\ell}, \quad w_0 = C \sin \frac{\pi x}{\ell}, \quad \alpha = A \sin \frac{\pi x}{\ell}.^{\dagger} \tag{3.6.23}$$

These expressions satisfy the boundary conditions (3.6.21) and (3.6.22). Substitution of these expressions into (3.6.20) yields

$$\begin{array}{llll} (N_y - N)\,B & & -Nz_0 A & = 0 \\ & (N_z - N)\,C & +Ny_0 A & = 0 \\ -Nz_0 B + & Ny_0 C & +i_0^2\,(N_t - N)\,A & = 0, \end{array} \tag{3.6.24}$$

where we have introduced the notation

$$\frac{\pi^2 EI_z}{\ell^2} = N_y, \quad \frac{\pi^2 EI_y}{\ell^2} = N_z, \quad \frac{1}{i_0^2}\left(\frac{\pi^2 E\Gamma}{\ell^2} + S_2\right) = N_t, \tag{3.6.25}$$

where N_t is the lowest buckling load due to torsion. The condition for a non-trivial solution is the vanishing of the determinant, i.e.,

$$i_0^2\,(N_y - N)\,(N_z - N)\,(N_t - N) + y_0^2 N^2\,(N - N_y) + z_0^2 N^2\,(N - N_z) = 0. \tag{3.6.26}$$

When y_0 and z_0 are zero, the roots of this equation are $N = (N_y, N_z, N_t)$. Now let at least y_0 or z_0 be nonzero. Then for $N = \min(N_y, N_z, N_t)$, the left-hand member of (3.6.26) is negative. The left-hand member can only become positive for $N < \min(N_y, N_z, N_t)$. This condition means that the smallest positive root of this equation is smaller than N_y, N_z, and N_t, i.e., the actual buckling load is smaller than the buckling load in bending and smaller than the buckling load in torsion.

Let us now consider the special case that the y-axis is an axis of symmetry, i.e., $z_0 = 0$. Then we have

$$N = N_y \quad \text{or} \quad i_0^2\,(N_z - N)\,(N_t - N) - y_0^2 N^2 = 0, \tag{3.6.27}$$

where for the first case there is no interaction, but there is interaction in the second case. Let us now investigate this second case more closely. Equation (3.6.27) may be rewritten in the form

$$N^2\left(1 - \frac{y_0^2}{i_0^2}\right) - (N_z + N_t)\,N + N_z N_t = 0. \tag{3.6.28}$$

This equation is symmetric in N_t and N_z; hence without loss of generality we may assume

$$N_z/N_t = \eta < 1. \tag{3.6.29}$$

† We are only interested in the lowest buckling mode $k = 1$. In general, one writes $v_0 = \sum_{k=1}^{\infty} B_k \sin \frac{k\pi x}{\ell}$, and so on.

Introducing the notation $N/N_z = \lambda$, we may rewrite the (3.6.28) to yield

$$\lambda^2\left(1-\frac{y_0^2}{i_0^2}\right)-\left(1+\frac{1}{\eta}\right)\lambda+\frac{1}{\eta}=0. \tag{3.6.30}$$

The smallest root is given by

$$\lambda=\frac{1}{2\left(1-y_0^2/i_0^2\right)}\left[(1+1/\eta)-\sqrt{(1-1/\eta)^2+4\eta y_0^2/i_0^2}\right]. \tag{3.6.31}$$

Let us now consider the two limiting cases

$$\eta \to 0 \quad \text{and} \quad \eta \to 1. \tag{3.6.32}$$

For $\eta \to 0$, we may write

$$\lambda=\lim_{\eta\to 0}\frac{1}{2\left(1-y_0^2/i_0^2\right)}\frac{1}{\eta}\left[1+\eta-\left(1-\eta+2\eta y_0^2/i_0^2\right)\right]=1, \tag{3.6.33}$$

as expected. For $\eta \to 1$, we find

$$\lambda=\frac{1}{1+y_0/i_0}<1. \tag{3.6.34}$$

Because $y_0/i_0 < 1$, this means

$$\frac{1}{2}<\lambda<1. \tag{3.6.35}$$

This result shows that under combined torsion and bending, the buckling load may be considerably smaller than N_z or N_t, especially when $y_0/i_0 \to 1$. An example of such a structure is the thin-walled circular tube with an open cross section (see Figure 3.6.2).

In this case, we have

$$S_t=\frac{2}{3}\pi \mathcal{G}Rt^3,\quad \Gamma=\pi\left(\frac{2}{3}\pi^3-4\right)R^5t,\quad y_0/i_0=\frac{2}{\sqrt{5}}. \tag{3.6.36}$$

For $\eta = 1$, we obtain $\lambda = 0.528$, i.e., a reduction of 47% of the buckling load. We emphasize here once again that the theory is only valid for $b^2/L^2 \ll 1$.

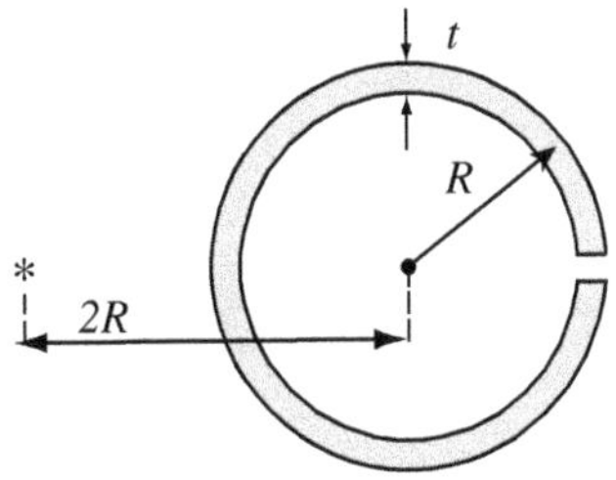

Figure 3.6.2

3.7 Infinite plate between flat smooth stamps

We consider a homogeneous isotropic elastic plate of thickness h, length ℓ, and infinite in width, compressed between flat smooth rigid stamps. The material is such that it can have finite deformations in the elastic range (a neo-Hookean material), and its elastic energy per unit volume of the undeformed body is given by

$$W = \frac{1}{2}\left[\lambda_1^2 + \lambda_2^2 + \lambda_3^2 - 3\right], \tag{3.7.1}$$

where λ_1, λ_2, and λ_3 are the extension ratios in the fundamental state I, i.e.,

$$\lambda_1 = 1 + \varepsilon_1, \quad \lambda_2 = 1 + \varepsilon_2, \quad \lambda_3 = 1 + \varepsilon_3, \tag{3.7.2}$$

where ε_1 $(i = 1, 2, 3)$ are the principal strains. In (3.7.1), the elastic constant is taken to be unity. We shall assume that the material is incompressible, i.e.,

$$\lambda_1\lambda_2\lambda_3 = 1, \tag{3.7.3}$$

and we consider the case of plane strain, i.e., $\lambda_3 = 1$, so that

$$\lambda_1\lambda_2 = 1. \tag{3.7.4}$$

We now consider the following states (see Figure 3.7.1), where $\sigma_1 h$ is the force per unit width of the plate. Let (x, y) be the coordinates of a material point in the undeformed state. The coordinates of this point in the fundamental state are denoted by $(\bar{x}, \bar{y})$, where

$$\bar{x} = \lambda_1 x, \quad \bar{y} = \lambda_2 y. \tag{3.7.5}$$

In the adjacent state, the coordinates are $(\bar{\bar{x}}, \bar{\bar{y}})$, where

$$\bar{\bar{x}} = \lambda_1 x + u, \quad \bar{\bar{y}} = \lambda_2 y + v. \tag{3.7.6}$$

The square of the length of an infinitely small material element in the undeformed state is given by

$$(ds)^2 = (dx)^2 + (dy)^2. \tag{3.7.7}$$

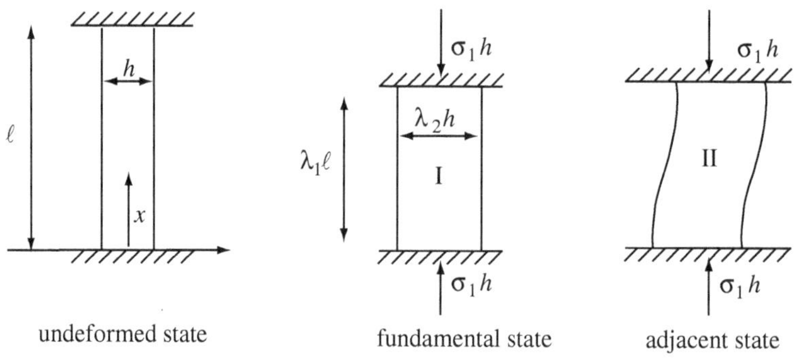

Figure 3.7.1

In the fundamental state I, it is given by

$$(d\bar{s})^2 = (d\bar{x})^2 + (d\bar{y})^2 = \lambda_1^2 (dx)^2 + \lambda_2^2 (dy)^2 , \tag{3.7.8}$$

and in the adjacent state II, it is given by

$$\begin{aligned}(d\bar{\bar{s}})^2 = (d\bar{\bar{x}})^2 + (d\bar{\bar{y}})^2 &= [(\lambda_1 + u_{,x})^2] (dx)^2 + [(\lambda_2 + v_{,x})^2 u_{,y}^2] (dy)^2 \\ &+ 2 [(\lambda_1 + u_{,x}) u_{,y} + (\lambda_2 + v_{,y}) v_{,y}] \, dx\, dy.\end{aligned} \tag{3.7.9}$$

The components of the metric tensor in the adjacent state are thus

$$\begin{aligned}\bar{\bar{g}}_{xx} &= (\lambda_1 + u_{,x})^2 + v_{,x}^2 \\ \bar{\bar{g}}_{yy} &= (\lambda_2 + v_{,y})^2 + u_{,y}^2 \\ \bar{\bar{g}}_{xy} &= (\lambda_1 + u_{,x}) u_{,y} + (\lambda_2 + v_{,y}) v_{,x}.\end{aligned} \tag{3.7.10}$$

The principal values of the metric tensor follow from

$$\begin{vmatrix} \bar{\bar{g}}_{xx} - \bar{\bar{g}} & \bar{\bar{g}}_{xy} \\ \bar{\bar{g}}_{xy} & \bar{\bar{g}}_{yy} - \bar{\bar{g}} \end{vmatrix} = 0, \tag{3.7.11}$$

and in the corresponding principal directions we have

$$\bar{\bar{g}}_{11} = \bar{\bar{\lambda}}_1^2, \quad \bar{\bar{g}}_{22} = \bar{\bar{\lambda}}_2^2, \tag{3.7.12}$$

where $\bar{\bar{\lambda}}_1$ and $\bar{\bar{\lambda}}_2$ are the extension ratios in those directions. The first invariant of the metric tensor is

$$\bar{\bar{g}}_{11} + \bar{\bar{g}}_{22} = \bar{\bar{\lambda}}_1^2 + \bar{\bar{\lambda}}_2^2 = \bar{\bar{g}}_{xx} + \bar{\bar{g}}_{yy}. \tag{3.7.13}$$

The elastic energy density in the adjacent state II is now given by

$$\begin{aligned}W_{\text{II}} &= \frac{1}{2} \left(\bar{\bar{g}}_{xx} + \bar{\bar{g}}_{yy} - 2\right) \\ &= \frac{1}{2} [(\lambda_1 + u_{,x})^2 + v_{,x}^2 + (\lambda_2 + v_{,y})^2 + u_{,y}^2 - 2].\end{aligned} \tag{3.7.14}$$

In the fundamental state I, the elastic energy density is

$$W_{\text{I}} = \frac{1}{2} \left(\bar{g}_{xx} + \bar{g}_{yy} - 2\right) = \frac{1}{2} \left(\lambda_1^2 + \lambda_2^2 - 2\right), \tag{3.7.15}$$

so that the increment of the elastic energy density in passing from the fundamental state I to the adjacent state II is

$$W_{\text{II}} - W_{\text{I}} = \lambda_1 u_{,x} + \lambda_2 v_{,y} + \frac{1}{2} \left(u_{,x}^2 + u_{,y}^2 + v_{,x}^2 + u_{,y}^2\right). \tag{3.7.16}$$

In addition to this energy expression, we shall need an explicit expression for the incompressibility condition, which with the aid of the second invariant of the metric tensor can be written as

$$1 = \lambda_1 \lambda_2 = \bar{\bar{\lambda}}_1 \bar{\bar{\lambda}}_2 = \sqrt{\bar{\bar{g}}_{11} \bar{\bar{g}}_{22}} = \sqrt{\bar{\bar{g}}_{xx} \bar{\bar{g}}_{yy} - \bar{\bar{g}}_{xy}}. \tag{3.7.17}$$

However, it is slightly simpler to derive the incompressibility condition from

$$\frac{d\overline{\overline{v}}}{dv} = \begin{vmatrix} \partial\overline{\overline{x}}/\partial x & \partial\overline{\overline{x}}/\partial y \\ \partial\overline{\overline{y}}/\partial x & \partial\overline{\overline{y}}/\partial y \end{vmatrix} = \begin{vmatrix} \lambda_1 + u_{,x} & u_{,y} \\ v_{,x} & \lambda_2 + v_{,y} \end{vmatrix} = \lambda_1\lambda_2,$$

from which follows

$$\lambda_2 u_{,x} + \lambda_1 v_{,y} + u_{,x}v_{,y} - u_{,y}v_{,x} = 0. \tag{3.7.18}$$

This incompressibility condition is a nonlinear side condition in our problem.

The total potential energy per unit width of the plate is now given by

$$\begin{aligned} P[\mathbf{u}] = &\int_0^{\ell}\int_{-h/2}^{h/2} \left[\lambda_1 u_{,x} + \lambda_2 v_{,y} + \frac{1}{2}\left(u_{,x}^2 + u_{,y}^2 + v_{,x}^2 + v_{,y}^2\right)\right] dx\, dy \\ &+ \sigma_1 \int_0^{\ell}\int_{-h/2}^{h/2} u_{,x}\, dx\, dy, \end{aligned} \tag{3.7.19}$$

where for the last term we have rewritten the term $\sigma_1 h \int u_{,x}\, dx$ as a surface integral. The stability condition is now that $P[\mathbf{u}] > 0$ under the nonlinear side condition (3.7.18). Using this side condition to eliminate $v_{,y}$, in the linear terms in (3.7.19), we find

$$\begin{aligned} P[\mathbf{u}] = &\int_0^{\ell}\int_{-h/2}^{h/2} \left[\left(\lambda_1 - \frac{\lambda_2^2}{\lambda_1} + \sigma_1\right) u_{,x} + \frac{1}{2}\left(u_{,x}^2 + u_{,y}^2 + v_{,x}^2 + v_{,y}^2\right)\right. \\ &\left. + \frac{\lambda_2}{\lambda_1}\left(u_{,y}v_{,x} - u_{,x}v_{,y}\right)\right] dx\, dy. \end{aligned} \tag{3.7.20}$$

The stability condition is still $P[\mathbf{u}] > 0$ under the side condition (3.7.18).

A necessary condition that $P[\mathbf{u}] > 0$ is the vanishing of the linear term in $P[\mathbf{u}]$, i.e.,

$$\sigma_1 = -\lambda_1 + \frac{\lambda_2^2}{\lambda_1} = -\lambda_1 + \frac{1}{\lambda_1^3}, \tag{3.7.21}$$

which is the equilibrium condition in the fundamental state. The discussion of the equilibrium in the fundamental state is now reduced to the discussion of

$$P[\mathbf{u}] = \int_0^{\ell}\int_{-h/2}^{h/2} \left[\frac{1}{2}\left(u_{,x}^2 + u_{,y}^2 + v_{,x}^2 + v_{,y}^2\right) + \frac{\lambda_2}{\lambda_1}\left(u_{,y}v_{,x} - u_{,x}v_{,y}\right)\right] dx\, dy \tag{3.7.22}$$

under the nonlinear side condition (3.7.18). Notice that $P[\mathbf{u}]$ is a homogeneous quadratic functional, i.e., $P[\mathbf{u}] = P_2[\mathbf{u}]$. A necessary condition for stability is that $P_2[\mathbf{u}] \geq 0$ for very small displacements from the fundamental state, which means that we may linearize the side condition to yield

$$\lambda_2 u_{,x} + \lambda_1 v_{,y} = 0. \tag{3.7.23}$$

A sufficient condition for stability is that $P_2[\mathbf{u}] > 0$ under the linear side condition (3.7.23).

Introducing a Lagrangian multiplier $p(x, y)$ to take into account the side condition, we now consider the functional

$$P[\mathbf{u}] + \int_0^{\ell} \int_{-h/2}^{h/2} p(x, y)(\lambda_2 u_{,x} + \lambda_1 v_{,y})\, dx\, dy. \tag{3.7.24}$$

A necessary condition for stability is now

$$\begin{aligned}\int_0^{\ell} \int_{-h/2}^{h/2} & [u_{,x}\eta_{,x} + u_{,y}\eta_{,y} + v_{,x}\zeta_{,x} + v_{,y}\zeta_{,y} \\ & + \frac{\lambda_2}{\lambda_1}(u_{,y}\zeta_{,x} + v_{,x}\eta_{,y} - u_{,x}\zeta_{,y} - v_{,y}\eta_{,x}) \\ & + (\lambda_2 u_{,x} + \lambda_1 v_{,y})\,\delta p + p(x, y)(\lambda_2\eta_{,x} + \lambda_1\zeta_{,y})]\, dx\, dy = 0,\end{aligned} \tag{3.7.25}$$

for all kinematically admissible displacements η, ζ. By integration by parts, we obtain

$$\begin{aligned}&\int_{-h/2}^{h/2} (u_{,x}\eta)\big|_0^{\ell}\, dy + \int_0^{\ell} (u_{,y}\eta)\big|_{-h/2}^{h/2}\, dx - \int_0^{\ell}\int_{-h/2}^{h/2} (u_{,xx} + u_{,yy})\,\eta\, dx\, dy \\ & + \int_{-h/2}^{h/2} (v_{,x}\zeta)\big|_0^{\ell}\, dy + \int_0^{\ell} (v_{,y}\zeta)\big|_{-h/2}^{h/2}\, dx - \int_0^{\ell}\int_{-h/2}^{h/2} (v_{,xx} + v_{,yy})\,\zeta\, dx\, dy \\ & + \frac{\lambda_2}{\lambda_1}\left[\int_{-h/2}^{h/2} (u_{,y}\zeta)\big|_0^{\ell}\, dy - \int_0^{\ell} (u_{,x}\zeta)\big|_{-h/2}^{h/2}\, dx + \int_0^{\ell} (v_{,x}\eta)\big|_{-h/2}^{h/2}\, dx - \int_{-h/2}^{h/2} (v_{,y}\eta)\big|_0^{\ell}\, dy\right] \\ & + \lambda_2 \int_{-h/2}^{h/2} (p\eta)\big|_0^{\ell}\, dy - \lambda_2 \int_0^{\ell}\int_{-h/2}^{h/2} p_{,x}\eta\, dx\, dy + \lambda_1 \int_0^{\ell} (p\zeta)\big|_{-h/2}^{h/2}\, dx - \lambda_1 \int_0^{\ell}\int_{-h/2}^{h/2} p_{,y}\zeta\, dx\, dy \\ & + \int_0^{\ell}\int_{-h/2}^{h/2} (\lambda_2 u_{,x} + \lambda_1 v_{,y})\delta p\, dx\, dy = 0.\end{aligned} \tag{3.7.26}$$

This yields the differential equations

$$\begin{aligned} u_{,xx} + u_{,yy} + \lambda_2 p_{,x} &= 0 \\ v_{,xx} + v_{,yy} + \lambda_1 p_{,y} &= 0 \\ \lambda_2 u_{,x} + \lambda_1 v_{,y} &= 0. \end{aligned} \tag{3.7.27}$$

At $x = 0$ and $x = \ell$, we have the kinematic boundary conditions

$$u_{,y}(0) = u_{,y}(\ell) = 0. \tag{3.7.28}$$

and at $x = 0$,

$$u(0) = 0, \tag{3.7.29}$$

which implies

$$\eta_{,y}(0) = \eta_{,y}(\ell) = 0, \quad \eta(0) = 0. \tag{3.7.30}$$

Taking into account these conditions, we find for the dynamic boundary conditions

$$\int_{-h/2}^{h/2} \left(u_{,x} - \frac{\lambda_2}{\lambda_1} v_{,y} + \lambda_2 p \right) dy = 0 \quad \text{at } x = \ell \tag{3.7.31}$$

$$v_{,x} + \frac{\lambda_2}{\lambda_1} u_{,y} = 0 \quad \text{at } x = 0 \text{ and } x = \ell \tag{3.7.32}$$

$$u_{,y} + \frac{\lambda_2}{\lambda_1} v_{,x} = 0 \quad \text{at } y = \pm h/2 \tag{3.7.33}$$

$$v_{,y} - \frac{\lambda_2}{\lambda_1} u_{,x} + \lambda_1 p = 0 \quad \text{at } y = \pm h/2. \tag{3.7.34}$$

The condition (3.7.32) may be simplified to

$$v_{,x} = 0 \quad \text{at } x = 0 \text{ and } x = \ell \tag{3.7.35}$$

due to (3.7.28). The system (3.7.27) may easily be shown to be of fourth order, and we have two boundary conditions on each part of the boundary.

We shall now restrict ourselves to the lowest buckling load, and taking into account the boundary conditions (3.7.28) and (3.7.29), we can try a solution for $u(x, y)$ of the form

$$u(x, y) = U(y) \sin \frac{\pi x}{\ell}. \tag{3.7.36}$$

From the first of the equations of (3.7.27), it then follows that $p(x, y)$ is of the form

$$p(x, y) = P(y) \cos \frac{\pi x}{\ell}, \tag{3.7.37}$$

and from the second or third of the equations of (3.7.27), we find

$$v(x, y) = V(y) \cos \frac{\pi x}{\ell}. \tag{3.7.38}$$

Substitution of these expressions into (3.7.27) yields

$$\begin{aligned} \ddot{U} - \frac{\pi^2}{\ell^2} U - \lambda_2 \frac{\pi}{\ell} P &= 0 \\ \ddot{V} - \frac{\pi^2}{\ell^2} V + \lambda_1 \dot{P} &= 0 \\ \lambda_1 \dot{V} + \lambda_2 \frac{\pi}{\ell} U &= 0, \end{aligned} \tag{3.7.39}$$

where $()^{\cdot} = d(\ )/dy$. This is a set of three homogeneous linear ordinary differential equations, which can be solved by the substitutions

$$U(y) = Ae^{\mu y}, \quad V(y) = Be^{\mu y}, \quad P(y) = Ce^{\mu y}, \tag{3.7.40}$$

which yield

$$\begin{aligned} \left(\mu^2 - \frac{\pi^2}{\ell^2}\right)A \qquad\qquad & - \lambda_2\frac{\pi}{\ell}C = 0 \\ \left(\mu^2 - \frac{\pi^2}{\ell^2}\right)B & + \lambda_1\mu C = 0 \\ \lambda_2\frac{\pi}{\ell}A + \lambda_1\mu B \qquad\qquad & = 0. \end{aligned} \tag{3.7.41}$$

The condition for a nontrivial solution is the vanishing of the determinant of the coefficients, which yields

$$\left(\mu^2 - \frac{\pi^2}{\ell^2}\right)\left(-\lambda_1^2\mu^2 + \lambda_2^2\frac{\pi^2}{\ell^2}\right) = 0. \tag{3.7.42}$$

The eigenvalues are thus

$$\mu_{1,2} = \pm\frac{\pi}{\ell}, \quad \mu_{3,4} = \pm\lambda\frac{\pi}{\ell}, \tag{3.7.43}$$

where $\lambda = \lambda_2/\lambda_1$.

Because we have two pairs of roots with opposite signs, we can also express the solutions as hyperbolic functions instead of exponential functions. Hence, we can write

$$\begin{aligned} U(y) &= \sum_{i=1}^{2} A_{i1}\sinh\mu_i y + \sum_{i=1}^{2} A_{i2}\cosh\mu_i y \\ V(y) &= \sum_{i=1}^{2} B_{i1}\cosh\mu_i y + \sum_{i=1}^{2} B_{i2}\sinh\mu_i y \\ P(y) &= \sum_{i=1}^{2} C_{i1}\sinh\mu_i y + \sum_{i=1}^{2} C_{i2}\cosh\mu_i y, \end{aligned} \tag{3.7.44}$$

where $\mu_1 = \pi/\ell$ and $\mu_2 = \lambda\pi/\ell$.

The solutions with the subscript 1 yield anti-symmetric deformations, and those with the subscript 2 symmetric deformations. Hence, we may split the problem into a symmetric and an anti-symmetric case.

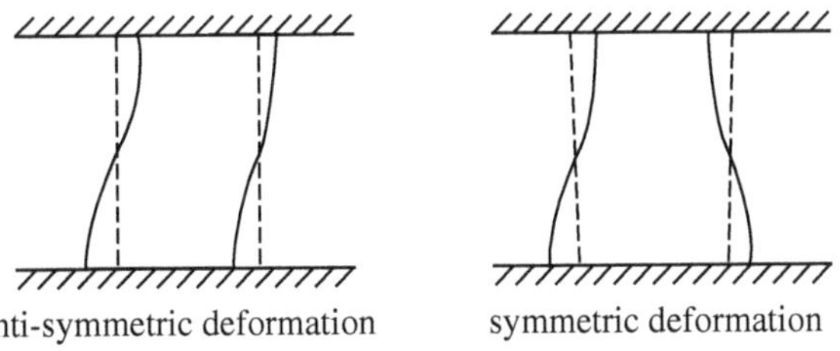

Figure 3.7.2

In the following, we shall discuss the case of anti-symmetric deformation in detail because calculations show that this type of buckling behavior occurs prior to that of symmetric deformation. The last type of deformation first occurs after a reduction of the length ℓ by about 50%. The calculations involved are entirely similar to those of the anti-symmetric case.

In the anti-symmetric case, we have

$$\begin{aligned} U(y) &= A_1 \sinh \frac{\pi y}{\ell} + A_2 \sinh \lambda \frac{\pi y}{\ell} \\ V(y) &= B_1 \cosh \frac{\pi y}{\ell} + B_2 \cosh \lambda \frac{\pi y}{\ell} \\ P(y) &= C_1 \sinh \frac{\pi y}{\ell} + C_2 \sinh \lambda \frac{\pi y}{\ell}. \end{aligned} \tag{3.7.45}$$

For $\mu = \pi/\ell$, we find from (3.7.41)

$$C_1 = 0, \quad B_1 = -\lambda A_1, \tag{3.7.46}$$

and for $\mu = \lambda\pi/\ell$, we find

$$B_2 = -A_2, \quad C_2 = \frac{\lambda^2 - 1}{\lambda_2} \frac{\pi}{\ell} A_2. \tag{3.7.47}$$

The boundary conditions (3.7.33) and (3.7.34), with the aid of (3.7.36) to (3.7.38), can be rewritten to yield

$$\dot{U} - \lambda \frac{\pi}{\ell} V = 0, \quad \dot{V} - \lambda \frac{\pi}{\ell} U + \lambda_1 P = 0, \tag{3.7.48}$$

for $y = \pm h/2$. Substitution of (3.7.45) into (3.7.48) and using (3.7.46) and (3.7.47) yields

$$\begin{aligned} l(1+\lambda^2) \cosh \frac{\pi h}{2\ell} A_1 + 2\lambda \cosh \lambda \frac{\pi h}{2\ell} A_2 = 0 \\ -2\lambda \sinh \frac{\pi h}{2\ell} A_1 - \left(\lambda + \frac{1}{\lambda}\right) \sinh \lambda \frac{\pi h}{2\ell} A_2 = 0. \end{aligned} \tag{3.7.49}$$

Notice that $\lambda = \lambda_2/\lambda_1$ is the single significant load parameter. Its value follows from the condition for a non-trivial solution of (3.7.49), which leads to

$$\left(\lambda^2 + 1\right)^2 \tanh \lambda\theta = 4\lambda^3 \tanh \theta, \tag{3.7.50}$$

where we have introduced the geometric parameter

$$\theta = \frac{\pi h}{2\ell}. \tag{3.7.51}$$

Apart from the solution $\lambda = 1$, which corresponds to the fundamental state only, (3.7.50) has one solution, $\lambda > 1$. Let us consider two limiting cases $\theta \to 0$ and $\theta \to \infty$.

For $\theta \ll 1$, we can use series expansions for the hyperbolic functions. Then (3.7.50) can be rewritten to yield

$$(\lambda^2 - 1)^2 = \frac{1}{3}\lambda^2\theta^2[(\lambda^2 + 1)^2 - 4] + O(\theta^3). \tag{3.7.52}$$

As for $\theta = 0$, $\lambda^2 = 1$ is a root of this equation; we try a solution of the form

$$\lambda^2 = 1 + \varepsilon, \quad 0 < \varepsilon \ll 1, \tag{3.7.53}$$

which yields

$$\varepsilon = \frac{4}{3}\theta^2 + O(\varepsilon\theta^2), \tag{3.7.54}$$

so that

$$\frac{\lambda_2}{\lambda_1} = 1 + \frac{2}{3}\theta^2. \tag{3.7.55}$$

Using the incompressibility condition $\lambda_1\lambda_2 = 1$, we find

$$\begin{aligned} \lambda_2 &= 1 + \frac{1}{3}\theta^2 = 1 + \frac{\pi^2 h^2}{12\ell^2} \\ \lambda_1 &= 1 - \frac{1}{3}\theta^2 = 1 - \frac{\pi^2 h^2}{12\ell^2}. \end{aligned} \tag{3.7.56}$$

This result for the thin plate ($h/\ell \ll 1$) is in complete agreement with the result for the Euler column, and as it is derived from a two-dimensional theory, it justifies the latter result.

For $\theta \to \infty$, (3.7.50) reduces to

$$(\lambda^2 + 1)^2 = 4\lambda^3, \tag{3.7.57}$$

from which follows $\lambda = 3.383$ (and of course, $\lambda = 1$). Hence, it follows that

$$\lambda_1 = \frac{1}{\sqrt{3.383}} \simeq 0.55, \tag{3.7.58}$$

which means a reduction in length of 45%. In this case, the deformations are concentrated near $y = \pm h/2$. In the case of symmetric deformation, the bifurcation equation becomes

$$4\lambda^3 \tanh \lambda\theta = (\lambda^2 + 1)^2 \tanh \theta, \tag{3.7.59}$$

which has a root $\lambda \geq 3.383$. Hence, it follows that bifurcation in an anti-symmetric mode ($\lambda \leq 3.383$) will occur prior to the occurrence of bifurcation in a symmetric mode.

We now continue with the discussion of the anti-symmetric case. From the first of the equations of (3.7.49), we obtain

$$\frac{A_2}{A_1} = -\frac{(1 + \lambda^2)}{2\lambda}\frac{\cosh \theta}{\cosh \lambda\theta}. \tag{3.7.60}$$

Using (3.7.36) to (3.7.38), (3.7.45) to (3.7.47), and (3.7.60), we find

$$
\begin{aligned}
u(x,y) &= A_1\left[\sinh\frac{\pi y}{\ell} - \frac{1+\lambda^2}{2\lambda}\frac{\cosh\theta}{\cosh\lambda\theta}\sinh\frac{\pi y}{\ell}\right]\sin\frac{\pi x}{\ell}\\
v(x,y) &= A_1\left[-\lambda\cosh\frac{\pi y}{\ell} + \frac{1+\lambda^2}{2\lambda}\frac{\cosh\theta}{\cosh\lambda\theta}\cosh\lambda\frac{\pi y}{\ell}\right]\cos\frac{\pi x}{\ell}\\
p(x,y) &= A_1\frac{1-\lambda^4}{2\lambda\lambda_2}\frac{\pi}{\ell}\frac{\cosh\theta}{\cosh\lambda\theta}\sinh\lambda\frac{\pi y}{\ell}\cos\frac{\pi x}{\ell}.
\end{aligned}
\tag{3.7.61}
$$

We now choose the constant A_1 so that $v(0,0)=1$, which yields

$$
A_1 = \frac{-2\lambda^{-1}\cosh\lambda\theta}{2\cosh\lambda\theta - (\lambda^{-2}+1)\cosh\theta}. \tag{3.7.62}
$$

Further, we introduce the shorthand

$$
2\cosh\lambda\theta - \left(\lambda^{-2}+1\right)\cosh\theta \equiv N. \tag{3.7.63}
$$

The expressions (3.7.61) can now be rewritten in the form

$$
\begin{aligned}
u_1(x,y) &= N^{-1}\sin\frac{\pi x}{\ell}\left[-2\frac{\cosh\lambda\theta}{\lambda}\sinh\frac{\pi y}{\ell} + (\lambda^{-2}+1)\cosh\theta\sinh\frac{\lambda\pi y}{\ell}\right]\\
v_1(x,y) &= N^{-1}\cos\frac{\pi x}{\ell}\left[2\cosh\lambda\theta\cosh\frac{\pi y}{\ell} - (\lambda^{-2}+1)\cosh\theta\cosh\frac{\lambda\pi y}{\ell}\right]\\
p_1(x,y) &= (\lambda_2 N)^{-1}\frac{\pi}{\ell}(\lambda^2-\lambda^{-2})\cosh\theta\sinh\frac{\lambda\pi y}{\ell}\cos\frac{\pi x}{\ell}.
\end{aligned}
\tag{3.7.64}
$$

Having determined the displacement fields for the critical case of neutral equilibrium, we now proceed with the discussion of its stability. Therefore, we consider small but finite displacements from the equilibrium configuration,

$$
\begin{aligned}
u(x,y) &= au_1(x,y) + \overline{u}(x,y)\\
v(x,y) &= av_1(x,y) + \overline{v}(x,y),
\end{aligned}
\tag{3.7.65}
$$

where $\overline{u}$ and $\overline{v}$ are orthogonal (in some suitable sense) to the buckling mode. Stability is now governed by the behavior of the functional (3.7.22) under the nonlinear side condition (3.7.18). To make full use of the properties of the functional in the critical case of neutral equilibrium, we consider the functional (3.7.22). Substituting (3.7.65) into this functional, we find

$$
\begin{aligned}
P_{\mathrm{II}} - P_{\mathrm{I}} = \int_0^{\ell}\int_{-h/2}^{h/2} \Big[&\frac{1}{2}a^2\left(u_{1,x}^2+u_{1,y}^2+v_{1,x}^2+v_{1,y}^2\right) + a\left(u_{1,x}\overline{u}_{,x}+u_{1,y}\overline{u}_{,y}+v_{1,x}\overline{v}_{,x}+v_{1,y}\overline{v}_{,y}\right)\\
&+\frac{1}{2}\left(\overline{u}_{,x}^2+\overline{u}_{,y}^2+\overline{v}_{,x}^2+\overline{v}_{,y}^2\right) + \lambda a^2\left(u_{1,y}v_{1,x}-u_{1,x}v_{1,y}\right)\\
&+\lambda a\left(u_{1,y}\overline{v}_{,x}+\overline{u}_{,y}\overline{v}_{1,x}-u_{1,x}\overline{v}_{,y}-\overline{u}_{,x}v_{1,y}\right) + \lambda\left(\overline{u}_{,y}\overline{v}_{,x}-\overline{u}_{,x}\overline{v}_{,y}\right)\Big]\,dx\,dy,
\end{aligned}
\tag{3.7.66}
$$

where the terms involving a^2 vanish because u_1 and v_1 satisfy the equations for the critical case of neutral equilibrium. Using (3.7.25) with $\eta = \overline{u}$, $\zeta = \overline{v}$, and $u = u_1, v = v_1, p = p_1$, we can rewrite the bilinear terms in (3.7.66) to yield

$$-\int_0^{\ell}\int_{-h/2}^{h/2} ap_1\left(\lambda_2\overline{u}_{,x} + \lambda_1\overline{v}_{,y}\right) dx\, dy, \tag{3.7.67}$$

so that our final result for the functional becomes

$$\begin{aligned} P_{\mathrm{II}} - P_{\mathrm{I}} = \int_0^{\ell}\int_{-h/2}^{h/2} \Big[&\frac{1}{2}\left(\overline{u}_{,x}^2 + \overline{u}_{,y}^2 + \overline{v}_{,x}^2 + \overline{v}_{,y}^2\right) + \lambda\left(\overline{u}_{,y}\overline{v}_{,x} - \overline{u}_{,x}\overline{v}_{,y}\right) \\ &- ap_1\left(\lambda_2\overline{u}_{,x} + \lambda_1\overline{v}_{,y}\right)\Big]\, dx\, dy. \end{aligned} \tag{3.7.68}$$

Substitution of (3.7.65) into the nonlinear side condition (3.7.18) yields

$$\begin{aligned} &\lambda_2\overline{u}_{,x} + \lambda_1\overline{v}_{,y} + a^2\left(u_{1,x}v_{1,y} - u_{1,y}v_{1,x}\right) \\ &\quad + a\left(u_{1,x}\overline{v}_{,y} + \overline{u}_{,x}v_{1,x} - u_{1,y}\overline{v}_{,x} - \overline{u}_{,y}v_{1,x}\right) + \overline{u}_{,x}\overline{v}_{,y} - \overline{u}_{,y}\overline{v}_{,x} = 0. \end{aligned} \tag{3.7.69}$$

We now use this nonlinear condition to rewrite the term involving p_1 in (3.7.68), which yields

$$\begin{aligned} &-\int_0^{\ell}\int_{-h/2}^{h/2} ap_1\left(\lambda_2\overline{u}_{,x} + \lambda_1\overline{v}_{,y}\right) dx\, dy \\ &= \int_0^{\ell}\int_{-h/2}^{h/2} \left[a^3p_1\left(u_{1,x}v_{1,y} - u_{1,y}v_{1,x}\right) + a^2p_1\left(u_{1,x}\overline{v}_{,y} + \overline{u}_{,x}v_{1,y} - u_{1,y}\overline{v}_{,x} - \overline{u}_{,y}v_{1,x}\right)\right. \\ &\quad \left. + ap_1\left(\overline{u}_{,x}\overline{v}_{,y} - \overline{u}_{,y}\overline{v}_{,x}\right)\right] dx\, dy, \end{aligned} \tag{3.7.70}$$

where the first term on the right-hand side vanishes because

$$\int_0^{\ell} \sin^2\frac{\pi x}{\ell}\cos\frac{\pi x}{\ell}\, dx = 0, \qquad \int_0^{\ell} \cos^3\frac{\pi x}{\ell}\, dx = 0.$$

The functional (3.7.68) can now be rewritten in the form

$$\begin{aligned} P_{\mathrm{II}} - P_{\mathrm{I}} = \int_0^{\ell}\int_{-h/2}^{h/2} \Big\{ &\frac{1}{2}\left(\overline{u}_{,x}^2 + \overline{u}_{,y}^2 + \overline{v}_{,x}^2 + \overline{v}_{,y}^2\right) + \lambda\left(\overline{u}_{,y}\overline{v}_{,x} - \overline{u}_{,x}\overline{v}_{,y}\right) \\ &+ a^2p_1\left(u_{1,x}\overline{v}_{,y} + \overline{u}_{,x}v_{1,y} - u_{1,y}\overline{v}_{,x} - \overline{u}_{,y}v_{1,x}\right) + ap_1\left(\overline{u}_{,x}\overline{v}_{,y} - \overline{u}_{,y}\overline{v}_{,x}\right) \\ &+ \overline{p}[\lambda_2\overline{u}_{,x} + \lambda_1\overline{v}_{,y} + a^2\left(u_{1,x}v_{1,y} - u_{1,y}v_{1,x}\right) \\ &+ a\left(u_{1,x}\overline{v}_{,y} + \overline{u}_{,x}v_{1,y} - u_{1,y}\overline{v}_{,x} - \overline{u}_{,y}v_{1,x}\right) + \overline{u}_{,x}\overline{v}_{,y} - \overline{u}_{,y}\overline{v}_{,x}] \\ &+ \mu\left(u_1\overline{u} + v_1\overline{v}\right)\Big\}\, dx\, dy, \end{aligned} \tag{3.7.71}$$

where we have taken into account the nonlinear side condition with the Lagrangian multiplier $\overline{p}\,(x, y)$ and the orthogonality condition

$$\int_0^{\ell}\int_{-h/2}^{h/2} (u_1\overline{u} + v_1\overline{v})\, dx\, dy = 0 \tag{3.7.72}$$

with the multiplier μ.† We now minimize this functional with respect to $\overline{u}$, $\overline{v}$, $\overline{p}$, and μ for a fixed value of the amplitude a. Variation with respect to $\overline{u}$ yields

$$\begin{aligned}\int_0^{\ell}\int_{-h/2}^{h/2} &\{\overline{u}_{,x}\delta\overline{u}_{,x} + \overline{u}_{,y}\delta\overline{u}_{,y} + \lambda\,(\overline{v}_{,x}\delta\overline{u}_{,y} - \overline{v}_{,y}\delta\overline{u}_{,x})\\ &+ a^2 p_1\,(v_{1,y}\delta\overline{u}_{,x} - v_{1,x}\delta\overline{u}_{,y}) + a p_1\,(\overline{v}_{,y}\delta\overline{u}_{,x} - \overline{v}_{,x}\delta\overline{u}_{,y})\\ &+ \overline{p}\,[\lambda_2\delta\overline{u}_{,x} + a\,(v_{1,y}\delta\overline{u}_{,x} - v_{1,x}\delta\overline{u}_{,y}) + \overline{v}_{,y}\delta\overline{u}_{,x} - \overline{v}_{,x}\delta\overline{u}_{,y}]\\ &+ \mu u_1\delta\overline{u}\}\, dx\, dy = 0.\end{aligned} \tag{3.7.73}$$

By integration by parts, we obtain

$$\begin{aligned}-\int_0^{\ell}\int_{-h/2}^{h/2} &[\overline{u}_{,xx} + \overline{u}_{,yy} + a^2\,(p_1 v_{1,y})_{,x} + a^2\,(p_1 v_{1,y})_{,x} - a^2\,(p_1 v_{1,x})_{,y}\\ &+ a\,(p_1\overline{v}_{,y})_{,x} - a\,(p_1\overline{v}_{,x})_{,y} + \lambda_2\overline{p}_{,x}\\ &+ a\,(\overline{p}v_{1,x})_{,x} - a\,(\overline{p}v_{1,x})_{,y} + (\overline{p}\overline{v}_{,y})_{,x} - (\overline{p}\overline{v}_{,x})_{,y} + \mu u_1]\,\delta\overline{u}\, dx\, dy\\ &+ \int_0^{\ell} (\overline{u}_{,y} + \lambda\overline{v}_{,x} - a^2 p_1 v_{1,x} - a p_1\overline{v}_{,x} - a\overline{p}v_{1,x} - \overline{p}\overline{v}_{,x})\,\delta\overline{u}\,\Big|_{-h/2}^{h/2}\, dx\\ &+ \int_{-h/2}^{h/2} (\overline{u}_{,x} - \lambda\overline{v}_{,y} + a^2 p_1 v_{1,y} + a p_1\overline{v}_{,y} + \lambda_2\overline{p} + a\overline{p}v_{1,y} + \overline{p}\overline{v}_{,y})\,\delta\overline{u}\,\Big|_0^{\ell}\, dy.\end{aligned} \tag{3.7.74}$$

Remembering the geometric boundary conditions $\overline{u} = 0$ at $x = 0$ and $\overline{u}_{,y} = 0$ at $x = \ell$, we obtain

$$\begin{aligned}&\overline{u}_{,xx} + \overline{u}_{,yy} + a^2(p_{1,x}v_{1,y} - p_{1,y}v_{1,x}) + a\,(p_{1,x}\overline{v}_{,y} - p_{1,y}\overline{v}_{,y})\\ &\quad + \lambda_2\overline{p}_{,x} + a(\overline{p}_{,x}v_{1,y} - \overline{p}_{,y}v_{1,y}) + \overline{p}_{,x}\overline{v}_{,y} - \overline{p}_{,y}\overline{v}_{,x} + \mu u_1 = 0,\end{aligned} \tag{3.7.75}$$

and the dynamic boundary conditions

$$\begin{aligned}&\overline{u}_{,y} + \lambda\overline{v}_{,x} - p_1(a^2 v_{1,x} + a\overline{v}_{,x}) - \overline{p}\,(a v_{1,x} + \overline{v}_{,x}) = 0 \quad \text{for} \quad y = \pm h/2\\ &\int_{-h/2}^{h/2} [\overline{u}_{,x} - \lambda\overline{v}_{,y} + (a^2 p_1 + a\overline{p})v_{1,y} + (a p_1 + \overline{p})\,\overline{v}_{,y} + \lambda_2\overline{p}]\, dy = 0 \quad \text{for} \quad x = \ell.\end{aligned} \tag{3.7.76}$$

† Notice that μ is independent of x and y.

Variation with respect to $\overline{v}$ yields

$$
\begin{aligned}
\int_0^{\ell}\int_{-h/2}^{h/2} & [\overline{v}_{,x}\,\delta\overline{v}_{,x} + \overline{v}_{,y}\delta\overline{v}_{,y} + \lambda\,(\overline{u}_{,y}\delta\overline{v}_{,x} - \overline{u}_{,x}\delta\overline{v}_{,y}) \\
& + a^2 p_1\,(u_{1,x}\delta\overline{v}_{,y} - u_{1,y}\delta\overline{v}_{,x}) + ap_1\,(\overline{u}_{,x}\delta\overline{v}_{,y} - \overline{u}_{,y}\delta\overline{v}_{,x}) \\
& + \overline{p}\,\{\lambda_1\delta\overline{v}_{,y} + a\,(u_{1,x}\delta\overline{v}_{,y} - u_{1,y}\delta\overline{v}_{,x}) + \overline{u}_{,x}\delta\overline{v}_{,y} - \overline{u}_{,y}\delta\overline{v}_{,x}\} \\
& + \mu v_1\delta\overline{v}]\; dx\, dy = 0.
\end{aligned}
\tag{3.7.77}
$$

Integration by parts yields

$$
\begin{aligned}
-\int_0^{\ell}\int_{-h/2}^{h/2} & [\overline{v}_{,xx} + \overline{v}_{,yy} + a^2\,(p_1 u_{1,x})_{,y} - a^2\,(p_1 u_{1,y})_{,x} \\
& + a\,(p_1\overline{u}_{,x})_{,y} - a\,(p_1\overline{u}_{,y})_{,x} + \lambda_1\overline{p}_{,y} \\
& + a\,(\overline{p}\,\overline{u}_{1,x})_{,y} - a\,(\overline{p}u_{,y})_{,x} + (\overline{p}\,\overline{u}_{,x})_{,y} - (\overline{p}\,\overline{u}_{,y})_{,x} + \mu v_1]\,\delta\overline{v}\, dx\, dy \\
& + \int_0^{\ell} (\overline{v}_{,y} - \lambda\overline{u}_{,x} + a^2 p_1 u_{1,x} + ap_1\overline{u}_{,x} + \lambda_1\overline{p} + a\overline{p}u_{1,x} + \overline{p}\overline{u}_{,x})\,\delta\overline{v}\,|_{-h/2}^{h/2}\, dx \\
& + \int_{-h/2}^{h/2} (\overline{v}_{,x} + \lambda\overline{u}_{,y} - a^2 p_1 u_{1,y} - ap_1\overline{u}_{,y} - a\overline{p}u_{1,y} - \overline{p}\,\overline{u}_{,y})\,\delta\overline{v}\,|_0^{\ell} dy = 0,
\end{aligned}
\tag{3.7.78}
$$

from which follows

$$
\begin{aligned}
& \overline{v}_{,xx} + \overline{v}_{,yy} + a^2(p_{1,y}u_{1,x} - p_{1,x}u_{1,y}) + a(p_{1,y}\overline{u}_{,x} - p_{1,x}\overline{u}_{,y}) \\
& \quad + \lambda_1\overline{p}_{,y} + a(\overline{p}_{,y}\overline{u}_{,x} - \overline{p}_{,x}u_{1,y}) + \overline{p}_{,y}\overline{u}_{,x} - \overline{p}_{,x}\overline{u}_{,y} + \mu v_1 = 0,
\end{aligned}
\tag{3.7.79}
$$

with the dynamic boundary conditions

$$
\overline{v}_{,y} - \lambda\overline{u}_{,x} + (a^2 p_1 + a\overline{p})u_{1,x} + (ap_1 + \overline{p})\overline{u}_{,x} + \lambda_1\overline{p} = 0 \quad \text{for} \quad y = \pm h/2 \tag{3.7.80}
$$

$$
\overline{v}_{,x} + \lambda\overline{u}_{,y} - (a^2 p_1 + a\overline{p})u_{1,y} - (ap_1 + \overline{p})\overline{u}_{,y} = 0 \quad \text{for} \quad x = 0 \quad \text{and} \quad x = \ell. \tag{3.7.81}
$$

Variation with respect to $\overline{p}$ and μ yields the nonlinear side condition and the orthogonality condition, respectively.

We now restrict ourselves to vanishingly small values of the amplitude a. Hence, we may write

$$
\begin{aligned}
\overline{u} &= \overline{u}^0 + a\overline{u}^1 + a^2\overline{u}^2 + \cdots \\
\overline{v} &= \overline{v}^0 + a\overline{v}^1 + a^2\overline{v}^2 + \cdots \\
\overline{p} &= \overline{p}^0 + a\overline{p}^1 + a^2\overline{p}^2 + \cdots \\
\mu &= \mu^0 + a\mu^1 + a^2\mu^2 + \cdots.
\end{aligned}
\tag{3.7.82}
$$

Introducing these expansions into the differential equations and the corresponding boundary conditions, we find to zeroth order

$$\begin{aligned}
&\overline{u}^0_{,xx} + \overline{u}^0_{,yy} + \lambda_2 \overline{p}^0_{,x} + \overline{p}^0_{,x}\overline{v}^0_{,y} - \overline{p}^0_{,y}\overline{v}^0_{,x} + \mu^0 u_1 = 0\\
&\overline{v}^0_{,xx} + \overline{v}^0_{,yy} + \lambda_1 \overline{p}^0_{,y} + \overline{p}^0_{,y}\overline{u}^0_{,x} - \overline{p}^0_{,x}\overline{u}^0_{,y} + \mu^0 v_1 = 0\\
&\lambda_2 \overline{u}^0_{,x} + \lambda_1 \overline{v}^0_{,y} + \overline{u}^0_{,x}\overline{v}^0_{,y} - \overline{u}^0_{,y}\overline{v}^0_{,x} = 0\\
&\int_0^{\ell}\int_{-h/2}^{h/2} \left(u_1 \overline{u}^0 + v_1 \overline{v}^0\right) dx\, dy = 0,
\end{aligned} \tag{3.7.83}$$

with the dynamic boundary conditions

$$\left.\begin{aligned}
\overline{u}^0_{,y} + \lambda \overline{v}^0_{,x} - \overline{p}^0 \overline{v}^0_{,x} = 0\\
\overline{v}^0_{,y} - \lambda \overline{u}^0_{,x} + \overline{p}^0 \overline{u}^0_{,x} + \lambda_1 \overline{p}^0 = 0
\end{aligned}\right\} \quad \text{for} \quad y = \pm h/2, \tag{3.7.84}$$

$$\int_{-h/2}^{h/2} \left(\overline{u}^0_{,x} - \lambda \overline{v}^0_{,y} + \overline{p}^0 \overline{v}^0_{,y} + \lambda_2 \overline{p}^0\right) dy = 0 \quad \text{for} \quad x = \ell \tag{3.7.85}$$

$$\overline{v}^0_{,x} + \lambda \overline{u}^0_{,y} - \overline{p}^0 \overline{u}^0_{,y} = 0 \quad \text{for} \quad x = 0 \quad \text{and} \quad x = \ell, \tag{3.7.86}$$

and the geometric boundary conditions

$$\overline{u}^0 = 0 \quad \text{for} \quad x = 0 \quad \text{and} \quad \overline{u}^0_{,y} = 0 \quad \text{for} \quad x = \ell. \tag{3.7.87}$$

Because the equations and the boundary conditions are homogeneous, the unique solution is

$$\overline{u}^0 = \overline{v}^0 = \overline{p}^0 = \overline{\mu}^0 = 0. \tag{3.7.88}$$

For term linear in a, we obtain

$$\begin{aligned}
&\overline{u}^1_{,xx} + \overline{u}^1_{,yy} + \lambda_2 \overline{p}^1_{,x} + \mu^1 u_1 = 0\\
&\overline{v}^1_{,xx} + \overline{v}^1_{,yy} + \lambda_1 \overline{p}^1_{,y} + \mu^1 v_1 = 0\\
&\lambda_2 \overline{u}^1_{,x} + \lambda_1 \overline{v}^1_{,y} = 0\\
&\int_0^{\ell}\int_{-h/2}^{h/2} \left(u_1 \overline{u}^1 + v_1 \overline{v}^1\right) dx\, dy = 0,
\end{aligned} \tag{3.7.89}$$

with the dynamic boundary conditions

$$\left.\begin{aligned}
\overline{u}^1_{,y} + \lambda \overline{v}^1_{,x} = 0\\
\overline{v}^1_{,y} - \lambda \overline{u}^1_{,x} + \lambda_1 \overline{p}^1 = 0
\end{aligned}\right\} \quad \text{for} \quad y = \pm h/2 \tag{3.7.90}$$

$$\int_{-h/2}^{h/2} \left(\overline{u}^1_{,x} - \lambda \overline{v}^1_{,y} + \lambda_2 \overline{p}^1\right) dy = 0 \quad \text{for} \quad x = \ell \tag{3.7.91}$$

$$\overline{v}^1_{,x} + \lambda \overline{u}^1_{,y} = 0 \quad \text{for} \quad x = 0 \quad \text{and} \quad x = \ell \tag{3.7.92}$$

and the geometric conditions

$$\overline{u}^1 = 0 \quad \text{for} \quad x = 0 \quad \text{and} \quad \overline{u}^1_{,y} = 0 \quad \text{for} \quad x = \ell. \tag{3.7.93}$$

Again, this is a set of homogeneous equations and boundary conditions, so the unique solution is given by

$$\overline{u}^1 = \overline{v}^1 = \overline{p}^1 = \mu^1 = 0. \tag{3.7.94}$$

For the terms quadratic in the amplitude a, we obtain

$$\begin{aligned}
&\overline{u}^2_{,xx} + \overline{u}^2_{,yy} + \lambda_2 \overline{p}^2_{,x} + \mu^2 u_1 = p_{1,y} v_{1,x} - p_{1,x} v_{1,y} \\
&\overline{v}^2_{,xx} + \overline{v}^2_{,yy} + \lambda_1 \overline{p}^2_{,y} + \mu^2 v_1 = p_{1,x} v_{1,y} - p_{1,y} v_{1,x} \\
&\lambda_2 \overline{u}^2_{,x} + \lambda_1 \overline{v}^2_{,y} = u_{1,y} v_{1,x} - u_{1,x} v_{1,y} \\
&\int_0^\ell \int_{-h/2}^{h/2} \left(u_1 \overline{u}^2 + v_1 \overline{v}^2\right) dx\, dy = 0,
\end{aligned} \tag{3.7.95}$$

with the dynamic boundary conditions

$$\left.\begin{aligned}
\overline{u}^2_{,y} + \lambda \overline{v}^2_{,x} &= p_1 v_{1,x} \\
\overline{v}^2_{,y} - \lambda \overline{u}^2_{,x} + \lambda_1 \overline{p}^2 &= p_1 u_{1,x}
\end{aligned}\right\} \quad \text{for} \quad y = \pm h/2 \tag{3.7.96}$$

$$\int_{-h/2}^{h/2} \left(\overline{u}^2_{,x} - \lambda \overline{v}^2_{,y} + \lambda_2 \overline{p}^2\right) dy = - \int_{-h/2}^{h/2} p_1 v_{1,y}\, dy \quad \text{for} \quad x = \ell \tag{3.7.97}$$

$$\overline{v}^2_{,x} + \lambda \overline{u}^2_{,y} = p_1 u_{1,y} \quad \text{for} \quad x = 0 \quad \text{and} \quad x = \ell, \tag{3.7.98}$$

and the geometric conditions

$$\overline{u}^2 = 0 \quad \text{at} \quad x = 0 \quad \text{and} \quad \overline{u}^2_{,y} = 0 \quad \text{at} \quad x = \ell. \tag{3.7.99}$$

The present equations and boundary conditions are the conditions for the stationary value of the functional

$$\begin{aligned}
(P_{\text{II}} - P_{\text{I}})^* = a^4 \int_0^\ell \int_{-h/2}^{h/2} \Bigg[&\frac{1}{2} \left(\overline{u}^{2^2}_{,x} + \overline{u}^{2^2}_{,y} + \overline{v}^{2^2}_{,x} + \overline{v}^{2^2}_{,y}\right) \\
&+ \lambda \left(\overline{u}^2_{,y} \overline{v}^2_{,x} - \overline{u}^2_{,x} \overline{u}^2_{,y}\right) \\
&+ p_1 \left(u_{1,x} \overline{v}^2_{,y} + \overline{u}^2_{,x} v_{1,y} - u_{1,y} \overline{v}^2_{,x} - \overline{u}^2_{,y} v_{1,x}\right) \Bigg] dx\, dy
\end{aligned} \tag{3.7.100}$$

under the linear side condition

$$\lambda_2 \overline{u}^2_{,x} + \lambda_1 \overline{v}^2_{,y} + u_{1,x} v_{1,y} - u_{1,y} v_{1,x} = 0 \tag{3.7.101}$$

and the orthogonality condition

$$\int_0^\ell \int_{-h/2}^{h/2} \left(u_1 \overline{u}^2 + v_1 \overline{v}^2\right) dx\, dy = 0. \tag{3.7.102}$$

This is the variational problem of a quadratic functional with the linear side condition, and has a second variation that is positive-definite under the requirement of orthogonality of the field $(\overline{u}^2, \overline{v}^2)$ with respect to the buckling mode $\mathbf{u}_1$. The solution of the variational problem thus represents a minimum of the functional $(P_{\text{II}} - P_{\text{I}})^*$. Writing

$$\text{Min}\,(P_{\text{II}} - P_{\text{I}})^* = a^4 A_4, \tag{3.7.103}$$

where A_4 is defined by the value of the integral in (3.7.100) evaluated for $\overline{u}^2, \overline{v}^2$ as solutions of the equations and boundary conditions (3.7.95) to (3.7.99), a necessary condition for stability is $A_4 \geq 0$, and a sufficient condition is $A_4 > 0$ in the case of neutral equilibrium at $\lambda = \lambda_{\text{cr}}$.

To avoid lengthy calculations, in the following we shall restrict ourselves to the case of a slender plate (i.e., $\theta \ll 1$). In that case, from (3.7.55) we have

$$\lambda_{\text{cr}} = 1 + \frac{2}{3}\theta^2 + O\left(\theta^4\right) = 1 + \frac{\pi^2 h^2}{6\ell^2} + O\left(\frac{h^4}{\ell^4}\right). \tag{3.7.104}$$

Because $\pi y/\ell$ is of order θ, we can write the buckling mode (3.7.64) in the form

$$\begin{aligned}
u_1 &= \left[\frac{\pi y}{\ell} + O\left(\theta^3\right)\right] \sin\frac{\pi x}{\ell} \\
v_1 &= \left[1 - \frac{\pi^2 y^2}{2\ell^2} + O\left(\theta^4\right)\right] \cos\frac{\pi x}{\ell} \\
p_1 &= \frac{\pi}{\ell}\left[\frac{2\pi y}{2\ell} + O\left(\theta^3\right)\right] \cos\frac{\pi x}{\ell},
\end{aligned} \tag{3.7.105}$$

where we must note that the derivative with respect to y of an error term θ^n is of order $\theta^{n-1}\ell^{-1}$.

We can now evaluate the expressions

$$\begin{aligned}
p_{1,x}v_{1,y} - p_{1,y}v_{1,x} &= \frac{\pi^3}{\ell^3}\left[1 + O\left(\theta^2\right)\right] \sin\frac{2\pi x}{\ell} \\
p_{1,y}u_{1,x} - p_{1,x}u_{1,y} &= \frac{2\pi^4 y}{\ell^4}\left[1 + O\left(\theta^2\right)\right] \sin\frac{2\pi x}{\ell} \\
u_{1,x}v_{1,y} - u_{1,y}v_{1,x} &= \frac{\pi^2}{2\ell^2}\left[1 + O\left(\theta^2\right)\right]\left(1 - \cos\frac{2\pi x}{\ell}\right) \\
-p_1 v_{1,x} &= \frac{\pi^3 y}{\ell^3}\left[1 + O\left(\theta^2\right)\right] \sin\frac{2\pi x}{\ell} \\
p_1 u_{1,x} &= \frac{\pi^4 y^2}{\ell^4}\left[1 + O\left(\theta^2\right)\right]\left(1 + \cos\frac{2\pi x}{\ell}\right).
\end{aligned} \tag{3.7.106}$$

$$p_1 v_{1,y} = -\frac{\pi^4 y^2}{\ell^4}\left[1 + O\left(\theta^2\right)\right]\left(1 + \cos\frac{2\pi x}{\ell}\right)$$

$$-p_1 u_{1,y} = -\frac{\pi^3 y}{\ell^3}\left[1 + O\left(\theta^2\right)\right]\sin\frac{2\pi x}{\ell}.$$

The equations (3.7.95) now become

$$\begin{aligned}
&\overline{u}^2_{,xx} + \overline{u}^2_{,yy} + \lambda_2 \overline{p}^2_{,x} = -\mu^2 \frac{\pi y}{\ell}\sin\frac{\pi x}{\ell} - \frac{\pi^3}{\ell^3}\sin\frac{2\pi x}{\ell} \\
&\overline{v}^2_{,xx} + \overline{v}^2_{,yy} + \lambda_1 \overline{p}^2_{,y} = -\mu^2\left(1 - \frac{\pi^2 y^2}{2\ell^2}\right)\cos\frac{\pi x}{\ell} - \frac{2\pi^4 y}{\ell^4} \\
&\lambda_2 \overline{u}^2_{,x} + \lambda_1 \overline{v}^2_{,y} = -\frac{\pi^2}{2\ell^2}\left(1 - \cos\frac{2\pi x}{\ell^2}\right) \\
&\int_0^\ell \int_{-h/2}^{h/2}\left[\frac{\pi y}{\ell}\sin\frac{\pi x}{\ell}\overline{u}^2 + \left(1 - \frac{\pi^2 y^2}{2\ell^2}\right)\cos\frac{\pi x}{\ell}\overline{v}^2\right]dx\,dy = 0,
\end{aligned} \tag{3.7.107}$$

with the dynamic boundary conditions

$$\left.\begin{aligned}
\overline{u}^2_{,y} + \lambda \overline{v}^2_{,x} &= \mp\frac{\pi^3 h}{2\ell^3}\sin\frac{2\pi x}{\ell} \\
\overline{v}^2_{,y} - \lambda \overline{u}^2_{,x} + \lambda_1 \overline{p}^2 &= \frac{\pi^4 h^2}{4\ell^4}\left(1 + \cos\frac{2\pi x}{\ell}\right)
\end{aligned}\right\} \quad \text{for} \quad y = \pm\frac{h}{2} \tag{3.7.108}$$

$$\int_{-h/2}^{h/2}\left(\overline{u}^2_{,x} - \lambda \overline{v}^2_{,y} + \lambda_2 \overline{p}^2\right)dy = \frac{\pi^4 h^3}{6\ell^4} \quad \text{for} \quad x = \ell \tag{3.7.109}$$

$$\overline{v}^2_{,x} + \lambda \overline{u}^2_{,y} = 0 \quad \text{for} \quad x = 0 \quad \text{and} \quad x = \ell, \tag{3.7.110}$$

and the geometric conditions

$$\overline{u}^2 = 0 \quad \text{at} \quad x = 0 \quad \text{and} \quad \overline{u}^2_{,y} = 0 \quad \text{at} \quad x = \ell. \tag{3.7.111}$$

The value μ^2 in (3.7.107) is still unknown. The terms in (3.7.107) involving μ^2 lead to terms in $\overline{u}^2$ and $\overline{v}^2$ proportional to $\sin \pi x/\ell$ and $\cos \pi x/\ell$, respectively, which do not satisfy the orthogonality condition, and hence we require

$$\mu^2 = 0. \tag{3.7.112}$$

We now try a solution in the form

$$\begin{aligned}
\overline{u}^2 &= \left(\frac{\pi}{8\ell} - \frac{\pi^3 y^2}{4\ell^3}\right)\sin\frac{2\pi x}{\ell} - \frac{\pi^2}{4\ell^2}x + C_1\frac{\pi^3 h^2}{\ell^3}\sin\frac{2\pi x}{\ell} + C_2\frac{\pi^4 h^2}{\ell^4}x \\
\overline{v}^2 &= \left(\frac{\pi^2\ y}{4\ell^2} - \frac{\pi^4 y^3}{2\ell^4}\right)\cos\frac{2\pi x}{\ell} - \frac{\pi^2}{4\ell^2}y + C_3\frac{\pi^4 h^2 y}{\ell^4}\cos\frac{2\pi x}{\ell} + C_4\frac{\pi^4 h^2}{\ell^4}y \\
\overline{p}^2 &= -\frac{\pi^4\ y^2}{\ell^4}\cos\frac{2\pi x}{\ell} - \frac{\pi^4\ y^2}{\ell^4} + C_5\frac{\pi^4 h^2}{\ell^4}\cos\frac{2\pi x}{\ell} + C_6\frac{\pi^4 h^2}{\ell^4},
\end{aligned} \tag{3.7.113}$$

where all numbers $C_1 - C_6$ are of order of magnitude unity. We have made use of the fact that λ_1 and λ_2 are of order unity with an error of order $\pi^2h^2/4\ell^2$, but in some places we had to allow for the more accurate relation $\lambda_2/\lambda_1 = \lambda = 1 + \pi^2h^2/6\ell^2$.

It is easily verified that the expressions (3.7.113) satisfy the equations (3.7.107) and the first boundary conditions (3.7.109) with relative errors of $O\left(h^2/L^2\right)$. For the left-hand member of the second of the boundary conditions (3.7.108), we obtain

$$\begin{aligned}
&\left(\frac{\pi^2}{4\ell^2} + \frac{3}{8}\frac{\pi^4h^2}{\ell^4}\right)\cos\frac{2\pi x}{\ell} - \frac{\pi^2}{4\ell^2} + C_3\frac{\pi^4h^2}{\ell^4}\cos\frac{2\pi x}{\ell} + C_4\frac{\pi^4h^2}{\ell^4} \\
&\quad - \lambda\left[\left(\frac{\pi^2}{4\ell^2} - \frac{\pi^4h^2}{8\ell^4}\right)\cos\frac{2\pi x}{\ell} - \frac{\pi^2}{4\ell^2} + C_1\frac{2\pi^4h^2}{\ell^4}\cos\frac{2\pi x}{\ell} + C_2\frac{\pi^4h^2}{\ell^4}\right] \\
&\quad + \lambda_1\left(-\frac{\pi^4h^2}{4\ell^4}\cos\frac{2\pi x}{\ell} - \frac{\pi^4h^2}{4\ell^4} + C_5\frac{\pi^4h^2}{\ell^4}\cos\frac{2\pi x}{\ell} + C_6\frac{\pi^4h^2}{\ell^4}\right) \\
&= \left(\frac{\pi^2}{4\ell^2} + \frac{3}{8}\frac{\pi^4h^2}{\ell^4} - \frac{\pi^2}{4\ell^2} + \frac{\pi^4h^2}{8\ell^4} - \frac{\pi^4h^2}{24\ell^4} - \frac{\pi^4h^2}{4\ell^4} + \cdots\right)\cos\frac{2\pi x}{\ell} \\
&\quad - \frac{\pi^2}{4\ell^2} + \frac{\pi^2}{4\ell^2} + \frac{\pi^4h^2}{24\ell^4} - \frac{\pi^4h^2}{4\ell^4} + C_4\frac{\pi^4h^2}{\ell^4} - C_2\frac{\pi^4h^2}{\ell^4} + C_6\frac{\pi^4h^2}{\ell^4} + \cdots \qquad (3.7.114) \\
&\quad + \left(C_3\frac{\pi^4h^2}{\ell^4} - 2C_1\frac{\pi^4h^2}{\ell^4} + C_5\frac{\pi^4h^2}{\ell^4}\right)\cos\frac{2\pi x}{\ell} \\
&= \left(\frac{5}{24} + C_3 + C_5 - 2C_1\right)\frac{\pi^4h^2}{\ell^4}\cos\frac{2\pi x}{\ell} + \left(-\frac{5}{24} + C_4 - C_2 + C_6\right)\frac{\pi^4h^2}{\ell^4} \\
&= -\frac{\pi^4h^2}{4\ell^4}\left(1 + \cos\frac{2\pi x}{\ell}\right) + \left(\frac{1}{24} + C_4 - C_2 + C_6\right)\frac{\pi^4h^2}{\ell^4} \\
&\quad + \left(\frac{11}{24} + C_3 + C_5 - 2C_1\right)\frac{\pi^4h^2}{\ell^4}\cos\frac{2\pi x}{\ell}.
\end{aligned}$$

It follows that

$$-C_2 + C_4 + C_6 + \frac{1}{24} = 0 \qquad (3.7.115)$$

$$-2C_1 + C_3 + C_5 + \frac{11}{24} = 0. \qquad (3.7.116)$$

For the left-hand member of (3.7.109), we obtain

$$\begin{aligned}
&2\left\{\left(\frac{\pi h}{16\ell} - \frac{\pi^3h^3}{96\ell^3}\right)\frac{2\pi}{\ell} - \frac{\pi^2h}{8\ell^2} + C_1\frac{\pi^3h^3}{2\ell^3}\frac{2\pi}{\ell} + C_2\frac{\pi^4h^3}{2\ell^4}\right. \\
&\quad - \lambda\left[\frac{\pi^2h}{8\ell^2} + \frac{\pi^4h^3}{16\ell^4} - \frac{\pi^2h}{8\ell^2} + C_3\frac{\pi^4h^3}{2\ell^4} + C_4\frac{\pi^4h^3}{2\ell^3}\right] \\
&\quad \left. + \lambda_2\left[-\frac{\pi^4h^3}{24\ell^4} - \frac{\pi^4h^3}{24\ell^4} + C_5\frac{\pi^4h^3}{2\ell^4} + C_6\frac{\pi^4h^3}{2\ell^4}\right]\right\} \qquad (3.7.117) \\
&= -\frac{\pi^4h^3}{3\ell^4} + \frac{\pi^4h^3}{\ell^4}(2C_1 + C_2 - C_3 - C_4 + C_5 + C_6) \\
&= \frac{\pi^4h^3}{6\ell^4} + \left(2C_1 + C_2 - C_3 - C_4 + C_5 + C_6 - \frac{1}{2}\right)\frac{\pi^4h^3}{\ell^4}.
\end{aligned}$$

Hence, it follows that

$$2C_1 + C_2 - C_3 - C_4 + C_5 + C_6 - \frac{1}{2} = 0. \tag{3.7.118}$$

When the constants C_1, C_2, and C_5 are chosen, the additional constants C_3, C_4, and C_6 can be calculated. For the evaluation of (3.7.103), we only need the dominating part of $\overline{u}^2$ and $\overline{v}^2$, i.e.,

$$\begin{aligned}\overline{u}^2 &= \left(\frac{\pi}{8\ell} - \frac{\pi^3 y^2}{4\ell^3}\right)\sin\frac{2\pi x}{\ell} - \frac{\pi^2}{4\ell^2}x \\ \overline{v}^2 &= \left(\frac{\pi^2}{4\ell^2}y + \frac{\pi^4 y^3}{2\ell^4}\right)\cos\frac{2\pi x}{\ell} - \frac{\pi^2}{4\ell^2}y.\end{aligned} \tag{3.7.119}$$

Substitution of these expressions into (3.7.100) yields

$$\text{Min}\,(P_{\text{II}} - P_{\text{I}})^* = \frac{\pi^6 h^2 a^4}{192\ell^6} h\ell\left[1 + O\left(h^2/\ell^2\right)\right]. \tag{3.7.120}$$

which shows that the critical state of neutral equilibrium is stable.

A comparison with Euler column theory requires the evaluation of

$$P_2\left[\mathbf{u}_1, \lambda = 1\right] = a^2 \int_0^{\ell}\int_{-h/2}^{h/2} \left[\frac{1}{2}\left(u_{1,x}^2 + u_{1,y}^2 + v_{1,x}^2 + v_{1,y}^2\right) + \left(u_{1,y}v_{1,x} - u_{1,x}v_{1,y}\right)\right] dx\, dy. \tag{3.7.121}$$

Using the expressions (3.7.105), we find

$$\begin{aligned}P_2\left[\mathbf{u}_1, \lambda = 1\right] &= a^2\left[\frac{1}{2}\frac{\pi^2 h}{\ell} + \frac{1}{48}\frac{\pi^4 h^3}{\ell^3} + O\left(\frac{h^5}{\ell^5}\right) - \frac{1}{2}\frac{\pi^2 h}{\ell} + \frac{1}{16}\frac{\pi^4 h^3}{\ell^3}\right] \\ &= \frac{1}{12}\frac{\pi^4 h^2 a^2}{\ell^4} h\ell\left[1 + O\left(\frac{h^2}{\ell^2}\right)\right].\end{aligned} \tag{3.7.122}$$

The ratio of the expressions (3.7.121) and (3.7.122) is $\pi^2 a^2/16\ell^2$ with a relative error of order h^2/ℓ^2. This is indeed in full agreement with the result of the Euler column in the initial post-buckling range, cf. (3.1.23).

3.8 Helical spring with a small pitch

The elastic stability of helical springs was first dealt with by HURLBRINK (1910) and GRAMMEL (1924). Both ignored the shear elasticity of the spring. Shortly thereafter, BIEZENO and KOCH took into account the effect of shear but they overestimated the effect, implying as a consequence that any spring, however short, would be liable to buckling, which is not confirmed by experiment. Later, HARINGX (1942) took into account the shear effect correctly. In the first part of this section we shall follow his approach.[†]

[†] Cf. J. A. HARINGX, On highly compressible helical springs and rubber rods, and their application for vibration-free mountings (Philips Research Laboratories, 1950).

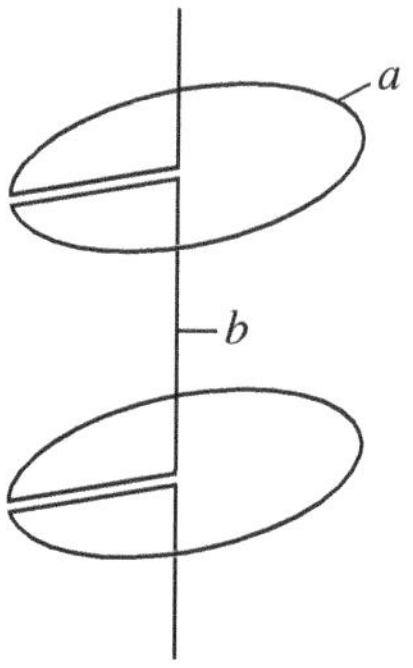

Figure 3.8.1

If the helical spring has a sufficiently small pitch when compressed, the deformation of one single coil under a certain load will differ little from that of an unclosed circular ring lying in a flat plane. Consequently, we may imagine the compressed helical spring as being replaced by a number of similar rings a connected by perfectly rigid elements b, as represented in Figure 3.8.1.

The influence of the axial force having been discounted as the successive coils are imagined as being flat after compression of the spring, we now must see what distortions occur as a result of the bending moments and transverse forces to be transmitted by these flat coils (see Figure 3.8.2).

In the following, we shall assume that the principal axes of inertia of the cross section of the ring are in the plane perpendicular to the plane of the ring. Let S_{b_1} and S_{b_2} be the bending stiffness of the cross section of the ring in the plane of the ring and perpendicular to it, respectively, and let S_t be the torsional stiffness of the cross section. Let u, v be the displacements due to N and D, respectively, and let φ be the rotation due to M. Elementary calculations show

$$N = \frac{S_t}{2\pi R^3} u, \quad D = \frac{S_{b_1}}{\pi R^3} v, \quad M = \frac{S_{b_2} \cdot S_t}{\pi R (S_{b_2} + S_t)} \varphi. \tag{3.8.1}$$

Haringx now smears the stiffness, i.e., he replaces the spring with n coils by a continuous rod. Let ℓ be the length of the spring; then the elongation of the spring is

$$\Delta\ell = \frac{2\pi R^3}{S_t} Nn \equiv \frac{N}{EA} \ell, \tag{3.8.2}$$

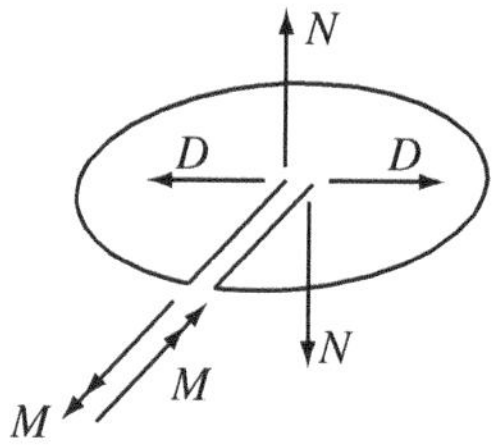

Figure 3.8.2

where the "stretching stiffness" EA is defined by

$$EA = \frac{S_t}{2\pi R^3 n}. \tag{3.8.3}$$

Further, for the total displacement due to D we have

$$v_{\text{tot}} = n\frac{\pi R^3 D}{S_{b_1}} \equiv \frac{D}{\beta EA}, \quad \beta = 2\frac{S_{b_1}}{S_t}. \tag{3.8.4}$$

The factor β becomes small for flat cross sections. If $t(y)$ is the thickness of the cross section, we then have

$$\beta = 2\frac{\int \frac{1}{12}Et^3(y)\,dy}{\int \frac{1}{3}Gt^3(y)\,dy} = 1 + \upsilon, \tag{3.8.5}$$

so that in general, $\beta \geq 1 + \upsilon$. Finally, the total angle of rotation is given by

$$\varphi_{\text{tot}} = n\frac{\pi R\,(S_{b_2} + S_t)}{S_{b_2} \cdot S_t} \equiv \frac{M}{B}\ell, \tag{3.8.6}$$

where B is the "bending stiffness,"

$$B = \frac{S_{b_2} \cdot S_t}{S_{b_2} + S_t}\frac{1}{\pi R}\frac{\ell}{n} \equiv \alpha\, EA\, R^2 \tag{3.8.7}$$

and

$$\alpha = \frac{2S_{b_2}}{S_{b_2} + S_t}. \tag{3.8.8}$$

For a circular cross section, the parameters α and β are given by

$$\alpha = \frac{2\,(1+\upsilon)}{2+\upsilon}, \quad \beta = 2\,(1+\upsilon)\,. \tag{3.8.9}$$

We now consider the deformation of an infinitesimally small element of the rod (see Figure 3.8.3).

Notice that in our model, the shear deformation is defined by the displacements of the transverse shear forces D. (This is not the case for ordinary beams.)

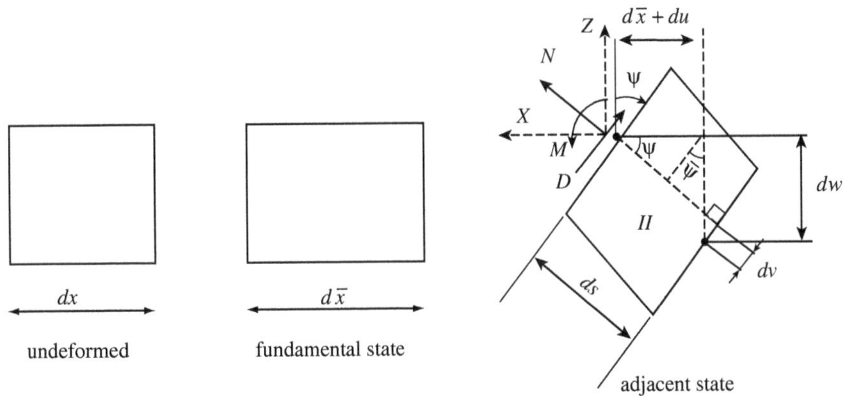

Figure 3.8.3

We now have the following relations,

$$\begin{aligned} ds &= (d\bar{x} + du)\cos\psi + dw\sin\psi \\ dv &= -(d\bar{x} + du)\sin\psi + dw\cos\psi. \end{aligned} \tag{3.8.10}$$

In the fundamental state, we have

$$N = -\lambda EA \quad \text{(which defines } \lambda\text{)} \tag{3.8.11}$$

$$\bar{x} = (1 - \lambda)x \tag{3.8.12}$$

so that with $()' = d(/dx)$, we have

$$\begin{aligned} \frac{ds}{dx} &= (1 - \lambda + u')\cos\psi + w'\sin\psi \\ \frac{dv}{dx} &= -(1 - \lambda + u')\sin\psi + w'\cos\psi. \end{aligned} \tag{3.8.13}$$

The elongation per unit length is thus

$$\varepsilon = (1 - \lambda + u')\cos\psi + w'\sin\psi - 1, \tag{3.8.14}$$

and the angle of shear is

$$\gamma = \frac{dv}{dx} = -(1 - \lambda + u')\sin\psi + w'\cos\psi. \tag{3.8.15}$$

Further, the curvature of the rod is given by

$$\kappa = \frac{d\psi}{dx} = \psi, \tag{3.8.16}$$

so we can write

$$N = EA\varepsilon, \quad D = \beta EA\gamma, \quad M = \alpha EAR^2\kappa. \tag{3.8.17}$$

In the fundamental state,

$$X = N = -\lambda EA. \tag{3.8.18}$$

In the adjacent state, we have

$$\begin{aligned} X &= N\cos\psi - D\sin\psi = \text{const.} \\ Z &= N\sin\psi + D\cos\psi = \text{const.} \equiv \zeta \text{EA} \end{aligned} \tag{3.8.19}$$

as conditions for the equilibrium of forces, and

$$M'dx + Dds - Ndv = 0 \tag{3.8.20}$$

for the equilibrium of moments. Solving N and D from (3.8.19), our results are

$$\begin{aligned} N &= X\cos\psi + Z\sin\psi = (-\lambda\cos\psi + \zeta\sin\psi)\,EA \\ D &= Z\cos\psi - X\sin\psi = (\lambda\sin\psi + \zeta\cos\psi)\,EA. \end{aligned} \tag{3.8.21}$$

Substituting these expressions into (3.8.20) and using the relations (3.8.13) to (3.8.17), we find

$$\begin{aligned} &\alpha R^2\psi'' + (\lambda\sin\psi + \zeta\cos\psi)\,[(1-\lambda+u')\cos\psi + w'\sin\psi] \\ &\quad - (-\lambda\cos\psi + \zeta\sin\psi)\,[-(1-\lambda+u')\sin\psi + w'\cos\psi] = 0, \end{aligned} \tag{3.8.22}$$

or, upon simplification,

$$\alpha R^2\psi'' + \lambda w' + \zeta(1-\lambda+u') = 0. \tag{3.8.23}$$

We shall now try to express w' and u' as functions of ψ, to obtain a differential equation in one variable. To this end, we solve $1-\lambda+u'$ and w' from (3.8.14) and (3.8.15) and make use of the relations (3.8.17) and (3.8.21). Our result is

$$\begin{aligned} 1-\lambda+u' &= (1+\varepsilon)\cos\psi - \gamma\sin\psi \\ &= (1-\lambda\cos\psi + \zeta\sin\psi)\cos\psi - \frac{1}{\beta}(\lambda\sin\psi + \zeta\cos\psi)\sin\psi \end{aligned} \tag{3.8.24}$$

$$\begin{aligned} w' &= \gamma\cos\psi + (1+\varepsilon)\sin\psi \\ &= \frac{1}{\beta}(\lambda\sin\psi + \zeta\cos\psi)\cos\psi + (1-\lambda\cos\psi + \zeta\sin\psi)\sin\psi. \end{aligned} \tag{3.8.25}$$

Substitution of these expressions into (3.8.23) and rearranging terms yields

$$\begin{aligned} &\alpha R^2\psi'' + \lambda\sin\psi - \lambda^2\left(1-\frac{1}{\beta}\right)\sin\psi\cos\psi \\ &\quad + \zeta\left[\cos\psi - \left(1-\frac{1}{\beta}\right)\lambda\cos 2\psi + \zeta\left(1-\frac{1}{\beta}\right)\sin\psi\cos\psi\right] = 0. \end{aligned} \tag{3.8.26}$$

When there are no transverse shear forces in the spring (e.g., for a spring hinged on both ends), $\zeta = 0$ and, in other cases, $\zeta \ll 1$ in the case of small deflections ψ, $\zeta = O(\psi)$.

For the determination of the buckling mode, (3.8.26) can be linearized, which yields

$$\alpha R^2\psi'' + \lambda\psi - \lambda^2\left(1-\frac{1}{\beta}\right)\psi + \zeta\left[1 - \left(1-\frac{1}{\beta}\right)\lambda\right] = 0, \tag{3.8.27}$$

where we have neglected the term in ζ^2. As mentioned previously, when the spring is hinged at both ends, $\zeta = 0$. Nonzero values of ζ occur, e.g., in the cases shown in Figure 3.8.4.

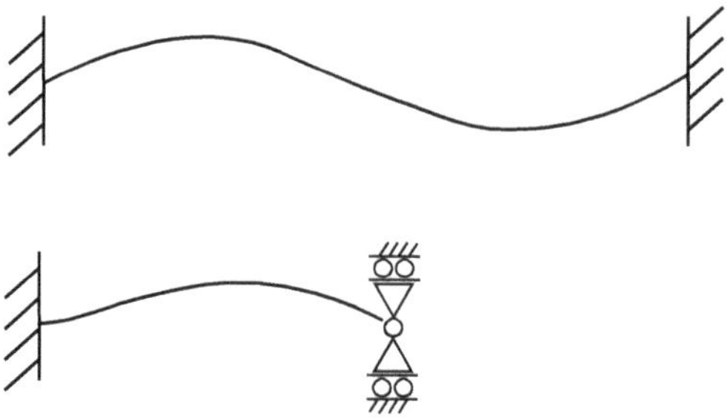

Figure 3.8.4

If ζ is nonzero, the value of ζ is determined from the condition

$$\int_0^\ell w'\,dx = 0, \tag{3.8.28}$$

which with (3.8.25) yields

$$\int_0^\ell \sin\psi - \lambda\left(1-\frac{1}{\beta}\right)\sin\psi\cos\psi + \zeta\left(\sin^2\psi + \frac{1}{\beta}\cos^2\psi\right)dx = 0, \tag{3.8.29}$$

or, linearized,

$$\int_0^\ell \left\{\psi\left[1-\lambda\left(1-\frac{1}{\beta}\right)\right]+\frac{\zeta}{\beta}\right\}dx = 0. \tag{3.8.30}$$

Let us now first consider the case that the spring is hinged at both ends. The lowest buckling mode is then given by

$$\psi = \cos\frac{\pi x}{\ell} \tag{3.8.31}$$

and the bifurcation condition becomes

$$\left(1-\frac{1}{\beta}\right)\lambda^2 - \lambda + \frac{\pi^2\alpha R^2}{\ell^2} = 0, \tag{3.8.32}$$

from which we obtain

$$\lambda_1 = \frac{\beta}{2(\beta-1)} - \left[\frac{\beta^2}{4(\beta-1)^2} - \frac{\pi^2\alpha\beta}{\beta-1}\frac{R^2}{\ell^2}\right]^{1/2}. \tag{3.8.33}$$

It follows that for sufficiently large values of R/ℓ (short spring), there is no critical load, i.e., no buckling can occur. We shall discuss this result in more detail later in this section.

When the spring is clamped at both ends, $\psi = 0$ for $x = 0$ and $x = \ell$, and for a symmetric buckling mode, there are no transverse shear forces, i.e., $\zeta = 0$. Then (3.8.30) is satisfied by

$$\psi = \sin\frac{2\pi x}{\ell} \tag{3.8.34}$$

and the bifurcation condition becomes

$$\left(1-\frac{1}{\beta}\right)\lambda^2-\lambda+\frac{4\pi^2\alpha R^2}{\ell^2}=0, \tag{3.8.35}$$

i.e., the last term is four times the corresponding term for the hinged-hinged spring.

We shall now treat the problem with the general theory of stability. The increment of the potential energy in passing from the undeformed state to the adjacent state is given by

$$\begin{aligned} P_{\text{II}}-P_{\text{I}}=\frac{1}{2}EA\int_0^\ell &\{\alpha R^2\psi'^2+[(1-\lambda+u')\cos\psi+w'\sin\psi-1]^2 \\ &+\beta[-(1-\lambda+u')\sin\psi+w'\cos\psi]^2\}\,dx \\ &-\frac{1}{2}EA\int_0^\ell \lambda^2\,dx+\lambda EA\int_0^\ell(-\lambda+u')\,dx. \end{aligned} \tag{3.8.36}$$

The equations for the equilibrium of forces are obtained by variations of this functional with respect to u' and w'. Variation with respect to u' yields

$$\begin{aligned} \int_0^\ell &\{[(1-\lambda+u')\cos\psi+w'\sin\psi-1]\cos\psi \\ &-\beta[-(1-\lambda+u')\sin\psi+w'\cos\psi]\sin\psi+\lambda\}\,\delta u'dx=0, \end{aligned} \tag{3.8.37}$$

and because there are no restrictions on $\delta u'$, it follows that

$$\begin{aligned} &[(1-\lambda+u')\cos\psi+w'\sin\psi-1]\cos\psi \\ &+\beta[-(1-\lambda+u')\sin\psi+w'\cos\psi]\sin\psi+\lambda=0. \end{aligned} \tag{3.8.38}$$

Variation with respect to w' yields

$$\begin{aligned} \int_0^\ell &\{[(1-\lambda+u')\cos\psi+w'\sin\psi-1]\sin\psi \\ &+\beta[-(1-\lambda+u')\sin\psi+w'\cos\psi]\cos\psi+c\}\delta w'dx=0, \end{aligned} \tag{3.8.39}$$

where we have introduced the Lagrangian multiplier c because $\delta w'$ is not arbitrary, as it must satisfy the kinematic condition $\int_0^\ell \delta w'dx=0$.

The second equilibrium equation now reads

$$\begin{aligned} &[(1-\lambda+u')\cos\psi+w'\sin\psi-1]\sin\psi \\ &+\beta[-(1-\lambda+u')\sin\psi+w'\cos\psi]\cos\psi+c=0. \end{aligned} \tag{3.8.40}$$

Multiplying (3.8.38) by $\cos\psi$ and (3.8.40) by $\sin\psi$ and adding the resulting equations, we obtain

$$(1-\lambda+u')\cos\psi+w'\sin\psi-1+\lambda\cos\psi+c\sin\psi=0. \tag{3.8.41}$$

Multiplying (3.8.38) by $\sin\psi$ and (3.8.40) by $\cos\psi$ and subtracting the resulting equations, we find

$$\beta[(1-\lambda+u')\sin\psi - w'\cos\psi] + \lambda\sin\psi - c\cos\psi = 0. \tag{3.8.42}$$

Using (3.8.41) and (3.8.42), we can rewrite the functional (3.8.36) to yield

$$\begin{aligned} P_{\mathrm{II}} - P_{\mathrm{I}} = \frac{1}{2}EA\int_0^\ell \Big[\alpha R^2\psi'^2 + (\lambda\cos\psi + c\sin\psi)^2 \\ + \frac{1}{\beta}(\lambda\sin\psi - c\cos\psi)^2 + 2\lambda u'\Big]\,dx, \end{aligned} \tag{3.8.43}$$

where we have omitted the constant terms in (3.8.36) because they are unimportant for our further discussion. We now still need an expression for u'. Multiplying (3.8.41) by $\beta\cos\psi$ and (3.8.42) by $\sin\psi$ and adding the resulting equations, we find

$$\begin{aligned} &\beta(1-\lambda+u') - \beta\cos\psi + \beta\lambda\cos^2\psi + \beta c\sin\psi\cos\psi \\ &+ \lambda\sin^2\psi - c\sin\psi\cos\psi = 0, \end{aligned}$$

from which

$$u' = -1 + \cos\psi + \lambda\left(1-\frac{1}{\beta}\right)\sin^2\psi - c\left(1-\frac{1}{\beta}\right)\sin\psi\cos\psi. \tag{3.8.44}$$

The functional (3.8.43) can now be rewritten to yield

$$\begin{aligned} P_{\mathrm{II}} - P_{\mathrm{I}} = \frac{1}{2}EA\int_0^\ell \Big\{\alpha R^2\psi'^2 - 2\lambda(1-\cos\psi) \\ + \lambda^2\left[1 + \left(1-\frac{1}{\beta}\right)\sin^2\psi\right] + c^2\left[\sin^2\psi + \frac{1}{\beta}\cos^2\psi\right]\Big\}\,dx. \end{aligned} \tag{3.8.45}$$

The second variation is now given by

$$P_2[\psi;\lambda] = \frac{1}{2}EA\int_0^\ell \left[\alpha R^2\psi'^2 - \lambda\psi^2 + \lambda^2\left(1-\frac{1}{\beta}\right)\psi^2 + \frac{c^2}{\beta}\right]dx, \tag{3.8.46}$$

where we have made use of the fact that $c \ll 1$. When there are no kinematic conditions, $c = 0$. In this case, the equation for neutral equilibrium is given by

$$\alpha R^2\psi'' + \lambda\left[1 - \lambda\left(1-\frac{1}{\beta}\right)\right]\psi = 0, \tag{3.8.47}$$

in full agreement with (3.8.27) with $\zeta = 0$. There are no cubic terms, so $P_{21}[\bar{a}\mathbf{u}, \bar{\mathbf{u}}] = 0$. The stability in the critical case of neutral equilibrium is thus governed by the fourth degree terms, i.e., by

$$P_4[\psi;\lambda] = \frac{1}{2}EA\int_0^\ell \left[\frac{1}{12}\lambda\psi^4 - \frac{1}{3}\lambda^2\left(1-\frac{1}{\beta}\right)\psi^4 + c^2\left(1-\frac{1}{\beta}\right)\psi^2\right]dx. \tag{3.8.48}$$

For $c = 0$, this becomes

$$P_4[\psi;\lambda] = \frac{1}{2}EA\int_0^{\ell}\frac{1}{12}\left[\lambda - 4\lambda^2\left(1-\frac{1}{\beta}\right)\right]\psi^4\,dx. \tag{3.8.49}$$

For a hinged-hinged spring, we have already discussed the buckling mode and the critical load, cf. (3.8.31) to (3.8.33). The critical load λ_1 given in (3.8.33) is real for

$$\frac{\pi^2R^2}{\ell^2} \leq \frac{\beta}{4\alpha(\beta-1)}. \tag{3.8.50}$$

The sign of $P_4[\psi;\lambda]$ is determined by the coefficient of ψ^4 in (3.8.49), i.e., by the sign of

$$f(\lambda_1) = \lambda_1 - 4\lambda_1^2\left(1-\frac{1}{\beta}\right). \tag{3.8.51}$$

Using (3.8.32), we obtain

$$f(\lambda_1) = \lambda_1 - 4\left(\lambda_1 - \frac{\pi^2\alpha R^2}{\ell^2}\right) = 4\pi^2\frac{\alpha R^2}{\ell^2} - 3\lambda_1. \tag{3.8.52}$$

Let us now first consider the critical case where the equality sign holds in (3.8.50), i.e.,

$$\lambda_1 = \frac{\beta}{2(\beta-1)}. \tag{3.8.53}$$

In this case,

$$f(\lambda_1) = \frac{\beta}{\beta-1} - \frac{3\beta}{2(\beta-1)} = \frac{3\beta}{2(1-\beta)} < 0, \tag{3.8.54}$$

so the equilibrium is unstable in that point.

Let us now investigate what happens in the points where $f(\lambda_1) = 0$, i.e., for

$$\lambda_1 = \frac{4\pi^2R^2}{3\ell^2}\alpha. \tag{3.8.55}$$

Substitution into the bifurcation condition (3.8.32) yields

$$\frac{\beta-1}{\beta}\left(\frac{4\pi^2R^2}{3\ell^2}\alpha\right)^2 - \frac{4\pi^2R^2}{3\ell^2}\alpha + \pi^2\frac{\alpha R^2}{\ell^2} = 0,$$

or

$$\frac{\pi^2R^2}{\ell^2}\alpha = \frac{3\beta}{16(\beta-1)} < \frac{\beta}{4(\beta-1)}, \tag{3.8.56}$$

i.e., the equilibrium is unstable for

$$\frac{\beta}{4(\beta-1)} \leq \lambda_1 \leq \frac{3\beta}{4(\beta-1)}. \tag{3.8.57}$$

We can now represent our results in the graph in Figure 3.8.5.

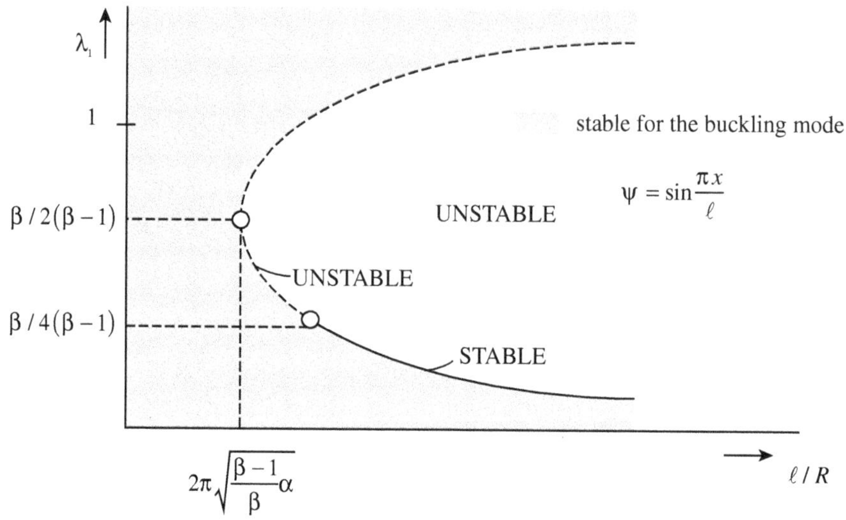

Figure 3.8.5

For a spring with a circular cross section, no buckling will occur for

$$\frac{\ell}{R} < 2\pi\sqrt{\frac{\beta-1}{\beta}}\alpha = 2\pi\sqrt{\frac{1+2\nu}{2+\nu}}. \tag{3.8.58}$$

It is still an open question whether the unstable domain on the boundary for values of λ_1 such that $\beta/4\,(\beta-1) < \lambda_1 < \beta/2\,(\beta-1)$ does actually exist, or that it stems from the approximation that the spring is built up from flat circular rings.

3.9 Torsion of a shaft

We consider a shaft of arbitrary (constant) cross section loaded by a torque W (see Figure 3.9.1).

Let, x, y, z be axes fixed in space, and let y_1, z_1 be (co-rotating) principle axes of inertia, which are rotated over an angle kx with respect to the fixed axes y, z. Here k is the torsion angle per unit length.

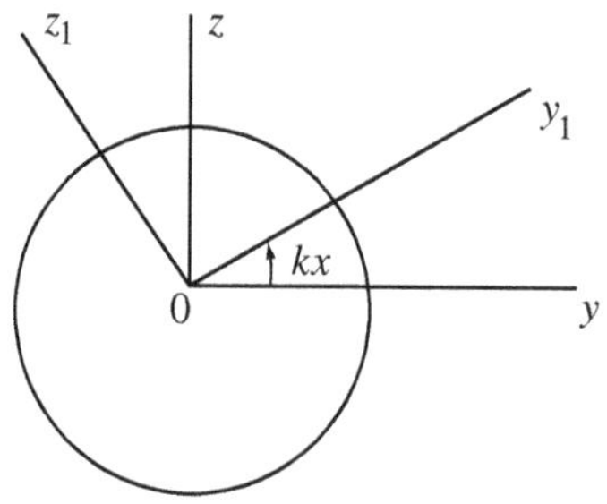

Figure 3.9.1

In the fundamental state, we have pure torsion, and the only non-vanishing components of the stress tensor are

$$\tau_{xy} = \frac{\partial F}{\partial z}, \quad \tau_{xz} = -\frac{\partial F}{\partial y}, \tag{3.9.1}$$

where F is the stress function of torsion. In the adjacent state, bending will occur, and the point 0 will have displacements $v_0(x)$, $w_0(x)$ along the y and z axes, respectively. Assuming that Bernoulli's hypothesis holds, the displacement u of a point (x, y, z) is given by

$$u = -yv_0'(x) - zw_0'(x), \tag{3.9.2}$$

where $()' = d()/dx$. To obtain an approximate uni-axial state of stress, we assume the following displacement field,

$$\begin{aligned} v(x, y, z) &= v_0(x) + \frac{1}{2}\nu v_0''(y^2 - z^2) + \nu w_0'' yz \\ w(x, y, z) &= w_0(x) + \nu v'' yz + \frac{1}{2}\nu w_0''(z^2 - y^2). \end{aligned} \tag{3.9.3}$$

This field satisfies

$$\begin{aligned} \frac{\partial v}{\partial y} &= \frac{\partial w}{\partial z} = -\nu\frac{\partial u}{\partial x} \\ \psi_{yz} &= \frac{\partial v}{\partial z} + \frac{\partial w}{\partial y} = 0. \end{aligned} \tag{3.9.4}$$

However, there are non-vanishing angles of shear,

$$\psi_{xy} = \frac{\partial u}{\partial y} + \frac{\partial v}{\partial x} = \frac{1}{2}\nu v_0''(y^2 - z^2) + \nu w_0'' yz. \tag{3.9.5}$$

Let ℓ be the wavelength of the deformation pattern, and let b be a characteristic length in the cross section. The contribution of ψ_{xy} in the energy function is then of order $b^4 v_0^2/\ell^6$, whereas the contributions due to $\partial u/\partial x$ are of order $b^2 v_0^2/\ell^4$. Under the assumption $b^2/\ell^2 \ll 1$, we can now neglect the contribution of ψ_{xy} in the energy, which results in a relative error of $O(b^2/\ell^2)$.

The energy density due to bending is given by

$$\int_0^\ell dx \iint_A \frac{1}{2} E\,(yv_0'' + zw_0'')^2\, dy\, dz. \tag{3.9.6}$$

It is now convenient to employ the principle axes of inertia. Using the relations

$$\begin{aligned} y &= y_1 \cos kx - z_1 \sin kx \\ z &= y_1 \sin kx + z_1 \cos kx, \end{aligned} \tag{3.9.7}$$

we obtain for the energy density

$$\int_0^\ell \left[\frac{1}{2} EI_{z_1}\,(v_0'' \cos kx + w_0'' \sin kx)^2 + \frac{1}{2} EI_{y_1}\,(w_0'' \cos kx - v_0'' \sin kx)^2\right] dx\, dy. \tag{3.9.8}$$

Because we are dealing with buckling, we cannot neglect the stresses in the fundamental state. Recalling our general expression for the second variation in the case of dead-weight loads,

$$P_2[\mathbf{u}] = \int \left[\frac{1}{2} S_{ij} u_{h,i} u_{h,j} + \frac{1}{2} E_{ijkl} \theta_{ij} \theta_{k\ell}\right] dV, \tag{3.9.9}$$

where we have already evaluated the second term (3.9.8), we are still in need of an explicit form of the first term. Because for torsion we have only two nonzero stress components, we find

$$\begin{aligned}
\frac{1}{2} S_{ij} u_{h,i} u_{h,j} &= \tau_{xy} \left(u_{,x} u_{,y} + v_{,x} v_{,y} + w_{,x} w_{,y}\right) + \tau_{xz}\left(u_{,x} u_{,z} + v_{,x} v_{,z} + w_{,x} w_{,z}\right) \\
&= \tau_{xy} \left\{(-y v_0'' - z w_0'')(-v_0') + \left[v_0' + \frac{1}{2} \nu v_0'''(y^2 - z^2) + \nu w_0''' yz\right]\right. \\
&\quad \left.\times (\nu v_0'' y + \nu w_0'' z) + \left[w_0' + \nu v_0''' yz + \frac{1}{2}\nu w_0'''(z^2 - y^2)\right](\nu v_0'' z - \nu w_0'' y)\right\} \\
&\quad + \tau_{xz}\left\{(-y v_0'' - z w_0'')(-w_0') + \left[v_0' + \frac{1}{2}\nu v_0'''(y^2 - z^2) + \nu w_0''' yz\right]\right. \\
&\quad \left.\times(-\nu v_0'' z + \nu w_0'' y) + \left[w_0' + \nu v_0'' yz + \frac{1}{2}\nu w_0'''(z^2 - y^2)\right](\nu v_0'' y + \nu w_0'' z)\right\}.
\end{aligned} \tag{3.9.10}$$

We shall now argue that the terms with third derivatives may be neglected. Consider the term

$$\iint_A \tau_{xy} \times \frac{1}{2}\nu v_0'''(y^2 - z^2) \times \nu v_0''' y \, dx\, dy = O\left(\frac{\tau v_0^2 b^5}{\ell^5}\right). \tag{3.9.11}$$

In our discussion of ψ_{xy}, we have already neglected terms of $O\left(E v_0^2 b^6/\ell^6\right)$, and have taken into account terms of $O\left(E v_0^2 b^4/\ell^4\right)$. Hence, it follows that we can neglect the contribution of (3.9.11) in the energy and, similarly, the other terms involving third derivatives with respect to x.

The expression (3.9.10) can now be rewritten to yield

$$\begin{aligned}
\frac{1}{2} S_{ij} u_{h,i} u_{h,j} &= v_0' v_0'' \left[y\tau_{xy} + \nu y \tau_{xy} - \nu z \tau_{xz}\right] + w_0' w_0'' \left[-\nu y \tau_{xy} + z\tau_{xz} + \nu z \tau_{xz}\right] \\
&\quad + v_0' w_0'' \left[z\tau_{xy} + \nu z \tau_{xy} + \nu y \tau_{xz}\right] + v_0'' w_0' \left[\nu z \tau_{xy} + y \tau_{xz} + \nu y \tau_{xz}\right].
\end{aligned} \tag{3.9.12}$$

Integrating this expression over the cross-sectional area, we must note that

$$\iint_A y \tau_{xy}\, dy\, dz = \iint y \frac{\partial F}{\partial z}\, dy\, dz = \oint_{\text{edge}} y F n_z\, ds = F_{\text{edge}} \oint y n_z\, ds = 0, \tag{3.9.13}$$

where F_{edge} is the constant value of F at the edge. Similarly, we have

$$\iint_A z \tau_{xz}\, dy\, dz = 0. \tag{3.9.14}$$

This implies that after integration, the first and the second term between brackets in (3.9.12) vanish and in the third and the fourth term between brackets the terms with v vanish because

$$\iint_A z\tau_{xy}\,dy\,dz = -\frac{1}{2}W \qquad \iint_A y\tau_{xz}\,dy\,dz = \frac{1}{2}W. \tag{3.9.15}$$

Further, we recall that the torque W is given by

$$W = \iint_A (y\tau_{xz} - z\tau_{xy})\,dy\,dz, \tag{3.9.16}$$

so the torsion energy per unit length is given by

$$\frac{1}{2}W\,(v_0''w_0' - v_0'\,w_0''). \tag{3.9.17}$$

The second variation now becomes

$$\begin{aligned} P_2[\mathbf{u}] = \int_0^\ell \Big[\frac{1}{2}EI_{z_1}\,(v_0''\cos kx + w_0''\sin kx)^2 \\ + \frac{1}{2}EI_{z_1}\,(w_0''\cos kx - v_0''\sin kx)^2 + \frac{1}{2}W\,(v_0w_0' - v_0'w_0)\Big]\,dx. \end{aligned} \tag{3.9.18}$$

For a shaft with a circular cross section $EI_{y_1} = EI_{z_1} = B$, the functional reduces to

$$P_2[\mathbf{u}] = \int_0^\ell \Big[\frac{1}{2}B\,(v_0''^2 + w_0''^2) + \frac{1}{2}W\,(v_0w_0' - v_0'w_0'')\Big]\,dx. \tag{3.9.19}$$

Our further discussion will be restricted to this functional.

The condition for neutral equilibrium is

$$\begin{aligned} P_{11}[\mathbf{u}, \zeta] = \int_0^\ell \Big[\frac{1}{2}B\,(v_0''^2 + w_0''^2) \\ + \frac{1}{2}W\,(v_0''\,\zeta' + w_0'\,\eta'' - v_0'\,\zeta'' - w_0''\,\eta')\Big]\,dx = 0. \end{aligned} \tag{3.9.20}$$

By integration by parts, we obtain

$$\begin{aligned} &\left(Bv_0'' + \frac{1}{2}Ww_0'\right)\eta'\Big|_0^\ell + \left(Bw_0'' - \frac{1}{2}Wv_0'\right)\zeta'\Big|_0^\ell \\ &- (Bv_0''' + Ww_0'')\,\eta'\Big|_0^\ell - (Bw_0''' - Wv_0'')\,\zeta\Big|_0^\ell \\ &+ \int_0^\ell [(Bv_0'''' + Ww_0''')\,\eta + (Bv_0'''' - Wv_0''')\,\zeta]\,dx = 0, \end{aligned} \tag{3.9.21}$$

from which we obtain

$$Bv_0'''' + Ww_0''' = 0, \quad Bw_0'''' - Wv_0''' = 0. \tag{3.9.22}$$

Let us now consider the case that the shaft is supported at its ends. It is now convenient to choose the origin of our coordinate system in the middle of the shaft, and let $-\ell \leq x \leq \ell$, i.e., we consider a shaft of length 2ℓ. The geometric conditions are then

$$v_0 = w_0 = 0 \quad \text{for} \quad x = \pm\ell. \tag{3.9.23}$$

Modifying the boundaries in (3.9.21) correspondingly, we obtain the dynamic boundary conditions

$$Bv_0'' + \frac{1}{2}Ww_0' = 0, \quad Bw_0'' - \frac{1}{2}Wv_0' = 0 \quad \text{for} \quad x = \pm\ell. \tag{3.9.24}$$

To solve this problem, we set

$$v_0 = Ce^{i\mu x}, \quad w_0 = De^{i\mu x}, \tag{3.9.25}$$

and substitution into the differential equations yields

$$\begin{aligned} \mu^4 BC - i\mu^3 WD &= 0 \\ \mu^4 BD + i\mu^3 WC &= 0. \end{aligned} \tag{3.9.26}$$

The condition for a non-trivial solution is now

$$\mu^8 - \mu^6 \frac{W^2}{B^2} = 0, \tag{3.9.27}$$

i.e.,

$$\mu_1, \ldots, \mu_6 = 0, \quad \mu_7 = \frac{W}{B}, \quad \mu_8 = -\frac{W}{B}. \tag{3.9.28}$$

Defining

$$\frac{W}{B} = \mu,$$

we can write the solution in the form

$$\begin{aligned} v_0 &= \underline{C_1 \cos \mu x} + C_2 \sin \mu x + \underline{C_3 x^2} + C_4 x + \underline{C_5} \\ w_0 &= \underline{C_1 \sin \mu x} - C_2 \cos \mu x + C_6 x^2 + \underline{C_7 x} + C_8 \end{aligned} \tag{3.9.29}$$

because the differential equations are satisfied identically by quadratic polynomials. Due to symmetry, we can split the solution into even and odd functions. Choosing an even function for v_0, w_0 must be odd due to (3.9.22). The problem with v_0 odd and w_0 even is readily found from the previous one by interchanging v_0 and w_0. The advantage of splitting up the problem is that we must now only satisfy four boundary conditions. From (3.9.23), we obtain

$$\begin{aligned} v_0\,(x = \ell) &= C_1 \cos \mu\ell + C_3 \ell^2 + C_5 = 0 \\ w_0\,(x = \ell) &= C_1 \sin \mu\ell + C_7 \ell^2 = 0 \end{aligned} \tag{3.9.30}$$

and from (3.9.24),

$$\begin{aligned} -\frac{1}{2}\mu^2 \cos \mu\ell\, BC_1 + 2C_3 B + \frac{1}{2}WC_7 &= 0 \\ \frac{1}{2}\mu^2 \sin \mu\ell\, BC_1 - C_3 B\ell &= 0. \end{aligned} \tag{3.9.31}$$

Because C_5 only occurs in the first of the equations of (3.9.31), the condition for a non-trivial solution is

$$\begin{vmatrix} \sin \mu\ell & 0 & \ell \\ -\frac{1}{2}\mu^2 \cos \mu\ell & 2 & \frac{1}{2}\mu \\ -\frac{1}{2}\mu^2 \sin \mu\ell & -\mu\ell & 0 \end{vmatrix} = 0, \tag{3.9.32}$$

from which follows

$$\tan \mu\ell = -\frac{1}{3}\mu\ell. \tag{3.9.33}$$

The smallest non-negative root of this transcendental equation is

$$\mu\ell = 1.566\frac{\pi}{2}, \tag{3.9.34}$$

which yields

$$W = 1.566\pi\frac{B}{2\ell}, \tag{3.9.35}$$

which is the critical value of the torque W.

Let us now examine the boundary conditions more closely. We consider the end cross section at $x = \ell$ (see Figure 3.9.2).

The axes y, z are fixed in space, and y_1, z_1 are co-rotating axes. The displacement $u(\ell, y_1, z_1)$ is given by

$$u(\ell, y_1, z_{,1}) = -y_1 v_0'(\ell) - z w_0'(\ell), \tag{3.9.36}$$

which implies that the shear stresses τ_{xy} cause a bending moment

$$\iint_A \tau_{xy} u \, dy \, dx = M_z \tag{3.9.37}$$

and, similarly for the shear stresses,

$$\iint_A \tau_{xz} u \, dy \, dz = -M_y. \tag{3.9.38}$$

This result means that the load on the end faces is not directed along the undeformed axis. Bending moments M_y and M_z must be added. Evaluating the integrals

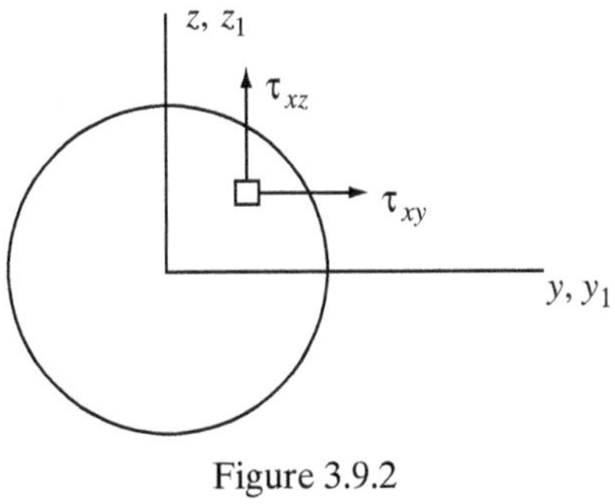

Figure 3.9.2

in (3.9.37) and (3.9.38), we find

$$
\begin{aligned}
M_y &= \iint_A \tau_{xz} \left(y_1 v_0^1 (\ell) + z w_0' (\ell) \right) dy_1 \, dz_1 = \frac{1}{2} W v_0' (\ell) \\
M_z &= - \iint_A \tau_{xz} \left(y_1 v_0' (\ell) + z w_0' (\ell) \right) dy_1 \, dz_1 = \frac{1}{2} W w_0' (\ell) .
\end{aligned}
\tag{3.9.39}
$$

These moments are called by ZIEGLER† *semi-tangential* moments, i.e., the moments are obtained by multiplying the torque by half the angle of rotation.

In the problem treated previously, the loading was conservative. GREENHILL‡ treated a similar problem, but in his case the bending moments were absent. The loading is then *non-conservative*. GREENHILL's boundary conditions are obtained from our previous problem by superimposing loads

$$
\Delta M_z = -\frac{1}{2} W w_0', \quad \Delta M_y = -\frac{1}{2} W v_0' \tag{3.9.40}
$$

for $x = \pm \ell$. The boundary conditions now become

$$
B w_0'' - W v_0' = 0, \quad B v_0'' + W w_0' = 0, \quad x = \pm \ell. \tag{3.9.41}
$$

The differential equations are still given by (3.9.22), and the general solution is

$$
\begin{aligned}
v_0 &= \underline{C_1 \cos \mu x} + C_2 \sin \mu x + \underline{C_3 x^2} + C_4 x + \underline{C_5} \\
w_0 &= \underline{C_1 \sin \mu x} - C_2 \cos \mu x + C_6 x^2 + \underline{C_7 x} + C_8,
\end{aligned}
\tag{3.9.42}
$$

where $\mu = W/B$. By arguments similar to those used in the previous problem, we may restrict ourselves to the underlined terms. The kinematic boundary conditions $v_0 = w_0 = 0$ at $x = \ell$ require

$$
\begin{aligned}
& C_1 \cos \mu\ell + C_3 \ell^2 + C_5 = 0 \\
& C_1 \sin \mu\ell + C_7 \ell = 0,
\end{aligned}
\tag{3.9.43}
$$

and the dynamic boundary conditions become

$$
\begin{aligned}
& -2 C_3 \ell W = 0 \\
& 2 B C_3 + W C_7 = 0,
\end{aligned}
\tag{3.9.44}
$$

which yields

$$
C_3 = C_7 = 0, \quad C_5 = -C_1 \cos \mu\ell, \quad C_1 \sin \mu\ell = 0, \tag{3.9.45}
$$

and hence, $\mu\ell = \pi$, so that the critical torque is given by

$$
W = 2\pi \frac{B}{2\ell}. \tag{3.9.46}
$$

The correctness of this result, obtained from the equations for neutral equilibrium for a conservative system, is in this case verified by the solution of the dynamic problem.

† Cf. H. ZIEGLER, *Principles of Structural Stability* (Blaisdell Publ. Comp.), p. 124.
‡ Cf. A. G. GREENHILL, *Proc. Inst. Mech. Engrs.*, 182 (1883).

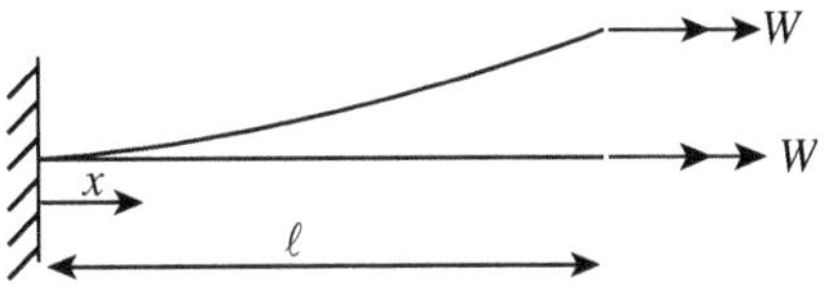

Figure 3.9.3

To show that this equivalent is not always justified, we consider the problem shown in Figure 3.9.3.

The boundary conditions in this case are

$$\begin{aligned} &v_0 = w_0 = 0, \quad v_0' = w_0' = 0, \quad \text{at } x = 0 \\ &\left.\begin{aligned} Bv_0'' + Ww_0' = 0, \quad Bw_0'' - Wv_0' = 0 \\ Bv_0''' + Ww_0'' = 0, \quad Bw_0''' - Wv_0'' = 0 \end{aligned}\right\} \quad \text{at } x = \ell. \end{aligned} \tag{3.9.47}$$

Integrating the differential equations (3.9.22) twice and making use of the boundary conditions at $x = \ell$, we find

$$Bv_0'' + Ww_0' = 0, \quad Bw_0'' - Wv_0' = 0. \tag{3.9.48}$$

The general solution to these equations reads

$$\begin{aligned} v_0 &= C_1 \cos \mu x + C_2 \sin \mu x + C_3 \\ w_0 &= C_1 \sin \mu x - C_2 \cos \mu x + C_4. \end{aligned} \tag{3.9.49}$$

From the kinematic conditions at $x = 0$, we find

$$\begin{aligned} C_1 + C_3 = 0, \quad & -C_2 + C_4 = 0 \\ C_2\mu = 0, \quad & C_1\mu = 0. \end{aligned} \tag{3.9.50}$$

These results mean that for $\mu \neq 0$, there is no neutral equilibrium (Ziegler, 1950). The dynamic problem yields increasing amplitudes for non-vanishing W (flutter).

We shall analyze this phenomenon in the following (simpler) problem, in which we consider a shaft without mass and a concentrated mass m at its end $x = \ell$ (see Figure 3.9.4).

Because the shaft has no mass, the differential equations are unaltered. The boundary conditions in this case are

$$\begin{aligned} &v_0 = w_0 = v_0' = w_0' = 0 \quad \text{at } x = 0 \\ &\left.\begin{aligned} Bv_0'' + Ww_0' = 0, \quad & Bw_0'' - Wv_0' = 0 \\ Bw_0''' + Ww_0'' = m\ddot{v}_0, \quad & Bw_0''' - Wv_0'' = m\ddot{w}_0 \end{aligned}\right\} \quad \text{at } x = \ell, \end{aligned} \tag{3.9.51}$$

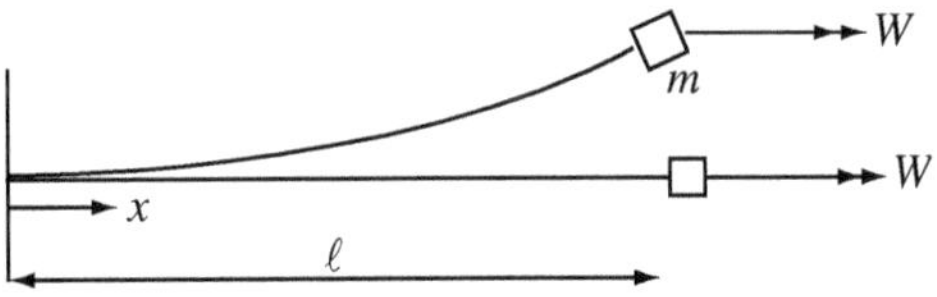

Figure 3.9.4

where $()' = \partial()/\partial x$ and $\dot{()} = \partial()/\partial t$. Introducing the complex function

$$V(x,t) = v_0(x,t) + iw_0(x,t), \tag{3.9.52}$$

we may combine the two differential equations to yield

$$V'''' - i\mu V''' = 0, \tag{3.9.53}$$

where $\mu = W/B$. The kinematic boundary conditions now become

$$V = V' = 0 \quad \text{at } x = 0, \tag{3.9.54}$$

and the dynamic boundary conditions now read

$$V'' - i\mu V' = 0, \quad V''' - i\mu V'' = \frac{m}{B}\ddot{V} \quad \text{at } x = \ell. \tag{3.9.55}$$

Introducing $V = U(x)e^{i\omega t}$ into the differential equation, we find

$$U = D_1 e^{i\mu x} + D_2 x^2 + D_3 x + D_4. \tag{3.9.56}$$

From the kinematic boundary conditions, we obtain

$$D_1 + D_4 = 0, \quad i\mu D_1 + D_3 = 0. \tag{3.9.57}$$

Hence, we can write

$$U = D_1\left(e^{i\mu x} - 1 - i\mu x\right) + D_2 x^2. \tag{3.9.58}$$

From the dynamic boundary conditions, we obtain

$$\begin{aligned} &-\mu^2 D_1 + 2(1 - i\mu\ell) D_2 = 0 \\ &-\frac{m\omega^2}{B}\left(e^{i\mu\ell} - 1 - i\mu\ell\right) D_1 + \left(-\frac{m\omega^2}{B}\ell^2 + 2i\mu\right) D_2 = 0. \end{aligned} \tag{3.9.59}$$

The condition for a non-trivial solution is

$$\begin{vmatrix} -\mu^2 & 2(1 - i\mu\ell) \\ -\dfrac{m\omega^2}{B}\left(e^{i\mu\ell} - 1 - i\mu\ell\right) & -\dfrac{m\omega^2}{B}\ell^2 + 2i\mu \end{vmatrix} = 0, \tag{3.9.60}$$

which yields

$$\frac{m\omega^2}{B}[\mu^2\ell^2 + 2(1 - i\mu\ell)(e^{i\mu\ell} - 1 - i\mu\ell)] - 2i\mu^3 = 0, \tag{3.9.61}$$

so that the square of the frequency is given by

$$\omega^2 = \frac{2i\mu^3 B/m}{\mu^2\ell^2 + 2(1 - i\mu\ell)(e^{i\mu\ell} - 1 - i\mu\ell)}. \tag{3.9.62}$$

It follows that ω^2 is complex unless the denominator is purely imaginary, which is not the case, as we shall show.

The fact that ω^2 is complex implies that there is one root with a negative imaginary part, which implies that V, and hence v_0 and w_0, increase exponentially with time. This fact means that for $\mu \neq 0$, i.e., when there is a torque, the system is always unstable. To show that ω^2 is complex, we consider small values of $\omega\ell$. The denominator can now be written as

$$\begin{aligned}
&\mu^2\ell^2 + 2(1 - i\mu\ell)\left[1 + i\mu\ell - \frac{1}{2}\mu^2\ell^2 - \frac{1}{6}i\mu^3\ell^3 + \frac{1}{24}\mu^4\ell^4 + O(\mu^5\ell^5) - i\mu\ell - 1\right] \\
&= \mu^2\ell^2\left[1 - 1 + i\mu\ell - \frac{1}{3}i\mu\ell - \frac{1}{3}\mu^2\ell^2 + \frac{1}{12}\mu^2\ell^2 + O(\mu^3\ell^3)\right] \\
&= \mu^3\ell^3\left[\frac{2}{3}i - \frac{1}{4}\mu\ell + O(\mu^2\ell^2)\right],
\end{aligned}$$

so that

$$\omega^2 = \frac{3B}{m\ell^3}\left[1 - \frac{3}{8}i\mu\ell + O(\mu^2\ell^2)\right]. \tag{3.9.63}$$

Although at first sight this result seems alarming, it is not important for actual constructions because it is virtually impossible to apply a torque the way it is assumed in this problem. To apply a torque, one usually makes use of the universal (Cardan) joint, and then the system is conservative. We shall treat this problem in the next section.

3.10 Torsion of a shaft with a Cardan (Hooke's) joint

We consider a shaft loaded in torsion with a Cardan (Hooke's) joint (see Figure 3.10.1).

Because large rotations may occur, we shall first derive an expression for the rotation vector for finite rotations.

Let $\mathbf{n}$ be a unit vector along the axis of rotation, and let α be the angle of rotation. The position vector of a point is denoted by $\mathbf{r}$ (see Figure 3.10.2).
The position of the point after rotation is denoted by $\mathbf{r}'$, and is given by

$$\mathbf{r}' = r\cos\theta\,\mathbf{e}_1 + r\sin\theta\cos\alpha\,\mathbf{e}_2 + r\sin\theta\sin\alpha\,\mathbf{e}_3, \tag{3.10.1}$$

which can be rewritten to yield

$$\begin{aligned}
\mathbf{r}' &= (\mathbf{r}\cdot\mathbf{n})\,\mathbf{n} + \{\mathbf{r} - (\mathbf{r}\cdot\mathbf{n})\,\mathbf{n}\}\cos\alpha + (\mathbf{n}\times\mathbf{r})\sin\alpha \\
&= \mathbf{r}\cos\alpha + (\mathbf{r}\cdot\mathbf{n})\,\mathbf{n}\,(1 - \cos\alpha) + (\mathbf{n}\times\mathbf{r})\sin\alpha.
\end{aligned} \tag{3.10.2}$$

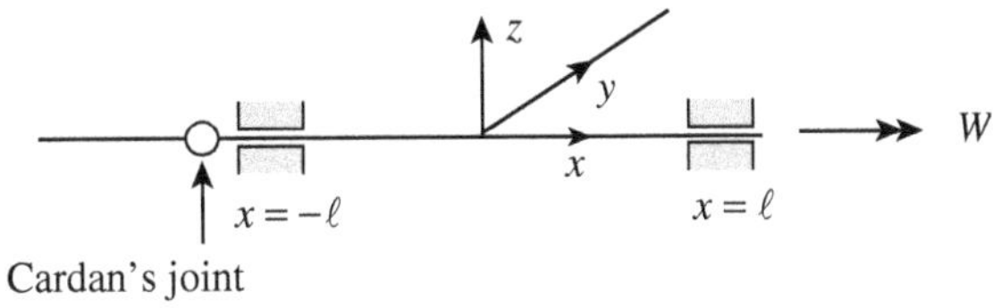

Figure 3.10.1

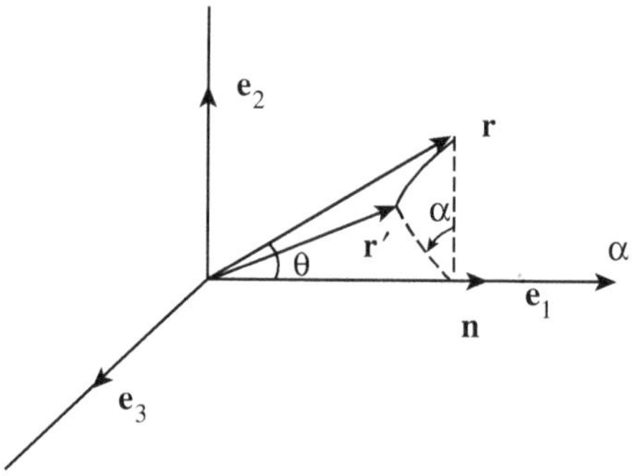

Figure 3.10.2

Without loss of generality, we may now restrict ourselves to values of α such that $-\pi \le \alpha \le \pi$. We now introduce a rotation vector ω defined by

$$\boldsymbol{\omega} = \mathbf{n} \cdot 2 \sin \alpha/2 \tag{3.10.3}$$

with components ω_i $(i \in \{1, 2, 3\})$. Notice that two subsequent rotations do not yield a rotation vector that is the sum of the two corresponding rotation vectors. We now consider the rotation of the triad $(\mathbf{e}_1, \mathbf{e}_2, \mathbf{e}_3)$ to obtain the rotation matrix. Let us denote the rotated triad by $(\mathbf{e}'_1, \mathbf{e}'_2, \mathbf{e}'_3)$. The triad $(\mathbf{e}_1, \mathbf{e}_2, \mathbf{e}_3)$ is fixed to the body under consideration (see Figure 3.10.3).

With $\mathbf{r} = \mathbf{e}_1$ and $\mathbf{n} = \boldsymbol{\omega}/(2 \sin \alpha/2)$, we obtain from (3.10.2),

$$\mathbf{e}'_1 = \mathbf{e}_1 \cos \alpha + \frac{\omega_1}{2 \sin \alpha/2} \frac{\boldsymbol{\omega}}{2 \sin \alpha/2} (1 - \cos \alpha) + \frac{\boldsymbol{\omega} \times \mathbf{e}_1}{2 \sin \alpha/2} \sin \alpha \tag{3.10.4}$$

with

$$\begin{aligned} |\boldsymbol{\omega}|^2 &= \omega^2 = \omega_1^2 + \omega_2^2 + \omega_3^2 = 4 \sin^2 \alpha/2 \\ \cos \alpha &= 1 - 2 \sin^2 \alpha/2 = 1 - \frac{1}{2}\omega^2 \\ \boldsymbol{\omega} \times \mathbf{e}_1 &= (\omega_1 \mathbf{e}_1 + \omega_2 \mathbf{e}_2 + \omega_3 \mathbf{e}_3) \times \mathbf{e}_1 \\ &= -\,\omega_2 \mathbf{e}_3 + \omega_3 \mathbf{e}_2 \end{aligned}$$

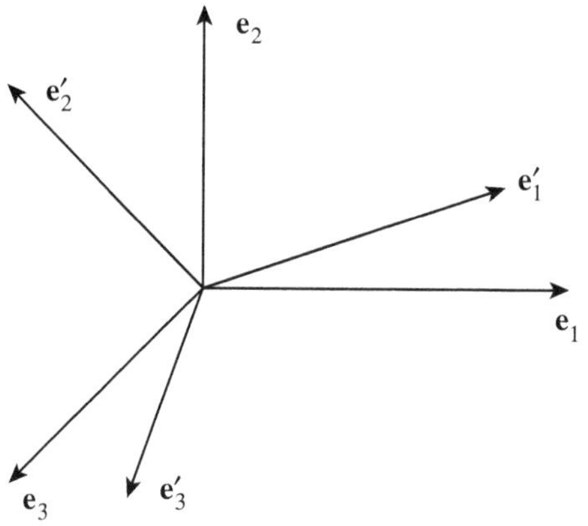

Figure 3.10.3

and

$$\cos\alpha/2 = \sqrt{\frac{1}{2}(1+\cos\alpha)} = \sqrt{1-\frac{1}{4}\omega^2}.$$

We can then rewrite this expression to yield

$$\begin{aligned}\mathbf{e}'_1 = &\left[1-\frac{1}{2}\left(\omega_2^2+\omega_3^2\right)\right]\mathbf{e}_1 + \left[\frac{1}{2}\omega_1\omega_2+\omega_3\left(1-\frac{1}{4}\omega^2\right)^{1/2}\right]\mathbf{e}_2 \\ &+\left[\frac{1}{2}\omega_1\omega_3-\omega_2\left(1-\frac{1}{4}\omega^2\right)^{1/2}\right]\mathbf{e}_3.\end{aligned} \tag{3.10.5}$$

By a cyclic interchanging of the subscript, we obtain

$$\begin{aligned}\mathbf{e}'_2 = &\left[\frac{1}{2}\omega_2\omega_1-\omega_3\left(1-\frac{1}{4}\omega^2\right)^{1/2}\right]\mathbf{e}_1 + \left[1-\frac{1}{2}\left(\omega_3^2+\omega_1^2\right)\right]\mathbf{e}_2 \\ &+\left[\frac{1}{2}\omega_2\omega_3+\omega_1\left(1-\frac{1}{4}\omega^2\right)^{1/2}\right]\mathbf{e}_3.\end{aligned} \tag{3.10.6}$$

$$\begin{aligned}\mathbf{e}'_3 = &\left[\frac{1}{2}\omega_3\omega_1+\omega_2\left(1-\frac{1}{4}\omega^2\right)^{1/2}\right]\mathbf{e}_1 + \left[\frac{1}{2}\omega_3\omega_2-\omega_1\left(1-\frac{1}{4}\omega^2\right)^{1/2}\right]\mathbf{e}_2 \\ &+\left[1-\frac{1}{2}\left(\omega_1^2+\omega_2^2\right)\right]\mathbf{e}_3.\end{aligned} \tag{3.10.7}$$

We can now write our results in the form

$$\mathbf{e}'_i = R_{ij}\,\mathbf{e}_j, \tag{3.10.8}$$

where the rotation matrix R_{ij} is given by

$$R_{ij} = \delta_{ij}\left(1-\frac{1}{2}\omega^2\right)+\frac{1}{2}\omega_i\omega_j+\varepsilon_{ijk}\omega_k\left(1-\frac{1}{4}\omega^2\right)^{1/2}, \tag{3.10.9}$$

where ε_{ijk} is the alternating tensor. Up to now, all the expressions were exact. For infinitesimal rotations, the rotation matrix can be linearized,

$$R_{ij} = \delta_{ij}+\varepsilon_{ijk}\omega_k+O(\omega^2). \tag{3.10.10}$$

For our purpose, this expression is insufficient because we need quadratic terms for the second variation of the elastic energy. Taking into account quadratic terms, we obtain

$$R_{ij} = \delta_{ij}\left(1-\frac{1}{2}\omega^2\right)+\frac{1}{2}\omega_i\omega_j+\varepsilon_{ijk}\omega_k+O(\omega^3). \tag{3.10.11}$$

Before we begin with the discussion of the buckling problem, let us make a few remarks. The Cardan Joint was invented by the physician and mathematician G. Cardan in Italy in the 16th century, but in most English-speaking countries it is referred to as Hooke's joint. The construction is sketched in Figure 3.10.4.

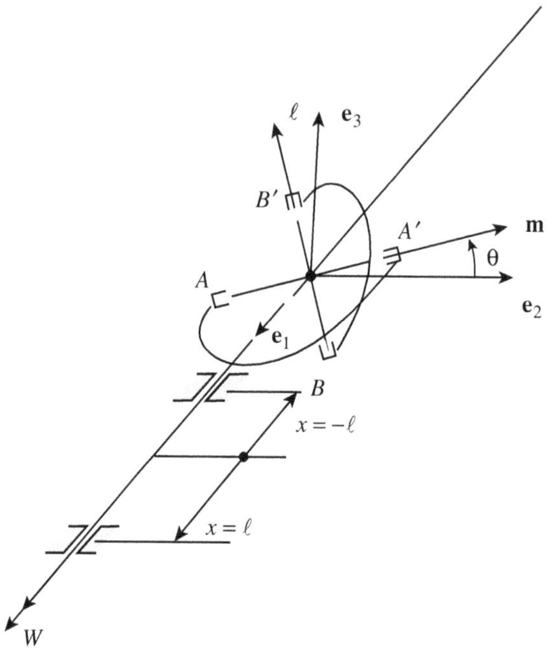

Figure 3.10.4

The bars AA′ and BB′ are rigid and are rigidly connected perpendicular to each other. The semi-rings are also rigid. The bearings in A, A′, B, B′ are frictionless.

Along the bars AA′ and BB′, we introduce the unit vectors $\mathbf{m}$ and ℓ, respectively. These vectors are expressed in terms of the triad $(\mathbf{e}_1, \mathbf{e}_2, \mathbf{e}_3)$, which is fixed in space,

$$\begin{aligned} \ell &= -\sin\theta\,\mathbf{e}_2 + \cos\theta\,\mathbf{e}_3 \\ \mathbf{m} &= \cos\theta\,\mathbf{e}_2 + \sin\theta\,\mathbf{e}_3. \end{aligned} \tag{3.10.12}$$

Let $\left(\mathbf{e}_1', \mathbf{e}_2', \mathbf{e}_3'\right)$ be connected to the deformable shaft; thus

$$\ell' = -\sin\theta\,\mathbf{e}_2' + \cos\theta\,\mathbf{e}_3'. \tag{3.10.13}$$

Let φ be the torsion angle at $x = -\ell$, so

$$\mathbf{m}' = \cos(\theta+\varphi)\,\mathbf{e}_2 + \sin(\theta+\varphi)\,\mathbf{e}_3. \tag{3.10.14}$$

Because ℓ' and $\mathbf{m}'$ are perpendicular to each other, we have

$$(-\sin\theta\,\mathbf{e}_2' + \cos\theta\,\mathbf{e}_3')\cdot[\cos(\theta+\varphi)\,\mathbf{e}_2 + \sin(\theta+\varphi)\,\mathbf{e}_3] = 0, \tag{3.10.15}$$

which yields

$$\begin{aligned} &-\sin\theta\cos(\theta+\varphi)\left[1-\frac{1}{2}\left(\omega_3^2+\omega_1^2\right)\right] - \sin\theta\sin(\theta+\varphi)\left[\omega_1\left(1-\frac{1}{4}\omega^2\right)^{1/2}+\frac{1}{2}\omega_2\omega_3\right] \\ &+\cos\theta\cos(\theta+\varphi)\left[-\omega_1\left(1-\frac{1}{4}\omega^2\right)^{1/2}+\frac{1}{2}\omega_3\omega_2\right] \end{aligned} \tag{3.10.16}$$

$$+ \cos\theta \sin(\theta+\varphi)\left[1-\frac{1}{2}\left(\omega_1^2+\omega_2^2\right)\right]=0,$$

or

$$\begin{aligned} &-\cos\varphi\,\omega_1\left(1-\frac{1}{4}\omega^2\right)^{1/2}+\frac{1}{2}\omega_2\omega_3\cos(2\theta+\varphi) \\ &+\sin\varphi\left(1-\frac{1}{2}\omega_1^2\right)+\frac{1}{2}\omega_3^2\sin\theta\cos(\theta-\varphi)-\frac{1}{2}\omega_2^2\cos\theta\sin(\theta+\varphi)=0. \end{aligned} \tag{3.10.17}$$

For known $\boldsymbol{\omega}$, the angle φ can be determined from this (exact) equation. Because we only need an approximate solution that is correct up to and including quadratic terms, we can simplify this equation to yield

$$-\omega_1+\frac{1}{2}\omega_2\omega_3\cos 2\theta+\varphi+\frac{1}{2}\omega_3^2\sin\theta\cos\theta-\frac{1}{2}\omega_2^2\cos\theta\sin\theta=0, \tag{3.10.18}$$

from which follows

$$\varphi=\omega_1-\frac{1}{2}\omega_2\omega_3\cos 2\theta-\frac{1}{4}\left(\omega_3^2-\omega_2^2\right)\sin 2\theta. \tag{3.10.19}$$

The second variation is now obtained by adding to the functional (3.9.20) the terms due to the rotations at $x=\pm\ell$,

$$\begin{aligned} P_2[\mathbf{u}] = &\int_{-\ell}^{\ell}\left[\frac{1}{2}B\left(v_0''^2+w_0''^2\right)-\frac{1}{2}W\left(v_0'w_0''-v_0''w_0'\right)\right]dx \\ &+\frac{1}{2}W\left[\omega_2\omega_3\cos 2\theta+\frac{1}{2}\left(\omega_3^2-\omega_2^2\right)\sin 2\theta\right]_{x=\ell} \\ &-\frac{1}{2}W\left[\omega_2\omega_3\cos 2\theta-\frac{1}{2}\left(\omega_3^2-\omega_2^2\right)\sin 2\theta\right]_{x=-\ell}. \end{aligned} \tag{3.10.20}$$

With

$$\omega_2=-w_0', \quad \omega_3=v_0', \tag{3.10.21}$$

we can rewrite this functional in the form

$$\begin{aligned} P_2[\mathbf{u}] = &\int_{-\ell}^{\ell}\left[\frac{1}{2}B\left(v_0''^2+w_0''^2\right)-\frac{1}{2}W\left(v_0'w_0''-v_0''w_0'\right)\right]dx \\ &+\frac{1}{2}W\left[-v_0'w_0'\cos 2\theta+\frac{1}{2}\left(v_0'^2-w_0'^2\right)\sin 2\theta\right]_{x=\ell} \\ &+\frac{1}{2}W\left[v_0'w_0'\cos 2\theta+\frac{1}{2}\left(v_0'^2-w_0'^2\right)\sin 2\theta\right]_{x=-\ell}. \end{aligned} \tag{3.10.22}$$

The condition for neutral equilibrium is

$$\begin{aligned} P_{11}[\mathbf{u},\zeta] = &\left(Bv_0''+\frac{1}{2}Ww_0''\right)\eta'\,\Big|_{-\ell}^{\ell}+\left(Bw_0''-\frac{1}{2}Wv_0'\right)\varsigma'\,\Big|_{-\ell}^{\ell} \\ &-\left(Bv_0'''+Ww_0''\right)\eta\,\Big|_{-\ell}^{\ell}-\left(Bw_0'''-Wv_0''\right)\varsigma\,\Big|_{-\ell}^{\ell} \\ &+\int_{-\ell}^{\ell}\left[\left(Bv_0''''+Ww_0'''\right)\eta+\left(Bw_0''''-Wv_0'''\right)\varsigma\right]dx \end{aligned}$$

$$+\frac{1}{2}W[-w_0'\cos 2\theta + v_0'\sin 2\theta]\,\eta'\,|_{x=\ell} \tag{3.10.23}$$
$$+\frac{1}{2}W[-v_0'\cos 2\theta - w_0'\sin 2\theta]\,\varsigma'\,|_{x=\ell}$$
$$+\frac{1}{2}W[w_0'\cos 2\theta + v_0'\sin 2\theta]\,\eta'\,|_{x=-\ell}$$
$$+\frac{1}{2}W[v_0'\cos 2\theta - w_0'\sin 2\theta]\,\varsigma'\,|_{x=-\ell} = 0,$$

where η and ς are arbitrary kinematically admissible displacement fields. Assuming that $x = \pm\ell$, we have the kinematic conditions

$$v_0(\pm\ell) = w_0(\pm\ell) = 0. \tag{3.10.24}$$

We arrive at the following dynamic boundary conditions at $x = \ell$,

$$\begin{aligned}\left[Bv_0'' + \frac{1}{2}Wv_0'\sin 2\theta + \frac{1}{2}Ww_0'(1-\cos 2\theta)\right]_{x=\ell} &= 0\\ \left[Bw_0'' - \frac{1}{2}Wv_0'(1+\cos 2\theta) - \frac{1}{2}Ww_0'\sin 2\theta\right]_{x=\ell} &= 0,\end{aligned} \tag{3.10.25}$$

and at $x = -\ell$, we have

$$\begin{aligned}\left[-Bv_0'' + \frac{1}{2}Wv_0'\sin 2\theta - \frac{1}{2}Ww_0'(1-\cos 2\theta)\right]_{x=-\ell} &= 0\\ \left[-Bw_0'' + \frac{1}{2}Wv_0'(1+\cos 2\theta) - \frac{1}{2}Ww_0'\sin 2\theta\right]_{x=-\ell} &= 0.^{\dagger}\end{aligned}$$

The differential equations are still given by

$$Bv_0'''' + Ww_0''' = 0, \quad Bw_0'''' - Wv_0''' = 0. \tag{3.10.26}$$

As already discussed in our previous examples, the general solution to these equations is

$$\begin{aligned}v_0 &= \underline{C_1\cos\mu x} + C_2\sin\mu x + \underline{C_3x^2} + C_4x + \underline{C_5}\\ w_0 &= \underline{C_1\sin\mu x} - C_2\cos\mu x + C_6x^2 + \underline{C_7x} + C_8\end{aligned} \tag{3.10.27}$$

where $\mu = W/B$.

We shall now first discuss the case that v_0 is an even function in x. Then w_0 is an odd function in x, as follows from the differential equation. This means that we must only consider the underlined terms in (3.10.27). Introduction of these expressions into the kinematic boundary conditions yields

$$\begin{aligned}&C_1\cos\mu\ell + C_3\ell^2 + C_5 = 0\\ &C_1\sin\mu\ell + C_7\ell = 0.\end{aligned} \tag{3.10.28}$$

† These boundary conditions can also be derived directly by requiring that the moments along ℓ and **m** vanish.

From the dynamic boundary conditions at $x = \ell$, we obtain

$$\begin{aligned} &\mu^2 [(1+\cos 2\theta)\cos\mu\ell + \sin 2\theta \sin\mu\ell]\, C_1 \\ &\quad -2\,(2+\mu\ell\sin 2\theta)\, C_3 - \mu\,(1-\cos 2\theta)\, C_7 = 0 \\ &\mu\,[(1-\cos 2\theta)\sin\mu\ell + \sin 2\theta\cos\mu\ell]\, C_1 \\ &\quad + 2\ell\,(1+\cos 2\theta)\, C_3 + \sin 2\theta\, C_7 = 0. \end{aligned} \tag{3.10.29}$$

Notice that due to symmetry and anti-symmetry, the boundary conditions at $x = -\ell$ are satisfied automatically. Because the constant C_5 only appears in the first of the kinematic conditions that does not contain C_7, we can first calculate C_1, C_3, and C_7 from the remaining equations and then determine C_5 from the first kinematic condition. The condition for a non-trivial solution of these three equations is

$$\begin{aligned} &\sin\mu\ell[-2(2+\mu\ell\sin 2\theta)\sin\theta + 2\mu\ell(1-\cos^2 2\theta)] \\ &\quad + \ell\{\mu^2[(1+\cos 2\theta)\cos\mu\ell + \sin 2\theta\sin\mu\ell]2\ell\,(1+\cos 2\theta) \\ &\quad + 2(2+\mu\ell\sin 2\theta)\mu[(1-\cos 2\theta)\sin\mu\ell + \sin 2\theta\cos\mu\ell]\} = 0, \end{aligned} \tag{3.10.30}$$

which can be reduced to

$$\lambda\sin\lambda\tan^2\theta + [(\lambda^2-1)\sin\lambda + \lambda\cos\lambda]\tan\theta + \lambda^2\cos\lambda = 0 \tag{3.10.31}$$

where $\lambda = \mu\ell$.

First, notice that there are no solutions for $\lambda \ll 1$. For $\theta = 0$, we have the solution $\lambda = \pi/2$, which is half the value of Greenhill's result. For $\theta = \pi/2$, we have $\lambda = \pi$ (Greenhill's value). Evaluating θ for given values of λ, we obtain the graph shown in Figure 3.10.5.

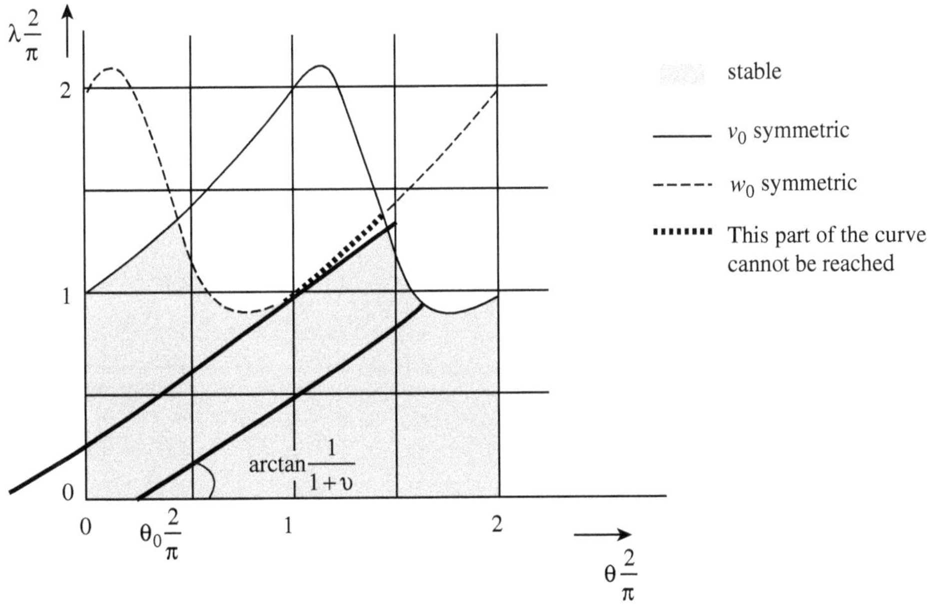

Figure 3.10.5

The case where w_0 is symmetric and v_0 is anti-symmetric is easily obtained by rotating the axes, which yield $v_0 \to w_0$, $w_0 \to -v_0$ for $\theta \to \theta + \pi/2$. Instead of (3.10.31), we obtain

$$\lambda \sin \lambda \text{cotan}^2 \theta - [(\lambda^2 - 1)\sin \lambda + \lambda \cos \lambda] \operatorname{cotan} \theta + \lambda^2 \cos \lambda = 0, \tag{3.10.32}$$

which has roots that are shifted over a distance $\pi/2$ compared to the roots of (3.10.31). The corresponding curve is drawn as a dashed curve in Figure 3.10.5.

The torsion angle follows from

$$\theta - \theta_0 = \frac{W\ell}{S_t}, \tag{3.10.33}$$

which yields

$$\theta_0 = \theta - \frac{W\ell}{S_t} = \theta - \lambda \frac{B}{S_t}. \tag{3.10.34}$$

This result is a straight line in the $\lambda - \theta$ graph. For a circular shaft, we have

$$\theta_0 = \theta - \lambda (1 + \upsilon), \tag{3.10.35}$$

and its tangent in the $\lambda - \theta$ plane is $\arctan(1/1 + \upsilon) < \pi/4$, which implies that this line may be tangent to the curve. The part of the curve above this line then cannot be reached, which implies that the critical torque is a discontinuous function of θ. From the practical point of view, these results are not as important because the critical value of the torque is very large.

3.11 Lateral buckling of a beam loaded in bending

We consider a beam with a slender cross section loaded by a bending moment M in the direction of a principle axis of inertia (see Figure 3.11.1).

The center of shear is (y_0, z_0). When M is sufficiently large, torsion and bending in the y-direction will occur besides the bending in the z-direction. The corresponding additional displacements in passing from the fundamental state to the adjacent state are (approximately) given by

$$\begin{aligned} u &= -y v_0'(x) + \psi_0(y,z)\alpha'(x) \\ v &= v_0(x) - \alpha(x)(z - z_0) + v^*(x,y,z) \\ w &= \alpha(x)(y - y_0) + w^*(x,y,z), \end{aligned} \tag{3.11.1}$$

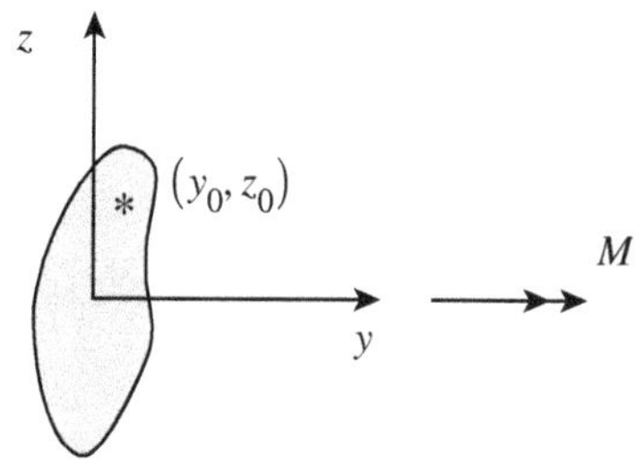

Figure 3.11.1

where the starred functions are added to obtain a uni-axial state of stress. However, the values of these functions are small compared to those of the other terms, and may for our purpose be neglected. In (3.11.1), $\psi_0(y, z)$ is the warping function with respect to the shear center and $\alpha(x)$ is the torsion angle per unit length. We know from our general theory that for dead-weight loads, the second variation is given by

$$P_2[\mathbf{u}] = \int_V \left[\frac{1}{2} S_{ij} u_{h,i} u_{h,j} + \frac{1}{2} E_{ijkl} \theta_{ij} \theta_{kl}\right] dV. \tag{3.11.2}$$

The second term in (3.11.2) is readily obtained in the form

$$\int_V \frac{1}{2} E_{ijkl} \theta_{ij} \theta_{kl}\, dV = \int_0^{\ell} \left[\frac{1}{2} E\mathcal{I}_z v_0''^2 + \frac{1}{2} S_t \alpha'^2 + \frac{1}{2} E\Gamma \alpha''^2\right] dx, \tag{3.11.3}$$

where Γ is the warping constant.

For the evaluation of the energy in the fundamental state, we notice that

$$S_{11} = \sigma_x = \frac{Mz}{\mathcal{I}_y} \tag{3.11.4}$$

is the only non-vanishing stress component. The contribution to the second variation is

$$\begin{aligned} \int_V \frac{1}{2} S_{11} \left(u_{1,1}^2 + u_{2,1}^2 + u_{3,1}^2\right) dV &= \int_V \frac{1}{2} \sigma_x \left(u_{,x}^2 + v_{,x}^2 + w_{,x}^2\right) dV \\ &= \int \frac{1}{2} \sigma_x \left\{(-y v_0 + \psi_0 \alpha'')^2 + \left[v_0' - \alpha'(z - z_0) + \frac{\partial v^*}{\partial x}\right]^2 \right. \\ &\quad \left. + \left[\alpha'(y - y_0) + \frac{\partial w^*}{\partial x}\right]^2\right\} dV. \end{aligned} \tag{3.11.5}$$

First, notice that the first term between the brackets can be neglected because in it is $\sigma_x u_{,x}^2$, which is always small compared to the term $E u_{,x}^2$, which is taken into account in (3.11.3). Second, the terms with an asterisk can be neglected as already discussed, so that we obtain

$$\begin{aligned} &\int_V \frac{1}{2} S_{ij} u_{h,i} u_{h,j}\, dV \\ &= \int dx \iint \frac{Mz}{2\mathcal{I}_y} \left[\{v_0' - \alpha'(z - z_0)\}^2 + \alpha'^2 (y - y_0)^2\right] dy\, dz \\ &= \int dx \iint \frac{Mz}{2\mathcal{I}_y} \left[(v_0' + \alpha' z_0)^2 + \alpha'^2 y_0^2 - 2\alpha'(v_0' + \alpha' z_0) z - 2\alpha'^2 y y_0 + \alpha'^2 z^2 + \alpha'^2 y^2\right] dy\, dz. \end{aligned} \tag{3.11.6}$$

The first two terms between the brackets vanish after integration because the origin of our (y, z) axes is the center of gravity, and the fourth term vanishes after integration because our axes are principle axes of inertia. Carrying out the integration, we obtain

$$\begin{aligned} &\int_V \frac{1}{2} S_{ij} u_{h,i} u_{h,j}\, dV \\ &= \int dx \left[-M\alpha'(v_0' + \alpha' z_0) + \frac{1}{2} \frac{M\alpha'^2}{\mathcal{I}_y} \iint z(y^2 + z^2)\, dy\, dz\right]. \end{aligned} \tag{3.11.7}$$

Introducing the shorthand

$$z_0 - \frac{1}{2\mathcal{I}_y}\iint z(y^2+z^2)\,dy\,dz = c, \tag{3.11.8}$$

for the second variation we finally obtain

$$P_2[\mathbf{u}] = \int\left[\frac{1}{2}E\mathcal{I}_z v_0''^2 + \frac{1}{2}S_t\alpha'^2 + \frac{1}{2}E\Gamma\alpha''^2 - M\alpha' v_0' - cM\alpha'^2\right]dx. \tag{3.11.9†}$$

Notice that when the y-axis is an axis of symmetry, $c = 0$.

When the beam is loaded by a shear force through the center of shear, the contributions of the shear stresses (according to Saint-Venant's theory for the bending of beams) to the elastic energy are

$$\begin{aligned}\int \frac{1}{2}S_{ij}u_{h,i}u_{h,j}\,dV &= \int [S_{12}(u_{1,1}u_{1,2}+u_{2,1}u_{2,2}+u_{3,1}u_{3,2}) \\ &\quad + S_{13}(u_{1,1}u_{1,3}+u_{2,1}u_{2,3}+u_{3,1}u_{3,3})]\,dV \\ &= \int\left\{\tau_{xy}\left[(-yv_0''+\psi_0\alpha'')\left(-v_0'+\alpha'\frac{\partial\psi_0}{\partial y}\right)+\alpha'(y-y_0)\alpha\right]\right. \\ &\quad \left. + \tau_{xz}\left[(-yv_0''+\psi_0\alpha'')\alpha'\frac{\partial\psi_0}{\partial z}+\{v_0'-\alpha'(z-z_0)\}(-\alpha)\right]\right\}dV\end{aligned} \tag{3.11.10}$$

We now notice that $|\psi_0(y,z)| = O(b^2)$, where b is the height of the beam, so that

$$\left|\psi_0\frac{\partial\psi_0}{\partial y}\alpha'\alpha''\right| = O\left(\frac{b^3\alpha^2}{\ell^3}\right) \tag{3.11.11}$$

and that

$$|\alpha\alpha'(y-y_0)| = O\left(\frac{b\alpha^2}{\ell}\right). \tag{3.11.12}$$

Further, we have the estimates

$$\begin{aligned}|yv_0''v_0'| &= O\left(\frac{bv_0^2}{\ell^3}\right) = O\left(\frac{b^3\alpha^2}{\ell^3}\right) \\ \left|\alpha' yv_0''\frac{\partial\psi_0}{\partial y}\right| &= O\left(\frac{b^2v_0\alpha}{\ell^3}\right) = O\left(\frac{b^3\alpha^2}{\ell^3}\right) \\ |\alpha''\psi_0 v_0'| &= O\left(\frac{b^2v_0\alpha}{\ell^3}\right) = O\left(\frac{b^3\alpha^2}{\ell^3}\right),\end{aligned} \tag{3.11.13}$$

so that by only taking into account the term $\alpha\alpha'(y-y_0)$ in the first line of (3.11.10), we make a relative error of order b^2/ℓ^2, which is admissible for $b^2/\ell^2 \ll 1$. A similar

† In the literature, the term with c is often missing. However, this is not always admissible.

argument holds for terms in the second line of (3.11.10), and our final result becomes

$$\begin{aligned}
&\int \frac{1}{2} S_{ij} u_{h,i} u_{h,j} \, dV \\
&\quad = \int dx \iint \left[(y\tau_{xy} + z\tau_{xz})\, \alpha\alpha' - \alpha\alpha' \tau_{xy} y_0 + \tau_{xz} \alpha \left(v_0' + \alpha' z_0 \right) \right] dy\, dz \\
&\quad = \int dx \iint \alpha\alpha' \left(y\tau_{xy} + z\tau_{xz} \right) dy\, dz + \int M' \alpha \left(v_0' + \alpha' z_0 \right) dx.
\end{aligned} \tag{3.11.14}$$

Here we have made use of the fact that

$$\iint \tau_{xy}\, dy\, dz = 0, \quad \int \tau_{xz}\, dy\, dz = D = M', \tag{3.11.15}$$

where $D(x)$ is the transverse shear force in a cross section. The first term in (3.11.14) can now be rewritten by noticing that

$$\begin{aligned}
&\iint (y\tau_{xy} + z\tau_{xz})\, dy\, dz \\
&\quad = \iint \left\{ \frac{1}{2} \frac{\partial}{\partial y} \left[(y^2 + z^2) \tau_{xy} \right] + \frac{1}{2} \frac{\partial}{\partial z} \left[(y^2 + z^2) \tau_{xz} \right] \right\} dy\, dz \\
&\quad\quad - \frac{1}{2} \iint (y^2 + z^2) \left(\frac{\partial \tau_{xy}}{\partial y} + \frac{\partial \tau_{xz}}{\partial z} \right) dy\, dz \\
&\quad = \frac{1}{2} \oint_{\text{edge}} (y^2 + z^2)(\tau_{xy} n_y + \tau_{xz} n_z)\, ds + \frac{1}{2} \iint (y^2 + z^2) \frac{\partial \sigma_x}{\partial x}\, dy\, dz.
\end{aligned} \tag{3.11.16}$$

The line integral vanishes because τ_{xy} and τ_{xz} vanish at the edge, and using (3.11.4) our final result is

$$\begin{aligned}
\iint (y\tau_{xy} + z\tau_{xz})\, dy\, dz &= \frac{1}{2} \frac{M'}{\mathcal{I}_y} \iint z \left(y^2 + z^2 \right) dy\, dz \\
&= M' (z_0 - c).
\end{aligned} \tag{3.11.17}$$

The second variation for a beam loaded by a transverse shear force (dead-weight load), applied in the center of shear-by-shear stresses according to Saint-Venant's theory, is now given by

$$\begin{aligned}
P_2[\mathbf{u}] &= \int_0^{\ell} \left[\frac{1}{2} E\mathcal{I}_z v_0''^2 + \frac{1}{2} S_t \alpha'^2 + \frac{1}{2} E\Gamma \alpha''^2 - M\alpha' v_0' \right. \\
&\quad \left. - cM\alpha'^2 - M'\alpha \left(v_0' + z_0 \alpha' \right) + M'\alpha\alpha' (z_0 - c) \right] dx \\
&= \int_0^{\ell} \left[\frac{1}{2} E\mathcal{I}_z v_0''^2 + \frac{1}{2} S_t \alpha'^2 + \frac{1}{2} E\Gamma \alpha''^2 - (M\alpha)' v_0' - c\,(M\alpha)' \alpha' \right] dx.
\end{aligned} \tag{3.11.18}$$

This expression is also (approximately) valid for distributed loads, provided that they are distributed conforming to the shear stress distribution of Saint-Venant's theory. When the transverse shear force is not applied in the center of shear, we must add the contribution of this force to the energy (see Figure 3.11.2).

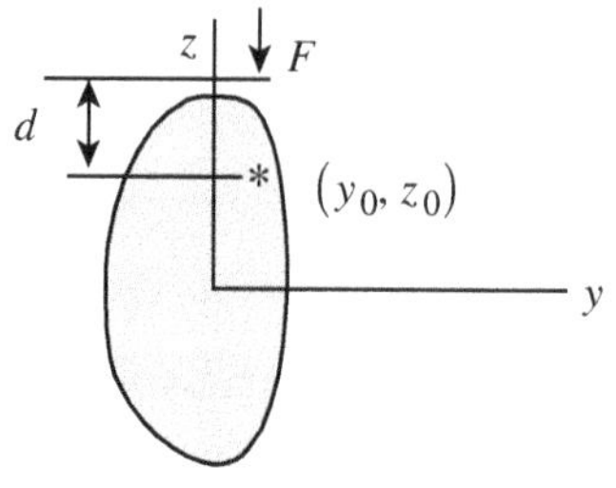

Figure 3.11.2

When the force is applied in a point $(y_0, z_0 + d)$, the contribution is

$$-\frac{1}{2}Fd\alpha^2. \qquad (3.11.19)$$

Because this contribution is negative, we have a destabilizing effect. When the force is applied in $(y_0, z_0 - d)$, this will be a stabilizing effect (see Figure 3.11.3).

For our further discussion, we shall now assume that the shear force is applied in the center of the shear. The necessary condition for neutral equilibrium is

$$\begin{aligned} P_{11}[\mathbf{u}, \xi] = \int_0^\ell & \left[E\mathcal{I}_z v_0''\eta'' + S_t\alpha'\varphi' + E\Gamma\alpha''\varphi'' \right. \\ & \left. - (M\alpha)'\eta' - (M\varphi)'v_0' - c(M\alpha)'\varphi' - c(M\varphi)'\alpha'\right] dx = 0, \end{aligned} \qquad (3.11.20)$$

where η and φ are kinematically admissible fields. By integration by parts, we obtain

$$\begin{aligned} & E\mathcal{I}_z v_0''\eta'\big|_0^\ell - \left[E\mathcal{I}_z v_0''' + (M\alpha)'\right]\eta\Big|_0^\ell + E\Gamma\alpha''\varphi'\big|_0^\ell \\ & + \left[S_t\alpha' - E\Gamma\alpha''' - Mv_0' - cM\alpha' - c(M\alpha)'\right]\varphi\big|_0^\ell \\ & + \int_0^\ell \left\{\left[E\mathcal{I}_z v_0'''' + (M\alpha)''\right]\eta + \left[-S_t\alpha'' + E\Gamma\alpha'''' + Mv_0'' + c(M\alpha)'' + cM\alpha''\right]\varphi\right\} dx = 0, \end{aligned} \qquad (3.11.21)$$

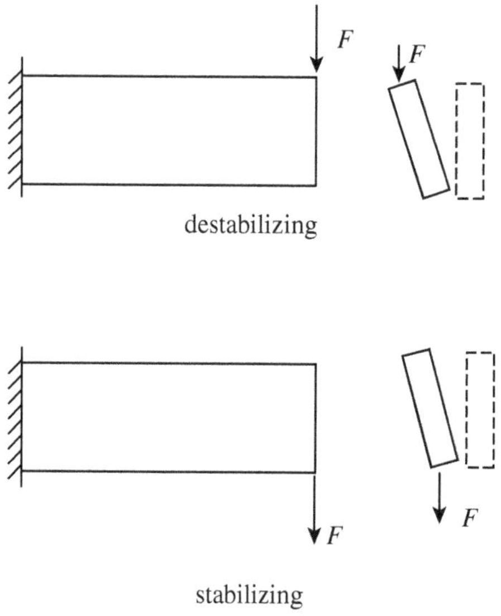

Figure 3.11.3

from which we obtain

$$\begin{aligned} E\mathcal{I}_z v_0'''' + (M\alpha)'' &= 0 \\ E\Gamma\alpha'''' - S_t\alpha'' + Mv_0'' + c\,(M\alpha)'' + cM\alpha'' &= 0. \end{aligned} \quad (3.11.22)$$

We now consider the following boundary conditions:

i) *Simply supported end.* In this case, the kinematic conditions are

$$v_0 = 0, \quad \alpha = 0. \quad (3.11.23)$$

It then follows that the dynamic boundary conditions are

$$E\mathcal{I}_z v_0'' = 0, \quad \underline{E\Gamma\alpha'' = 0}. \quad (3.11.24)$$

ii) *Clamped end.* Here we have only the kinematic conditions

$$v_0 = 0, \quad \alpha = 0, \quad v_0' = 0, \quad \underline{\alpha' = 0}. \quad (3.11.25)$$

iii) *Free end.* In this case, we have only dynamic boundary conditions

$$\begin{aligned} &E\mathcal{I}_z v_0'' = 0, \quad E\mathcal{I}_z v_0''' + (M\alpha)' = 0, \\ &\underline{E\Gamma\alpha'' = 0}, \quad S_t\alpha' - \underline{E\Gamma\alpha'''} - Mv_0' - cM\alpha' - c\,(M\alpha)' = 0. \end{aligned} \quad (3.11.26)$$

For narrow cross sections, $|\psi_0| = O\,(ht)$ (see Figure 3.11.4), and the warping constant Γ is of order $|\Gamma| = O(h^3t^3)$, which means the warping is not important, so that the elementary theory ($\Gamma = 0$) may be used. This is also true for a T profile (see Figure 3.11.5).

However, for other profiles we have different conditions (see Figure 3.11.6).

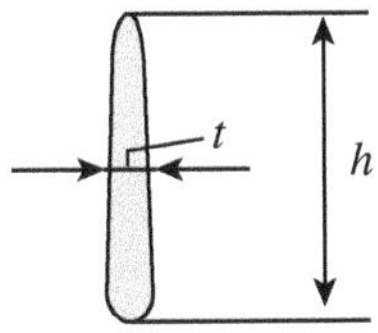

Figure 3.11.4

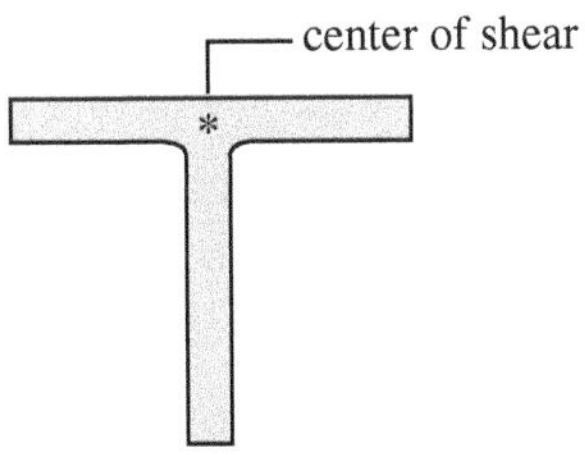

Figure 3.11.5

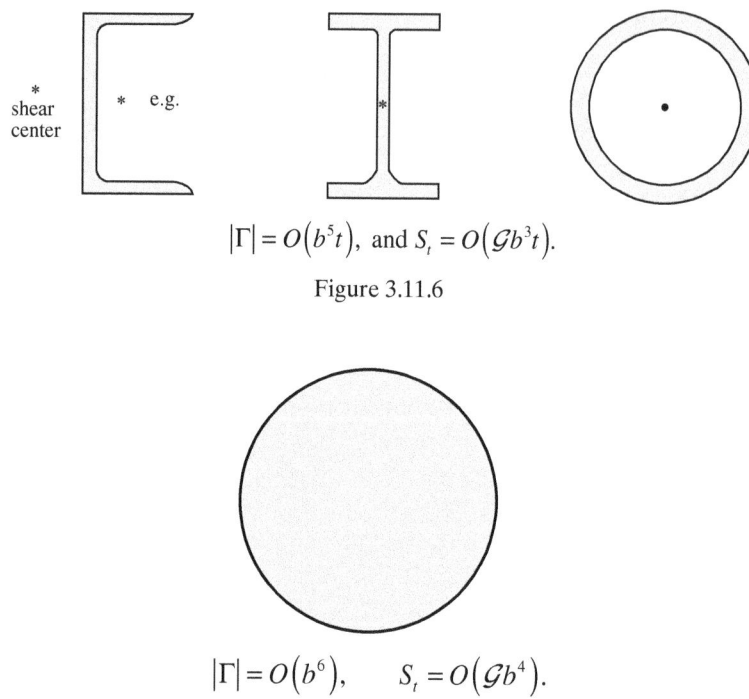

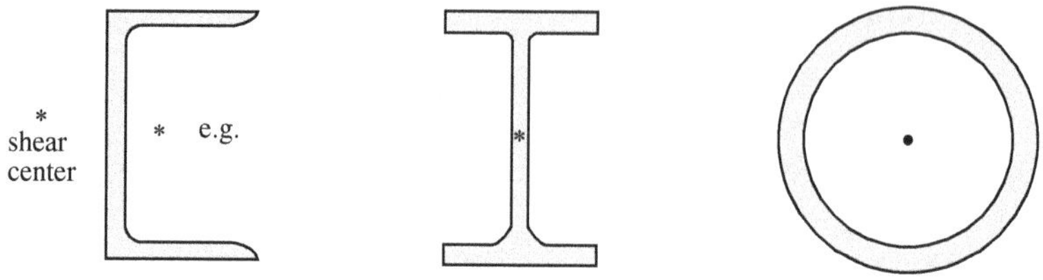

$|\Gamma| = O(b^5 t)$, and $S_t = O(\mathcal{G} b^3 t)$.

Figure 3.11.6

$|\Gamma| = O(b^6)$, $\quad S_t = O(\mathcal{G} b^4)$.

Figure 3.11.7

For a massive cross section, we have Figure 3.11.7.

For $\Gamma = 0$, the order of the system of differential equations (3.11.22) is reduced by two, and therefore the number of boundary conditions at each end of the beam is reduced by one. When $\Gamma = 0$, the underlined terms in (3.11.23) to (3.11.26) must be omitted. In the following examples, we shall assume that $\Gamma = 0$.

Consider a simply supported beam loaded by a constant bending couple M. The differential equations then are

$$E\mathcal{I}_z v_0'''' + M\alpha'' = 0, \quad -S_t\alpha'' + Mv_0'' + 2Mc\alpha'' = 0. \tag{3.11.27}$$

By choosing

$$\alpha = A\sin\frac{k\pi x}{\ell} \quad v_0 = B\sin\frac{k\pi x}{\ell}, \tag{3.11.28}$$

the boundary conditions (3.11.23) and (3.11.24) are satisfied identically. Substitution into the differential equations yield

$$\begin{aligned} &\frac{k^4\pi^4}{\ell^4}E\mathcal{I}_z B - \frac{k^2\pi^2}{\ell^2}MA = 0 \\ &-\frac{k^2\pi^2}{\ell^2}MB + \frac{k^2\pi^2}{\ell^2}(S_t - 2Mc)A = 0. \end{aligned} \tag{3.11.29}$$

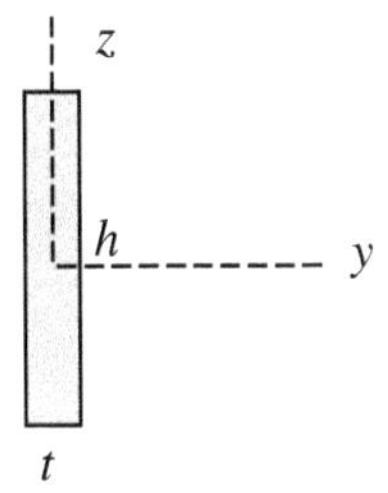

$$S_t = \frac{1}{3}\mathcal{G}t^3h,\quad \mathcal{I}_y = \frac{1}{12}th^3,\quad \mathcal{I}_y = \frac{1}{12}t^3h$$

$$M_{cr} = \frac{\pi}{6\ell}\frac{Et^3h}{\sqrt{2(1+\upsilon)}}$$

$$\sigma_{cr} = \frac{\pi}{\ell h}\frac{Et^2}{\sqrt{2(1+\upsilon)}} \cong \frac{5\pi}{8}\frac{t^2E}{h\ell} \quad (\upsilon = 0.28)$$

Figure 3.11.8. Example of slender rectangular section.

The condition for a non-trivial solution is

$$M^2 = \frac{k^2\pi^2}{\ell^3}E\mathcal{I}_zS_t\left(1 - \frac{2Mc}{S_t}\right). \tag{3.11.30}$$

When $c = 0$, we have

$$M = \pm\frac{k\pi}{\ell}\sqrt{E\mathcal{I}_zS_t}. \tag{3.11.31}$$

In general, we have $2Mc/S_t \ll 1$, and then

$$M \approx \pm\frac{k\pi}{\ell}\sqrt{E\mathcal{I}_zS_t}\left[1 \mp \frac{k\pi c}{\ell}\sqrt{\frac{E\mathcal{I}_z}{S_t}}\right]. \tag{3.11.32}$$

For example, if $t/h = 0.1, h/\ell = 0.1$, then $\sigma_{cr} \approx 0.002E$, which for most construction materials is still within the elastic range.

$$M_{cr} = \frac{\pi}{24\ell}\frac{Eht^3}{\sqrt{2(1+\upsilon)}}\left[1 - \frac{\pi h}{5\ell}\sqrt{\frac{1}{2}(1+\upsilon)}\right]$$

$$\sigma_{cr} = \pi\frac{t^2}{h^2}\frac{h}{\ell}\frac{E}{\sqrt{2(1+\upsilon)}}\left[1 - \frac{\pi h}{5\ell}\sqrt{\frac{1}{2}(1+\upsilon)}\right]$$

$$\approx \frac{5\pi}{8}E\frac{t^2}{h^2}\frac{h}{\ell} \quad \text{for} \quad \upsilon = 0.28 \quad \text{and} \quad h/\ell \ll 1.$$

Notice that this is (approximately) the same result as for the rectangular cross section.

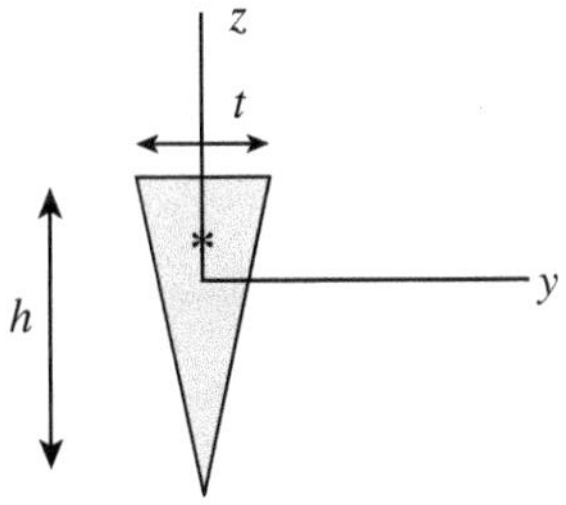

$$z_0 = \frac{2}{15}h,\quad c = \frac{1}{5}h$$

$$E\mathcal{I}_{z_0} = \frac{1}{48}Eht^3,\quad E\mathcal{I}_y = \frac{1}{36}Eth^3$$

$$S_t = \frac{Eht^3}{24(1+\upsilon)},\quad \frac{E\mathcal{I}_{z_0}}{S_t} = \frac{1}{2}(1+\upsilon)$$

Figure 3.11.9. Example of slender triangular section.

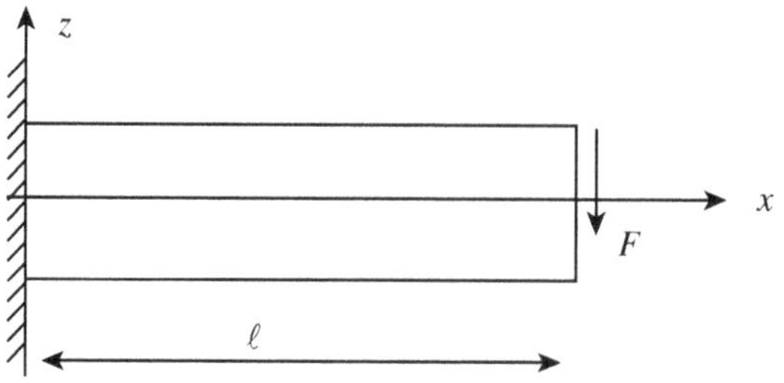

Figure 3.11.10

Let us now consider a beam clamped at one end and loaded by a transverse shear force (applied through the shear center) at the other end (see Figure 3.11.10).

We further assume that the cross section is such that $\Gamma = 0$, $c = 0$. In this case, the differential equations are

$$\begin{aligned} &E\mathcal{I}_z v_0'''' + [F(\ell - x)\alpha]'' = 0, \\ &F(\ell - x) v_0'' - S_t \alpha'' = 0. \end{aligned} \tag{3.11.33}$$

The boundary conditions are

$$v_0 = 0, \quad \alpha = 0, \quad v_0' = 0 \quad \text{at } x = 0 \tag{3.11.34}$$

$$\begin{aligned} &E\mathcal{I}_z v_0'' = 0, \quad E\mathcal{I}_z v_0''' + [F(\ell - x)\alpha]' = 0, \\ &S_t \alpha' = 0 \quad \text{at } x = \ell \end{aligned} \tag{3.11.35}$$

Integrating the first equation and taking into account the second condition of (3.11.35), we obtain

$$E\mathcal{I}_z v_0''' + [F(\ell - x)\alpha]' = 0. \tag{3.11.36}$$

Integrating once again and taking into account the first and the third conditions of (3.11.35), we obtain

$$E\mathcal{I}_z v_0'' + F(\ell - x)\alpha = 0. \tag{3.11.37}$$

Eliminating v_0'' from (3.11.37) and the second of the equations (3.11.33), we obtain an equation in α,

$$\alpha'' + \frac{F^2(\ell - x)^2}{E\mathcal{I}_z S_t}\alpha = 0 \tag{3.11.38}$$

to be solved under the conditions

$$\alpha = 0 \quad \text{at } x = 0, \quad \alpha' = 0 \quad \text{at } x = \ell. \tag{3.11.39}$$

Let us now introduce a new variable η

$$\eta = \frac{\ell - x}{\ell}, \tag{3.11.40}$$

the equation becomes then

$$\frac{d^2\alpha}{d\eta^2} + \frac{F^2\ell^4}{E\mathcal{I}_z S_t}\eta^2\alpha = 0, \tag{3.11.41}$$

or

$$\frac{d^2\alpha}{d\eta^2} + \lambda\eta^2\alpha = 0, \tag{3.11.42}$$

where λ is defined by

$$\lambda = \frac{F^2\ell^2}{E\mathcal{I}_z S_t}. \tag{3.11.43}$$

The equation must be solved under the conditions

$$\alpha = 0 \quad \text{at } \eta = 1, \quad \frac{d\alpha}{d\eta} = 0 \quad \text{at } \eta = 0. \tag{3.11.44}$$

The solutions to (3.11.42) are Bessel functions, but for our purpose it is more convenient to try a solution of the form

$$\alpha = \sum_{n=0}^{\infty} A_n\eta^n. \tag{3.11.45}$$

Substitution into the differential equation yields

$$\sum_{n=0}^{\infty} A_n (n-1)\,\eta^{n-2} + \lambda \sum_{n=0}^{\infty} A_n\eta^{n+2} = 0. \tag{3.11.46}$$

The coefficient of η^k is

$$A_{k+2}\,(k+2)\,(k+1) + \lambda A_{k-2} = 0, \tag{3.11.47}$$

so that

$$A_{k+2} = -\frac{\lambda}{(k+1)\,(k+2)}A_{k-2}. \tag{3.11.48}$$

The solution then becomes

$$\alpha = A_0\left(1 - \frac{\lambda}{3.4}\eta^4 + \frac{\lambda}{3.4}\frac{\lambda}{7.8}\eta^8 - \frac{\lambda}{3.4}\frac{\lambda}{7.8}\frac{\lambda}{11.12}\eta^{12} + \cdots\right). \tag{3.11.49}$$

This solution satisfies the boundary condition at $\eta = 0$. The value of λ follows from $\alpha = 0$ for $\eta = 1$, i.e., the series between the brackets must vanish for $\eta = 1$, which yields

$$\lambda = 16.104, \tag{3.11.50}$$

so the critical load is

$$F_{cr} = 4.013\frac{\sqrt{E\mathcal{I}_z S_t}}{\ell^2}. \tag{3.11.51}$$

For a beam with a rectangular cross section (t, h) the critical stress becomes

$$\sigma_{cr} = 4.013\frac{E}{\sqrt{2\,(1+\upsilon)}}\frac{t^2}{h^2}\frac{h}{\ell}. \tag{3.11.52}$$

As we have seen, the problem in this case can be reduced to a single second-order differential equation. This reduction is always possible when we have a beam with at least one free edge. In this case, the first of the equations of (3.11.33) can be integrated twice to yield

$$EI_z v_0'' + M\alpha = \text{const.} \tag{3.11.53}$$

When the bending moment at $x = 0$ or $x = \ell$ vanishes, the constant is equal to zero. Solving v_0'' from this equation (for simplicity, we shall assume that the constant is equal to zero), we find

$$v_0'' = -\frac{M}{EI_z}\alpha, \tag{3.11.54}$$

and substitution into the second of the equations of (3.11.33) yields

$$\alpha'' + \frac{M^2}{EI_z S_t}\alpha = 0. \tag{3.11.55}$$

This equation also holds for a distributed load.

An approach to get an approximate solution follows from the fact that (3.11.55) minimizes the functional

$$\int_0^\ell \left[\frac{1}{2}S_t\alpha'^2 - \frac{M^2}{2EI_z}\alpha^2\right] dx, \tag{3.11.56}$$

as follows from the variation of the functional, which yields

$$S_t\alpha'\delta\alpha\Big|_0^\ell - \int_0^\ell \left(S_t\alpha'' + \frac{M^2}{EI_z}\alpha\right)\delta\alpha\, dx = 0. \tag{3.11.57}$$

Let the boundary conditions be

$$\alpha = 0 \quad \text{at } x = 0, \quad \alpha' = 0 \quad \text{at } x = \ell, \tag{3.11.58}$$

in agreement with (3.11.57). We now assume

$$\alpha = \xi - \frac{1}{2}\xi^2, \quad \xi = x/\ell, \tag{3.11.59}$$

which satisfies both boundary conditions. Introducing this approximation into the functional (3.11.56) and setting the result equal to zero, we find when $M = F(\ell - x)$,

$$\int_0^1 \left[\frac{1}{2}\frac{S_t}{\ell^2}(1-\xi)^2 - \frac{F^2\ell^2}{2EI_z}(1-\xi)^2\left(\xi - \frac{1}{2}\xi^2\right)^2\right]\ell\, d\xi = 0, \tag{3.11.60}$$

which yields

$$\frac{F^2\ell^4}{EI_z S_t} = \frac{105}{6} = 17.5, \tag{3.11.61}$$

so that

$$F_{cr} = 4.18\frac{\sqrt{EI_z S_t}}{\ell^2}, \tag{3.11.62}$$

which gives an error of about 4% compared to the exact value.

3.12 Buckling of plates loaded in their plane

We consider a homogeneous, isotropic elastic flat plate, loaded in its plane. Without going into details, we note that a plate theory can be derived under the following assumptions:

i) Normals to the undeformed middle surface remain normal to the deformed middle surface.
ii) Changes in length of these normals may be neglected.
iii) The state of stress is approximately plane and parallel to the middle surface.

Further, we note that the stiffness of the plate in its plane is considerably larger than the bending stiffness, which means that after buckling, the displacements perpendicular to the middle plane are considerably larger than the displacement in the middle plane.

We now recall that stability in the fundamental state is governed by the functional

$$P[\mathbf{u}] = \int_V \left(\frac{1}{2}S_{ij}u_{h,i}u_{h,j} + \frac{1}{2}E_{ijkl}\gamma_{ij}\gamma_{k\ell}\right)dV, \tag{3.12.1}$$

where γ is the nonlinear strain tensor. For plates, the non-vanishing components of the strain tensor are

$$\begin{aligned}\gamma_{11}(z) &= u_{,x} + \frac{1}{2}w_{,x}^2 - zw_{,xx}\\ \gamma_{22}(z) &= v_{,y} + \frac{1}{2}w_{,y}^2 - zw_{,yy}\\ 2\gamma_{12}(z) &= u_{,y} + v_{,x} + w_{,x}w_{,y} - 2zw_{,xy},\end{aligned} \tag{3.12.2}$$

where $u(x, y)$, $v(x, y)$ are the displacements in the mid-plane, $w(x, y)$ is the displacement perpendicular to the mid-plane, and z is the distance to the mid-plane $(-h/2 \leq z \leq h/2)$. The contribution of the second term in (3.12.1) to the energy density can now be written as

$$\frac{1}{2}E_{ijkl}\gamma_{ij}\gamma_{k\ell} = \frac{E}{2(1-\upsilon^2)}\left[\gamma_{11}^2(z) + \gamma_{22}^2(z) + 2\upsilon\gamma_{11}(z)\gamma_{22}(z) + 2(1-\upsilon)\gamma_{12}^2(z)\right]. \tag{3.12.3}$$

Integrating (3.12.3) over the thickness of the plate, we obtain the energy per unit area,

$$\begin{aligned}V = \frac{Eh}{2(1-\upsilon^2)}&\left\{\left(u_{,x} + \frac{1}{2}w_{,x}^2\right)^2 + \left(v_{,y} + \frac{1}{2}w_{,y}^2\right)^2\right.\\ &+ 2\upsilon\left(u_{,x} + \frac{1}{2}w_{,x}^2\right)\left(v_{,y} + \frac{1}{2}w_{,y}^2\right) + \frac{1}{2}(1-\upsilon)\left(u_{,y} + v_{,x} + w_{,x}w_{,y}\right)^2\\ &\left.+ \frac{h^2}{12}\left[w_{,xx}^2 + w_{,yy}^2 + 2\upsilon w_{,xx}w_{,yy} + 2(1-\upsilon)w_{,xy}^2\right]\right\},\end{aligned} \tag{3.12.4}$$

where the last term represents the bending energy.

Assuming that in the fundamental state the only non-vanishing stress components are

$$S_{11} = S_x, \quad S_{22} = S_y, \quad S_{12} = S_{xy},$$

the corresponding energy per unit area of the middle plane is

$$\left[\frac{1}{2}S_x w_{,x}^2 + \frac{1}{2}S_y w_{,y}^2 + S_{xy} w_{,x} w_{,y}\right] h. \tag{3.12.5}$$

The energy functional for a plate loaded in its plane now becomes

$$\begin{aligned} P[\mathbf{u}] = \frac{Eh}{12\,(1-\upsilon^2)} \iint \Bigg\{ &\left(u_{,x} + \frac{1}{2}w_{,x}^2\right)^2 + \left(v_{,y} + \frac{1}{2}w_{,y}^2\right)^2 \\ &+ 2\upsilon\left(u_{,x} + \frac{1}{2}w_{,x}^2\right)\left(v_{,y} + \frac{1}{2}w_{,y}^2\right) + \frac{1}{2}(1-\upsilon)\,(u_{,y} + v_{,x} + w_{,x}w_{,y})^2 \\ &+ \frac{h^2}{12}\left[w_{,xx}^2 + w_{,yy}^2 + 2\upsilon w_{,xx}w_{,yy} + 2(1-\upsilon)w_{,xy}^2\right]\Bigg\}\,dx\,dy \\ &+ \frac{1}{2}h \iint \left[S_x w_{,x}^2 + S_y w_{,y}^2 + 2S_{xy} w_{,x} w_{,y}\right] dx\,dy. \end{aligned} \tag{3.12.6}$$

We now assume that the fundamental state is linear, i.e., that the stresses are proportional to the loads. This assumption is valid when the rotations are of the same order of magnitude as the strains, which are small. This is the case for a property-supported plate (no rotations, as in a rigid body). Under these assumptions, we can consider the x-, y-coordinates are the same as the coordinates in the undeformed state.

Stability is primarily determined by the second variation, which is given by

$$\begin{aligned} P_2[\mathbf{u}] = \frac{Eh}{2\,(1-\upsilon^2)} \iint \Bigg\{ &u_{,x}^2 + v_{,y}^2 + 2\upsilon u_{,x} v_{,y} + 2(1-\upsilon)\,(u_{,y} + v_{,x})^2 \\ &+ \frac{h^2}{12}\left[w_{,xx}^2 + w_{,yy}^2 + 2\upsilon w_{,xx}w_{,yy} + 2(1-\upsilon)w_{,xy}^2\right] \\ &+ \frac{1-\upsilon^2}{E}\left(S_x w_{,x}^2 + S_y w_{,y}^2 + 2S_{xy} w_{,x} w_{,y}\right)\Bigg\}\,dx\,dy. \end{aligned} \tag{3.12.7}$$

We may now draw some important conclusions from the structure of this second variation:

i) The first line in the integrand is positive-definite.† At neutral equilibrium, $P_2[\mathbf{u}]$ is semi-positive definite, which implies that the second and the third line together, which only depend on w, become semi-positive-definite, and that the first line must vanish at neutral equilibrium. This means that

$$u_1(x, y) = v_1(x, y) \equiv 0 \tag{3.12.8}$$

at neutral equilibrium.

† To show this we recall that $a^2 + b^2 + 2\upsilon ab + 2(1-\upsilon)c^2 = (a + \upsilon b)^2 + (1-\upsilon)b^2 + 2(1-\upsilon)c^2 > 0$ for $\forall a, b, c$, $|\upsilon| < 1$, and a, b, c not vanishing simultaneously.

ii) Because in (3.12.6) the third and fourth line together are semi-positive-definite and the first two lines together are positive-definite, it follows that in the critical state of neutral equilibrium $P[\mathbf{u}] > 0$, so that for flat plates loaded in their plane, the critical state of neutral equilibrium is stable. This conclusion does not hold for curved plates and shells.

Let us now derive the conditions for neutral equilibrium, which follow from

$$P_{11}[\mathbf{u},\zeta] = 0 = \frac{Eh^2}{12(1-\upsilon^2)} \iint \Big\{ w_{,xx}\zeta_{,xx} + w_{,yy}\zeta_{,yy} + \upsilon w_{,xx}\zeta_{,yy} + \upsilon w_{,yy}\zeta_{,xx} + 2(1-\upsilon)\, w_{,xy}\zeta_{,xy} + \frac{12(1-\upsilon^2)}{Eh^2}\left[S_x w_{,x}\zeta_{,x} + S_y w_{,y}\zeta_{,y} + S_{xy}\left(w_{,x}\zeta_{,y} + w_{,y}\zeta_{,x}\right)\right]\Big\}\, dx\, dy. \quad (3.12.9)$$

Using the divergence theorem once, with υ as the unit normal vector to the edge, we obtain positive in the outward direction

$$\frac{Eh^3}{12(1-\upsilon^2)} \oint_{\text{edge}} \left[w_{,xx}\zeta_{,x}\upsilon_x + w_{,yy}\zeta_{,y}\upsilon_y + \upsilon w_{,xx}\zeta_{,y}\upsilon_y + \upsilon w_{,yy}\zeta_{,x}\upsilon_x + (1-\upsilon)w_{,xy}\zeta_{,x}\upsilon_y + (1-\upsilon)w_{,xy}\zeta_{,y}\upsilon_x\right] ds + \frac{Eh^3}{12(1-\upsilon^2)} \iint \Big\{ -\left(w_{,xx}+\upsilon w_{,yy}\right)_{,x}\zeta_{,x} - \left(w_{,yy}+\upsilon w_{,xx}\right)_{,y}\zeta_{,y} - (1-\upsilon)w_{,xyy}\zeta_{,x} - (1-\upsilon)w_{,xyx}\zeta_{,y} + \frac{12(1-\upsilon^2)}{Eh^2}\left[S_x w_{,x}\zeta_{,x} + S_y w_{,y}\zeta_{,y} + S_{xy}\left(w_{,x}\zeta_{,y} + w_{,y}\zeta_{,x}\right)\right]\Big\}\, dx\, dy = 0. \quad (3.12.10)$$

Repeated application of the divergence theorem yields

$$\frac{Eh^3}{12(1-\upsilon^2)} \oint_{\text{edge}} \Big\{\left[\left(w_{,xx}+\upsilon w_{,yy}\right)\upsilon_x + (1-\upsilon)w_{,xy}\upsilon_y\right]\Big\}\zeta_{,x} + \left[\left(w_{,yy}+\upsilon w_{,xx}\right)\upsilon_y + (1-\upsilon)w_{,xy}\upsilon_x\right]\zeta_{,y} - \Big\{\left[\left(w_{,xx}+\upsilon w_{,yy}\right)_{,x} + (1-\upsilon)w_{,xyy}\right]\upsilon_x + \left[\left(w_{,yy}+\upsilon w_{,xx}\right)_{,y} + (1-\upsilon)w_{,xxy}\right]\upsilon_y\Big\}\zeta + \frac{12(1-\upsilon^2)}{Eh^2}\left[\left(S_x w_{,x} + S_{xy}w_{,y}\right)\upsilon_x + \left(S_y w_{,y} + S_{xy}w_{,x}\right)\upsilon_y\right]\zeta\Big\}\, dx\, dy + \frac{Eh^3}{12(1-\upsilon^2)} \iint \Big\{\Delta\Delta w - \frac{12(1-\upsilon^2)}{Eh}\left[S_x w_{,xx} + S_y w_{,yy} + 2S_{xy}w_{,xy} + \left(S_{x,x}+S_{xy,y}\right)w_{,x} + \left(S_{y,y}+S_{xy,x}\right)w_{,y}\right]\Big\}\zeta\, dx\, dy = 0, \quad (3.12.11)$$

where Δ is the Laplacian operator. The differential equation for neutral equilibrium is now given by

$$\Delta\Delta w - \frac{12(1-\upsilon^2)}{Eh^2}\left(S_x w_{,xx} + S_y w_{,yy} + 2S_{xy}w_{,xy} - Xw_{,x} - Yw_{,y}\right) = 0, \quad (3.12.12)$$

where X and Y are the mass forces per unit area. Here we have used the equilibrium equations

$$\begin{aligned} S_{x,x} + S_{xy,y} + X &= 0 \\ S_{xy,y} + S_{y,y} + Y &= 0. \end{aligned} \tag{3.12.13}$$

Using the relations

$$\zeta_{,x} = \zeta_{,s} t_x + \zeta_{,\upsilon} \upsilon_x, \qquad \zeta_{,y} = \zeta_{,s} t_y + \zeta_{,\upsilon} \upsilon_y, \tag{3.12.14}$$

where $\mathbf{t}$ is the unit tangent vector to the edge, we can rewrite the line integral in (3.12.11) to yield

$$\begin{aligned} \oint_{\text{edge}} \Big\{ & \left[(w_{,xx} + \upsilon w_{,yy}) \upsilon_x^2 + 2(1-\upsilon) w_{,xy} \upsilon_x \upsilon_y + (w_{,yy} + \upsilon w_{,xx}) \upsilon_y^2 \right] \zeta_{,\upsilon} \\ & + \left[(w_{,xx} + \upsilon w_{,yy}) \upsilon_x t_x + (w_{,yy} + \upsilon w_{,xx}) \upsilon_y t_y \right. \\ & \left. + (1-\upsilon) w_{,xy} (\upsilon_y t_x + \upsilon_x t_y) \right] \zeta_{,s} \\ & - \Big\{ \left[(w_{,xx} + \upsilon w_{,yy})_{,x} + (1-\upsilon) w_{,xyy} \right] \upsilon_x + \left[(w_{,yy} + \upsilon w_{,xx})_{,y} + (1-\upsilon) w_{,xxy} \right] \upsilon_y \\ & - \frac{12(1-\upsilon^2)}{Eh} \left[(S_x w_{,x} + S_{xy} w_{,y}) \upsilon_x + (S_y w_{,y} + S_{xy} w_{,x}) \upsilon_y \right] \Big\} \zeta \Big\} ds = 0. \end{aligned} \tag{3.12.15}$$

Let us now consider a part of the edge curve (see Figure 3.12.1).
It then follows that

$$t_y = \upsilon_x, \quad t_x = -\upsilon_y. \tag{3.12.16}$$

Further, we shall make use of the relation

$$\int_{s_1}^{s_2} F(s) \zeta_{,s} ds = F(s) \zeta \Big|_{s_1}^{s_2} - \int_{s_1}^{s_2} F_{,s} \zeta \, ds, \tag{3.12.17}$$

which holds when $F(s)$ is a differentiable function. This implies that the edge curve has no corners in $s_1 < s < s_2$. In the following, we shall assume that the edge curve is a smooth curve. From (3.12.15), we now obtain

$$\left[(w_{,xx} + \upsilon w_{,yy}) \upsilon_x^2 + 2(1-\upsilon) w_{,xy} \upsilon_x \upsilon_y + (w_{,yy} + \upsilon w_{,xx}) \upsilon_y^2 \right] \zeta_{,\upsilon} \Big|_{\text{edge}} = 0 \tag{3.12.18}$$

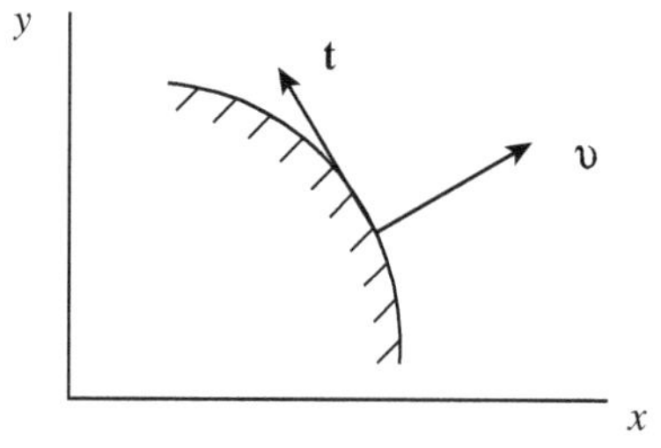

Figure 3.12.1

$$\Big\{(1-\upsilon)\left[(w_{,yy}-w_{,xx})\,\upsilon_x\upsilon_y+\left(\upsilon_x^2-\upsilon_y^2\right)w_{,xy}\right]_{,s}$$
$$+\left(w_{,xx}+w_{,yy}\right)_{,x}\upsilon_x+\left(w_{,xx}+w_{,yy}\right)_{,y}\upsilon_y \tag{3.12.19}$$
$$-\frac{12\left(1-\upsilon^2\right)}{Eh}\left[\left(S_xw_{,x}+S_{xy}w_{,y}\right)\upsilon_x+\left(S_yw_{,y}+S_{xy}w_{,x}\right)\upsilon_y\right]\Big\}\,\zeta\Big|_{\text{edge}}=0$$

$$(1-\upsilon)\left[(w_{,yy}-w_{,xx})\,\upsilon_x\upsilon_y+\left(\upsilon_x^2-\upsilon_y^2\right)w_{,xy}\right]\zeta\Big|_{s_1}^{s_2}=0. \tag{3.12.20}$$

First, we notice that the left-hand side of (3.12.20) vanishes identically along a closed curve. Further, it vanishes along parts of the edge that are supported such that $w = 0$, and it also vanishes along a free edge provided that the edge in the parts of the edge curve adjacent to the free part is properly supported.

In the following discussion, we shall assume that (3.12.20) is satisfied. Let us now consider the various boundary conditions.

i) *The free edge.* Because there are no restrictions imposed on ζ and $\zeta_{,\upsilon}$, their coefficients must vanish, i.e.,

$$\left(w_{,xx}+\upsilon w_{,yy}\right)\upsilon_x^2+2(1-\upsilon)w_{,xy}\upsilon_x\upsilon_y+\left(w_{,yy}+\upsilon w_{,xx}\right)\upsilon_y^2=0, \tag{3.12.21}$$

which means that the bending moment along the edge vanishes, and

$$(1-\upsilon)\left[(w_{,yy}-w_{,xx})\,\upsilon_x\upsilon_y+\left(\upsilon_x^2-\upsilon_y^2\right)w_{,xy}\right]_{,s}$$
$$+\left(w_{,xx}-w_{,yy}\right)_{,x}\upsilon_x+\left(w_{,xx}+w_{,yy}\right)_{,y}\upsilon_y \tag{3.12.22}$$
$$-\frac{12(1-\upsilon^2)}{Eh}\left[\left(S_xw_{,x}+S_{xy}w_{,y}\right)\upsilon_x+\left(S_yw_{,y}+S_{xy}w_{,x}\right)\upsilon_y\right]=0,$$

which means that the reduced transverse shear force vanishes at the edge.

ii) *The simply supported edge.* Here $\zeta \equiv 0$ along the edge, so (3.12.19) is satisfied, so that we have

$$(w_{,xx}+\upsilon w_{,yy})\upsilon_x^2+2(1-\upsilon)w_{,xy}\upsilon_x\upsilon_y+(w_{,yy}+\upsilon w_{,xx})\upsilon_y^2=0 \tag{3.12.23}$$

in addition to the kinematic condition $w = 0$.

iii) *The clamped edge.* Here we have the kinematic conditions

$$w=\frac{\partial w}{\partial \upsilon}=0 \quad \text{at the edge} \tag{3.12.24}$$

so that $\zeta = \zeta_{,\upsilon} = 0$ along the edge, which implies that the conditions (3.12.18) and (3.12.19) are satisfied automatically.

Let us now apply our result to a square plate, simply supported at its edges, loaded in compression by forces σ per unit length at its edges at $x = 0$ and $x = a$ (see Figure 3.12.2).

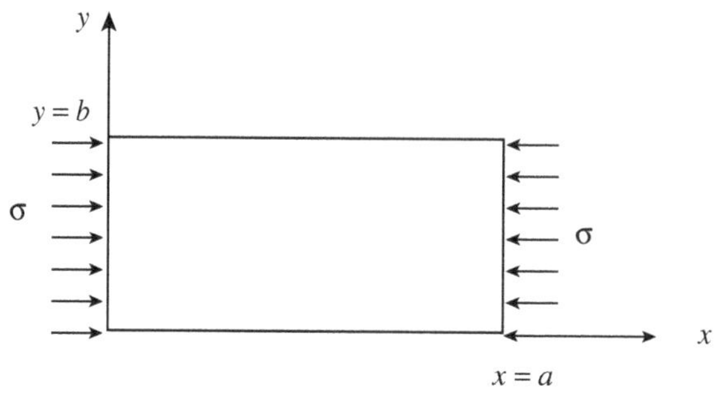

Figure 3.12.2

In this case, the only non-vanishing stress component in the fundamental state is $S_x = -\sigma$. Further, there are no mass forces, so the differential equation for neutral equilibrium is given by

$$\frac{Eh^3}{12(1-\upsilon^2)}\Delta\Delta w + h\sigma w_{,xx} = 0. \tag{3.12.25}$$

In this case, we have

$$\begin{array}{lll} \upsilon_x = 1 & \upsilon_y = 0 & \text{at } x = a \\ \upsilon_x = -1 & \upsilon_y = 0 & \text{at } x = 0 \\ \upsilon_y = 1 & \upsilon_x = 0 & \text{at } y = b \\ \upsilon_y = -1 & \upsilon_x = 0 & \text{at } y = 0 \end{array}$$

so that the boundary conditions $x = a$ are

$$w_{,xx} + \upsilon w_{,yy} = 0, \quad w = 0. \tag{3.12.26}$$

Because $w \equiv 0$ at $x = a$, $w_{,yy} = 0$, so that we can rewrite the boundary conditions as

$$w_{,xx} = 0 \quad w = 0 \quad \text{at } x = a, \quad x = 0. \tag{3.12.27}$$

Similarly,

$$w_{,yy} = 0 \quad w = 0 \quad \text{at } y = 0, \quad y = b. \tag{3.12.28}$$

Because both the differential equation and the boundary conditions contain only even derivatives, we can attempt a solution of the form

$$w(x, y) = C_{mn} \sin\frac{m\pi x}{a} \sin\frac{m\pi y}{b}, \tag{3.12.29}$$

which satisfies all the boundary conditions.† Substitution into the differential equation yields

$$\frac{Eh^3}{12(1-\upsilon^2)}\left(\frac{m^2\pi^2}{a^2} + \frac{n^2\pi^2}{b^2}\right)^2 - h\sigma\frac{m^2\pi^2}{a^2} = 0. \tag{3.12.30}$$

† Here m, n are integers.

For given values of m and n, (3.12.20) is satisfied for

$$\sigma = \frac{\pi^2 E h^2}{12(1-\upsilon^2)b^2}\left(m^2\frac{b}{a}+n^2\frac{a}{b}\right)^2\frac{1}{m^2}. \tag{3.12.31}$$

The critical load is obtained for the minimum value of the right-hand term, which for fixed m has a minimum value for $n = 1$, so that we must minimize the expression

$$\sigma = \frac{\pi^2 E h^2}{12(1-\upsilon^2)b^2}\left(m^2\frac{b^2}{a^2}+2+\frac{a^2}{b^2m^2}\right) \tag{3.12.32}$$

with respect to m. To this end, we assume that m is continuous. Differentiating the expression between the brackets with respect to m^2 and setting the result equal to zero, we find

$$m = \frac{a}{b}. \tag{3.12.33}$$

This is the exact result when a/b is an integer. In other cases, m is the integer that is closest to a/b. Hence, it follows that

$$\sigma_{\mathrm{cr}} \geq \frac{\pi^2 E h^2}{3(1-\upsilon^2)b^2} \tag{3.12.34}$$

where the equality sign holds when a/b is an integer. For sufficiently large values of a/b, (3.12.34) is a good approximation,† which means that for a sufficiently long plate we can always approximate a/b by an integer. The critical load for other boundary conditions is often expressed as

$$\sigma_{\mathrm{cr}} = \frac{\pi^2 E h^2}{3(1-\upsilon^2)}k, \tag{3.12.35}$$

i.e., it is expressed as a multiple of the critical load for a simply supported load.

When $a/b < 1$, the smallest wave number is $m = 1$, so we can write the critical load in the form

$$\sigma_{\mathrm{cr}} = \frac{\pi^2 E h^2}{12(1-\upsilon^2)a^2}\left(1+2\frac{a^2}{b^2}+\frac{a^4}{b^4}\right). \tag{3.12.36}$$

In the limiting case $a/b \to 0$, we have

$$\sigma_{\mathrm{cr}} = \frac{\pi^2 E h^2}{12(1-\upsilon^2)a^2}, \tag{3.12.37}$$

which also follows from the result for the Euler bar when we take into account that there is no anti-elastic bending, which accounts for the factor $(1-\upsilon^2)^{-1}$.

Let us now consider the solution of our buckling problem in more detail. We have attempted a solution in the form (3.12.29), which means that we have silently

† E.g., $a/b = 3.5$ yields for the factor between the brackets in (3.12.32), 4.072 for $m = 4$ and 4.095 when $m = 3$, which means an error of 2.5% when (3.12.34) is used.

assumed that the infinite series

$$w(x, y) = \sum_{m=1}^{\infty}\sum_{n=1}^{\infty} C_{mn} \sin\frac{m\pi x}{a} \sin\frac{n\pi y}{b} \tag{3.12.38}$$

can be differentiated term-wise, which is only admissible when the differentiated series is uniformly convergent for $0 \leq x \leq a$, $0 \leq y \leq b$. We shall now show that for a plate with simply supported edges, this series can be differentiated term-wise. To do this, we apply a Fourier transform to (3.12.25), which yields

$$\int_0^a\int_0^b \left[\frac{Eh^3}{12(1-\upsilon^2)}\Delta\Delta w + h\sigma w_{,xx}\right] \sin\frac{m\pi x}{a} \sin\frac{n\pi y}{b} dx\, dy = 0. \tag{3.12.39}$$

Let us first consider the term

$$\begin{aligned}\int_0^a w_{,xxxx} \sin\frac{m\pi x}{a} dx &= w_{,xxx} \sin\frac{m\pi x}{a}\Big|_0^a - \int_0^a w_{,xxx}\frac{m\pi}{a} \cos\frac{m\pi x}{a} dx \\ &= -w_{,xx}\frac{m\pi}{a} \cos\frac{m\pi x}{a}\Big|_0^a - \int_0^a w_{,xx}\frac{m^2\pi^2}{a^2} \sin\frac{m\pi x}{a} dx.\end{aligned} \tag{3.12.40}$$

Here the stock term vanishes because $w_{,xx} = 0$ at $x = a$ and $x = 0$. This stock term does not vanish in the case of a clamped edge. Continuing our partial integration, we obtain

$$\begin{aligned}\int_0^a w_{,xxxx} \sin\frac{m\pi x}{a} dx &= -w_{,x}\frac{m^2\pi^2}{a^2} \sin\frac{m\pi x}{a}\Big|_0^a \\ &\quad + \int_0^a w_{,x}\frac{m^3\pi^3}{a^3} \cos\frac{m\pi x}{a} dx \\ &= w\frac{m^3\pi^3}{a^3} \cos\frac{m\pi x}{a}\Big|_0^a + \frac{m^4\pi^4}{a^4}\int_0^a w \sin\frac{m\pi x}{a} dx.\end{aligned} \tag{3.12.41}$$

Here the stock term vanishes for both simply supported edges and clamped edges because then $w = 0$ at $x = 0$ and $x = a$. Similarly, we obtain for simply supported edges

$$\begin{aligned}\int_0^a w_{,xx} \sin\frac{m\pi x}{a} dx &= -\frac{m^2\pi^2}{a^2}\int_0^a w \sin\frac{m\pi x}{a} dx \\ \int_0^b w_{,yyyy} \sin\frac{n\pi y}{a} dy &= \frac{n^4\pi^4}{b^4}\int_0^b w \sin\frac{n\pi y}{a} dy\end{aligned} \tag{3.12.42}$$

$$\begin{aligned}&\int_0^a\int_0^b w_{,xx}w_{,yy} \sin\frac{m\pi x}{a} \sin\frac{n\pi y}{a} dx\, dy \\ &= \frac{m^2n^2\pi^4}{a^2b^2}\int_0^a\int_0^b w \sin\frac{m\pi x}{a} \sin\frac{n\pi y}{b} dx\, dy.\end{aligned}$$

Our final result for the Fourier transform of the differential equation can now be written as

$$\left[\frac{Eh^3}{12(1-\upsilon^2)}\left(\frac{m^2\pi^2}{a^2} + \frac{n^2\pi^2}{b^2}\right)^2 - h\sigma\frac{m^2\pi^2}{a^2}\right] W_{mn} = 0 \tag{3.12.43}$$

where

$$W_{mn} = \int_0^a \int_0^b w(x, y) \sin \frac{m\pi x}{a} \sin \frac{m\pi y}{b} dx\, dy \tag{3.12.44}$$

is the Fourier transform of $w(x, y)$. The equation (3.12.43) is valid for all integer values of m and n, and because $W_{mn} \not\equiv 0$ the expression between the brackets must vanish, which leads to our earlier result (3.12.31).

Introducing the series (3.12.38) into (3.12.44), we find the following relation between C_{mn} and the Fourier transform of $w(x, y)$:

$$W_{mn} = \frac{1}{4} C_{mn} ab. \tag{3.12.45}$$

To obtain this result, we have interchanged integration and summation, which is admissible because the series (3.12.38) is a uniformly convergent series. We see that only one term of the series is left, which justifies our earlier approach.

Let us now consider the influence of a free edge at $y = 0$ for the case that $a/b \ll 1$, when the other edges are simply supported (see Figure 3.12.3).
In this case, we shall only apply a Fourier transform in the x direction because $w = w'' = 0$ at $x = 0$ and $x = a$, but $w \neq 0$ at $y = 0$. With

$$W_m(y) = \int_0^a w(x, y) \sin \frac{m\pi}{a} dx, \tag{3.12.46}$$

we find

$$\frac{Eh^3}{12(1 - \upsilon^2)} \left(\frac{d^4 W_m}{dy^4} - 2 \frac{m^2\pi^2}{a^2} \frac{d^2 W_m}{dy^2} + \frac{m^4\pi^4}{a^4} W_m \right) - h\sigma \frac{m^2\pi^2}{a^2} W_m = 0, \tag{3.12.47}$$

where we have used the relation

$$\int_0^a \frac{d^n w}{dy^n} \sin \frac{m\pi x}{a} dx = \frac{d^n W_m}{dy^n}. \tag{3.12.48}$$

As discussed previously for a short plate ($a/b \ll 1$), the wave number is $m = 1$. Introducing the notation

$$y = \frac{a}{\pi} \eta, \quad d()/d\eta = ()^{\cdot}, \quad W_1 = W, \tag{3.12.49}$$

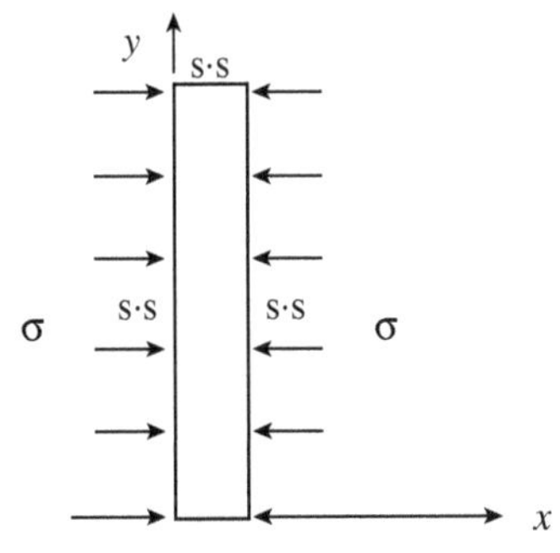

Figure 3.12.3

we may rewrite (3.12.47) in the form

$$\frac{\pi^2 Eh^2}{12(1-\upsilon^2)a^2}(W^{\cdot\cdot\cdot\cdot} - 2W^{\cdot\cdot} + W) - \sigma W = 0. \tag{3.12.50}$$

Further, it is convenient to write

$$\sigma = \lambda \frac{\pi^2 Eh2}{12(1-\upsilon^2)a^2} \tag{3.12.51}$$

so that the differential equation becomes

$$W^{\cdot\cdot\cdot\cdot} - 2W^{\cdot\cdot} + W(1-\lambda) = 0. \tag{3.12.52}$$

The boundary conditions are

$$\begin{aligned} &w_{,yy} + \upsilon w_{,xx} = 0, \quad w_{,yyy} + (2-\upsilon)w_{,xyy} = 0 \quad \text{at } y = 0 \\ &w = 0, \quad w_{,yy} = 0 \quad \text{at } y = b, \end{aligned} \tag{3.12.53}$$

and

$$w = 0, \quad w_{,xx} = 0 \quad \text{at } x = 0, \quad x = a. \tag{3.12.54}$$

We have already used these last conditions in the derivation of the differential equation for the Fourier transform of $w(x, y)$. Writing

$$w(x, y) = Y(y) \sin \frac{\pi x}{a}, \tag{3.12.55}$$

we obtain from (3.12.47)

$$W(y) = \frac{a}{2} Y(y). \tag{3.12.56}$$

Because the boundary conditions are homogeneous, we can now express them as

$$\begin{aligned} &W^{\cdot\cdot} - \upsilon W = 0, \quad W^{\cdot\cdot\cdot} - (2-\upsilon)W^{\cdot} = 0 \quad \text{at } \eta = 0 \\ &W = 0, \quad W^{\cdot\cdot} - \upsilon W = 0 \quad \text{at } \eta = \pi \frac{b}{a}. \end{aligned} \tag{3.12.57}$$

The boundary conditions at $x = 0$, $x = a$ are satisfied by our choice (3.12.55). To solve (3.12.52), we set

$$W = Ce^{\mu\eta}. \tag{3.12.58}$$

Introducing this expression into the differential equation, we obtain the characteristic equation

$$\mu^4 - 2\mu^2 + 1 - \lambda = 0, \tag{3.12.59}$$

from which

$$\mu = \pm(1 \pm \sqrt{\lambda})^{1/2}. \tag{3.12.60}$$

These roots are real when $0 \le \lambda \le 1$. Because we want solutions that decay with increasing distance from $y = 0$, we must only consider the solution

$$W = C_1 e^{-\sqrt{1+\sqrt{\lambda}}\eta} + C_2 e^{-\sqrt{1-\sqrt{\lambda}}\eta}. \tag{3.12.61}$$

Substitution into the boundary conditions at $\eta = 0$ yields

$$\begin{aligned}(1+\sqrt{\lambda}-\upsilon)C_1 + (1-\sqrt{\lambda}-\upsilon)C_2 &= 0\\ C_1(1+\sqrt{\lambda})^{1/2}(1-\upsilon-\sqrt{\lambda}) + C_2(1-\sqrt{\lambda})^{1/2}(1-\upsilon+\sqrt{\lambda}) &= 0.\end{aligned} \tag{3.12.62}$$

The condition for a non-trivial solution is now

$$f(\lambda,\upsilon) \equiv (1-\sqrt{\lambda})^{1/2}(1-\upsilon+\sqrt{\lambda})^2 - (1+\sqrt{\lambda})^{1/2}(1-\upsilon-\sqrt{\lambda})^2 = 0. \tag{3.12.63}$$

For $\lambda = 0$, we have $f(\lambda,\upsilon) = 0$, and for $\lambda = 1$ we have $f(\lambda,\upsilon) = -\sqrt{2}\upsilon^2$. To get a better picture of the curve $f(\lambda,\upsilon) = 0$, we also determine the derivative

$$\begin{aligned}\frac{df}{d\sqrt{\lambda}} &= 2(1-\sqrt{\lambda})^{1/2}(1-\upsilon+\sqrt{\lambda}) - \frac{1}{2}(1-\sqrt{\lambda})^{-1/2}(1-\upsilon+\sqrt{\lambda})^2\\ &\quad + 2(1+\sqrt{\lambda})^{1/2}(1-\upsilon-\sqrt{\lambda}) - \frac{1}{2}(1+\sqrt{\lambda})^{-1/2}(1-\upsilon-\sqrt{\lambda})^2.\end{aligned} \tag{3.12.64}$$

At $\sqrt{\lambda} = 1$, $df/d\sqrt{\lambda} = -\infty$, and for $\sqrt{\lambda} = 0$ we have $df/d\sqrt{\lambda} = (1-\upsilon)(3+\upsilon)$, so that we get the picture shown in Figure 3.12.4.
It follows that for $\upsilon > 0$, the root $\sqrt{\lambda}$ will be close to 1, so we set

$$\sqrt{\lambda} = 1 - \varepsilon^2. \tag{3.12.65}$$

Substitution into (3.12.63) yields

$$\varepsilon(2-\upsilon-\varepsilon^2)^2 - (2+\varepsilon^2)^{1/2}(\varepsilon^2-\upsilon)^2,$$

from which

$$\varepsilon = \frac{\sqrt{2}\upsilon^2}{(2-\upsilon)^2}\left[1 + O(\varepsilon^2)\right], \tag{3.12.66}$$

so that

$$\lambda_{cr} \approx 1 - \frac{4\upsilon^4}{(2-\upsilon)^4}. \tag{3.12.67}$$

The largest reduction is obtained for $\upsilon = 1/2$, where $\varepsilon \approx 0.157$ and

$$\lambda_{cr} = \frac{77}{81} = 0.9506. \tag{3.12.68}$$

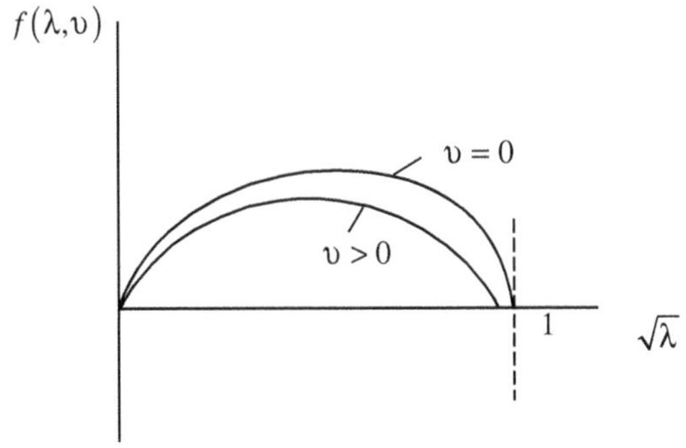

Figure 3.12.4

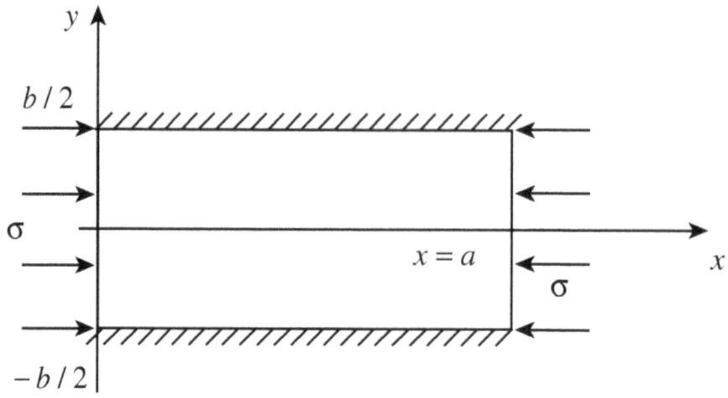

Figure 3.12.5

To a second-order approximation, ε is given by

$$\varepsilon = \frac{\sqrt{2}(2-\upsilon)^2}{\upsilon(8-\upsilon)}\left[1+\frac{2\upsilon^3(8-\upsilon)}{(2-\upsilon)^4}\right]^{1/2}, \tag{3.12.69}$$

which for $\upsilon = 1/2$ yields $\varepsilon = 0.1444$, and $\lambda_{cr} = 0.9585$. The "exact" root is $\varepsilon = 0.1486$, so that

$$\lambda_{cr} = 0.9563 \quad \text{(exact)}, \tag{3.12.70}$$

which means a reduction of the critical load of about 5% compared to the plate that is simply supported at $y = 0$.

Let us now consider a rectangular plate with clamped edges at $y = \pm b/2$, and simply supported edges at $x = 0$ and $x = a$, loaded in compression by forces σh per unit length at $x = 0$ and $x = a$ (see Figure 3.12.5).

However, the clamped edges are allowed to slide in the x-direction. We consider the case that the plate is sufficiently long, and that 2ℓ is the wavelength of the deformation pattern in the buckled state. The differential equation for neutral equilibrium is

$$\frac{Eh^3}{12(1-\upsilon^2)}\Delta\Delta w + \sigma h w_{,xx} = 0, \tag{3.12.71}$$

and the boundary conditions are

$$\begin{aligned} &w = 0, \quad w_{,xx} = 0 \quad \text{at } x = 0, \quad x = m\ell(\leq a) \\ &w = 0, \quad w_{,y} = 0 \quad \text{at } y = \pm b/2. \end{aligned} \tag{3.12.72}$$

Because the lowest bifurcation load will occur when $m = 1$, we try a solution of the form

$$w(x, y) = W(y)\sin\frac{\pi x}{\ell}. \tag{3.12.73}$$

Substitution into the differential equation yields

$$\frac{Eh^3}{12(1-\upsilon^2)}\left(\frac{d^4W}{dy^4} - \frac{2\pi^2}{\ell^2}\frac{d^2W}{dy^2} + \frac{\pi^4}{\ell^4}W\right) - \sigma h\frac{\pi^2}{\ell^2}W = 0. \tag{3.12.74}$$

It is now convenient to write

$$\sigma = \lambda \frac{\pi^2 E h^2}{3(1-\upsilon^2)b^2}, \tag{3.12.75}$$

i.e., the load is expressed as a multiple of the critical load for a sufficiently long plate, simply supported at its edges.

The differential equation can now be written as

$$\frac{d^4W}{dy^4} - \frac{2\pi^2}{\ell^2}\frac{d^2W}{dy^2} + \left(\frac{\pi^4}{\ell^4} - \frac{4\pi^4\lambda}{\ell^2 b^2}\right) W = 0. \tag{3.12.76}$$

The boundary conditions at $x = 0$ and $x = \ell$ are satisfied automatically by our choice for $w(x, y)$, and the conditions at $y = \pm b/2$ read

$$W = 0, \quad \frac{dW}{dy} = 0 \quad \text{at } y = \pm b/2. \tag{3.12.77}$$

To solve (3.12.76), we set

$$W = Ce^{\frac{\pi y}{\ell}\mu}. \tag{3.12.78}$$

Introduction of this expression into the differential equation yields the characteristic equation

$$\mu^4 - 2\mu^2 + 1 - 4\lambda\frac{\ell^2}{b^2} = 0, \tag{3.12.79}$$

from which

$$\mu = \pm\sqrt{1 \pm 2\frac{\ell}{b}\sqrt{\lambda}}. \tag{3.12.80}$$

As the plate with two clamped edges has a larger stiffness than the simply supported plate, $\lambda > 1$. This means that for $2\ell\sqrt{\lambda}/b > 1$, there will be two imaginary roots. Hence, we write

$$\begin{aligned} \mu_{1,2} &= \pm\alpha \quad \alpha = \left(1 + 2\frac{\ell}{b}\sqrt{\lambda}\right)^{1/2} \\ \mu_{3,4} &= \pm i\beta \quad \beta = \left(2\frac{\ell}{b}\sqrt{\lambda} - 1\right)^{1/2}. \end{aligned} \tag{3.12.81}$$

Because the construction and its loading are symmetric with respect to the x-axis, we can split the solution into a part that is symmetric with respect to $y = 0$ and an anti-symmetric part. Guided by the result for the simply supported plate, we expect the lowest bifurcation load for the symmetric solution; thus we consider the solution

$$W(y) = C_1 \cosh \alpha\pi y/\ell + C_2 \cos \beta\pi y/\ell. \tag{3.12.82}$$

Substitution of the boundary conditions (3.12.77) yields

$$\begin{aligned} C_1 \cosh \alpha\frac{\pi b}{2\ell} + C_2 \cos \beta\frac{\pi b}{2\ell} &= 0 \\ \alpha C_1 \sinh \alpha\frac{\pi b}{2\ell} - \beta C_2 \sin \beta\frac{\pi b}{2\ell} &= 0. \end{aligned} \tag{3.12.83}$$

The condition for a non-trivial solution is

$$-\beta\cosh\alpha\frac{\pi b}{2\ell}\sin\beta\frac{\pi b}{2\ell}-\alpha\sinh\alpha\frac{\pi b}{2\ell}\cos\beta\frac{\pi b}{2\ell}=0,$$

or rewritten,

$$\beta\tan\beta\frac{\pi b}{2\ell}+\alpha\tanh\alpha\frac{\pi b}{2\ell}=0. \tag{3.12.84}$$

This is an equation for λ with b/ℓ as a parameter. Minimizing with respect to b/ℓ yields (numerical calculations)

$$(\ell/b)_{\text{minimizing}}=0.66, \quad \lambda_{\text{cr}}=1.7425.^{\dagger} \tag{3.12.85}$$

To avoid the rather tedious calculations involved with the solution of the transcendental equation (3.12.83), we now try an approximate solution to our problem. Introducing a dimensionless coordinate η defined by

$$y=\eta b, \quad -\frac{1}{2}\leq\eta\leq\frac{1}{2}, \tag{3.12.86}$$

we assume the following symmetric polynomial for W,

$$W(\eta)=1-8\eta^2+16\eta^4, \tag{3.12.87}$$

which satisfies the boundary conditions

$$W=dW/d\eta=0 \quad \text{for} \quad \eta=\pm 1/2.$$

Because u and v are zero at bifurcation, it follows from (3.12.7) that we can write

$$P_2[\mathbf{u}_1]=\frac{Eh^3}{24(1-\upsilon^2)}\iint\left[(\Delta w)^2-2(1-\upsilon)\left(w_{,xx}w_{,yy}-w_{,xy}^2\right)-\frac{12(1-\upsilon^2)}{h^2}\frac{\sigma}{E}w_{,x}^2\right]dx\,dy. \tag{3.12.88}$$

The second variation is semi-positive definite for the exact solution. For the exact buckling load and an assumed (approximate) displacement, the field $P_2[\mathbf{u}]>0$, so that the application of Rayleigh's method (also called Rayleigh-Ritz method) where we put $P_2[\mathbf{u}]=0$ for an assumed displacement field yields an upper bound for the critical load.

For the evaluation of (3.12.88), it will be convenient to make use of the following property:

$$\iint_S\left[w_{,xx}w_{,yy}-w_{,xy}^2\right]dx\,dy=0 \tag{3.12.89}$$

for a domain bounded by straight lines (polygonal edge curve) with $w=$ const. at the boundary ∂S, and $\partial w/\partial\upsilon=0$ in the corner points at the boundary where $\boldsymbol{\upsilon}$ is the unit normal to ∂S, positive in the outward direction. To show this property, we apply

† This result is valid for $\ell/b\geq 1/2\sqrt{\lambda_{\text{cr}}}=0.37878$.

the divergence theorem to (3.12.89), which yields

$$
\begin{aligned}
&\iint_S \left(w_{,xx}w_{,yy} - w_{,xy}^2\right)dx\,dy = 0\\
&= \int_{\partial S} \left(w_{,xx}w_{,y}\upsilon_y - w_{,xy}w_{,y}\upsilon_x\right)ds - \iint_S \left(w_{,xxy}w_{,y} - w_{,xyx}w_{,y}\right)dx\,dy\\
&= \int_{\partial S} -w_{,y}\left(w_{,xx}t_x + w_{,xy}t_y\right)ds \qquad (3.12.90)\\
&= \int_{\partial S} -w_{,y}\left(w_{,xx}dx + w_{,xy}dy\right) = -\int_{\partial S} w_{,y}d\left(w_{,x}\right)\\
&= -\int_{\partial S} \left(w_{,\upsilon}\upsilon_y + w_{,s}t_y\right)d\left(w_{,\upsilon}\upsilon_x + w_{,s}t_x\right).
\end{aligned}
$$

So far, our result is fully general. When w is constant along ∂S, we have $w_{,s} = 0$ along ∂S, and along a straight line υ_x and υ_y are constant, so that for a polygonal edge curve where $w =$ const., we can write

$$
\iint_S \left(w_{,xx}w_{,yy} - w_{,xy}^2\right)dx\,dy = -\frac{1}{2}\sum_{i=1}^{n}\left(\upsilon_x\upsilon_y w_{,\upsilon}^2\right)\Big|_{s_i}^{s_{i+1}} \qquad (3.12.91)
$$

for a polygonal edge curve with n corner points, where s_i denotes the ith corner point and $s_{n+1} \equiv s_0$. When $w_{,\upsilon} = 0$, in the corner points we obtain the result (3.12.89). Notice that this result is valid for $\forall w \,|\, w \in C^3$, $w = 0$ on ∂S, $\partial w/\partial\upsilon = 0$ in the corner points on ∂S, where ∂S is a polygonal curve.

For simply supported and clamped straight edges, we have $w = 0$, $\partial w/\partial\upsilon = 0$ in the corner points, and thus the result (3.12.89) applies,† so that for these plates we can write

$$
P_2\left[\mathbf{u}_1\right] = \frac{Eh^3}{24\left(1-\upsilon^2\right)}\iint \left[(\Delta w)^2 - \frac{12\left(1-\upsilon^2\right)}{h^2}\frac{\sigma}{E}w_{,x}^2\right]dx\,dy. \qquad (3.12.92)
$$

Using the results

$$
\begin{aligned}
\int_0^{\ell}\int_{-\frac{1}{2}b}^{\frac{1}{2}b}(\Delta w)^2\,dx\,dy &= \int_0^{\ell}\int_{-\frac{1}{2}}^{\frac{1}{2}} b\left[-\frac{\pi^2}{\ell^2}\left(1 - 8\eta^2 + 16\eta^4\right) + \frac{1}{b^2}\left(-16 + 108\eta^2\right)\right]^2 \sin^2\frac{\pi x}{\ell}\,dx d\eta \qquad (3.12.93)\\
&= \frac{128}{b^3}\left(\frac{4}{5} + \frac{64}{105}C + \frac{128}{315}C^2\right),
\end{aligned}
$$

† Notice that the limiting case that $n \to \infty$ for a simply supported plate bounded by a regular polygonal curve with n corner points is not a simply supported but a *clamped* circular plate, since now $\partial w/\partial\upsilon = 0$ along the edge.

where

$$C = \frac{\pi^2 b^2}{16\ell^2}, \tag{3.12.94}$$

and

$$\int_0^{\ell}\int_{-\frac{1}{2}b}^{\frac{1}{2}b} w_{,x}^2\, dx\, dy = \int_0^{\ell}\int_{\frac{1}{2}}^{\frac{1}{2}} b\frac{\pi^2}{\ell^2}\cos^2\frac{\pi x}{\ell}\left(1 - 8\eta^2 + 16\eta^4\right)^2 dx\, d\eta$$

$$= \frac{64}{315}\frac{16C}{b}, \tag{3.12.95}$$

we obtain from the condition $P_2[\mathbf{u}_1] = 0$,

$$\sigma = \frac{Eh^2}{12(1-\upsilon^2)b^2}\frac{128\left(\frac{4}{5} + \frac{64}{105}C + \frac{128}{315}C^2\right)}{\frac{64}{315}\cdot 16C}. \tag{3.12.96}$$

Minimizing with respect to C, we obtain

$$C_{\min} = \sqrt{\frac{63}{32}} = 1.403 \tag{3.12.97}$$

and from (3.12.94),

$$(\ell/b)_{\min} = 0.663. \tag{3.12.98}$$

The critical load is now

$$\sigma_{cr} = \frac{\pi^2 Eh^2}{3(1-\upsilon^2)b^2}\, 1.7454, \tag{3.12.99}$$

which is about 0.2% higher than the exact value.

As a last example, we shall treat a rectangular plate, simply supported at its edges and loaded in shear by shear forces τh per unit length along its edges (see Figure 3.12.6).

From (3.12.12), the governing differential equation is now

$$\frac{Eh^3}{12(1-\upsilon^2)}\Delta\Delta w - 2\tau h w_{,xy} = 0. \tag{3.12.100}$$

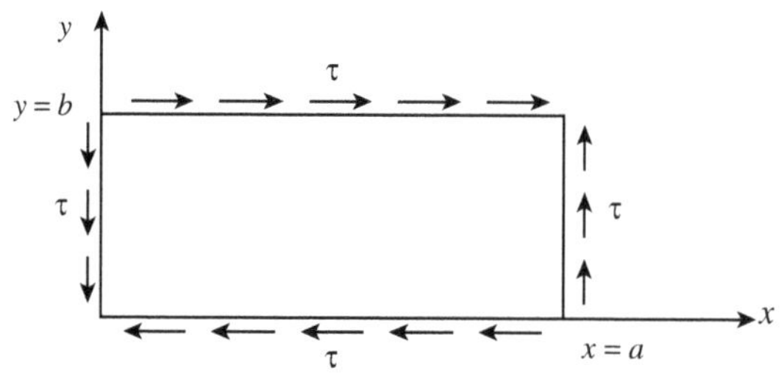

Figure 3.12.6

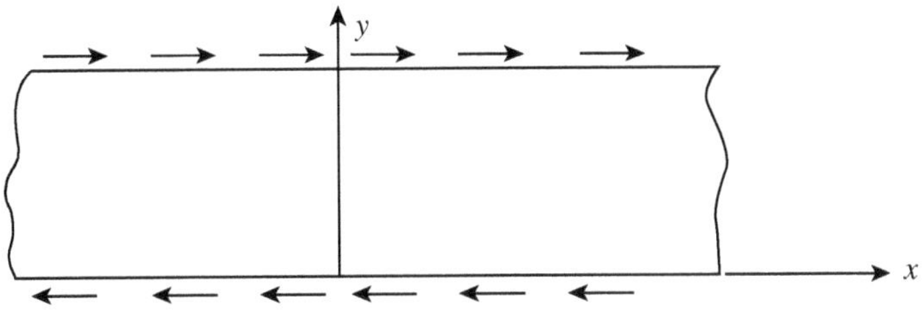

Figure 3.12.7

Notice that the differential equation now contains even and odd derivatives with respect to x and y. In the following, we shall assume that the plate is infinitely long, so that we do not take into account the boundary conditions on the vertical ends. The boundary conditions at $y = 0$, $y = b$ read

$$w = 0, \quad w_{,yy} = 0 \quad \text{at } y = 0, \quad y = b. \tag{3.12.101}$$

The exact solution to this problem was given by SOUTHWELL,† and his result was

$$\tau_{cr} = 5.35 \frac{\pi^2 E h^2}{12(1 - \upsilon^2) b^2}. \tag{3.12.102}$$

However, the numerical factor is not very accurate. Later, we shall give an upper bound that is slightly smaller.

The exact solution of this problem requires rather tedious calculations, and therefore we shall try to construct an approximate solution. We try a solution of the form

$$w(x, y) = f \sin \frac{\pi}{\ell}(x - my) \sin \frac{\pi y}{b}, \tag{3.12.103}$$

which satisfies the boundary condition $w = 0$ at $y = 0$, $y = b$. The deflection vanishes at the nodal lines $x - my = k\ell$ $(k = 0, \pm 1, \pm 2, \ldots)$.

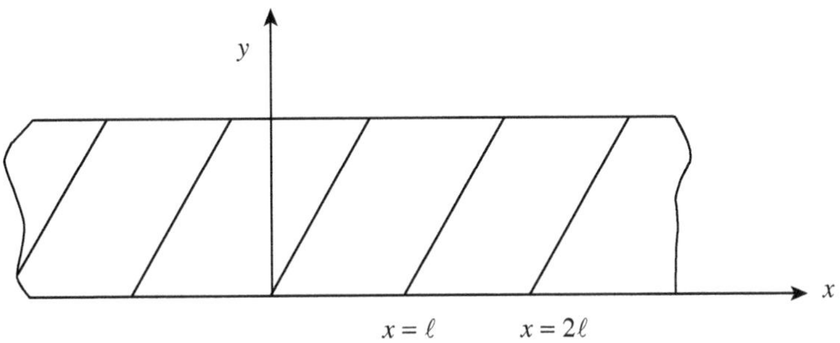

Figure 3.12.8

† Cf. R.V. SOUTHWELL, *Proc. Roy. Soc., London, Series A*, **105**, 582.

The energy functional is

$$\int_0^\ell \int_0^b \left\{ \frac{Eh^3}{12(1-\upsilon^2)} \left[(\Delta w)^2 - 2(1-\upsilon)\left(w_{,xx}w_{,yy} - w_{,xy}^2\right) \right] + \tau h w_{,x} w_{,y} \right\} dx\, dy, \tag{3.12.104}$$

where the second term vanishes because of (3.12.87). Applying the Rayleigh-Ritz method, we obtain

$$\tau = \frac{\pi^2 Eh^2}{24(1-\upsilon^2)b^2} \left[\frac{1}{m}\left(m^2+1\right)^2 \frac{b^2}{\ell^2} + \frac{1}{m}\frac{\ell^2}{b^2} + 6m + \frac{2}{m} \right], \tag{3.12.105}$$

where m and b/ℓ are still unknown parameters. Minimizing with respect to (b/ℓ^2), we obtain

$$\frac{1}{m}(m^2+1)^2 - \frac{1}{m}\frac{\ell^4}{b^4} = 0, \tag{3.12.106}$$

which yields

$$\frac{\ell^2}{b^2} = m^2 + 1. \tag{3.12.107}$$

Substitution into (3.12.104) yields

$$\tau = \frac{\pi^2 Eh^2}{24(1-\upsilon^2)b^2}\left(8m + \frac{4}{m}\right). \tag{3.12.108}$$

Minimizing with respect to m, we obtain

$$m = \frac{1}{2}\sqrt{2}, \tag{3.12.109}$$

so that the (upper bound for the) critical load is given by

$$\tau_{cr} = \frac{\pi^2 Eh^3}{3(1-\upsilon^2)b^2}\sqrt{2}, \tag{3.12.110}$$

which means a factor 1.414 instead of 1.3337 (Southwell). This result is a rather crude approximation, and is due to the fact that the boundary condition $w_{,yy} = 0$ is not satisfied at $y = 0$, $y = b$.

Let us now analyze the consequences of our assumption for the displacement field shown in Figure 3.12.9.

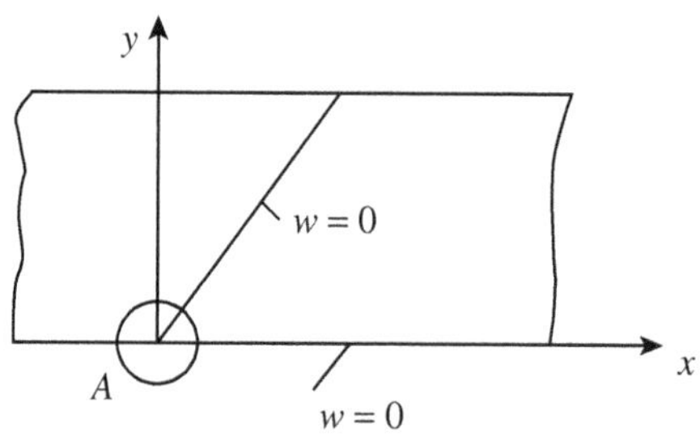

Figure 3.12.9

Because $w = 0$ along the edge and along the nodal line, we must have $w_{,x} = w_{,y} = 0$ in A but not $w_{,yy} = 0$. To satisfy this condition, we try a form for $w(x, y)$ that is slightly more general than (3.12.102):

$$w(x, y) = \sin\frac{\pi}{\ell}(x - \varphi(y))W(y), \quad W(0) = W(b) = 0. \tag{3.12.111}$$

We then obtain

$$w_{,y} = -\frac{\pi}{\ell}W(y)\varphi'(y)\cos\frac{\pi}{\ell}[x - \varphi(y)] + W'(y)\sin\frac{\pi}{\ell}[x - \varphi(y)] \tag{3.12.112}$$

$$\begin{aligned} w_{,yy} = {} & \left[W(y) - \frac{\pi^2}{\ell^2}W(y)\varphi'^2(y)\right]\sin[x - \varphi(y)] \\ & - [2W'(y)\varphi'(y) + W(y)\varphi''(y)]\cos\frac{\pi}{\ell}[x - \varphi(y)] \end{aligned} \tag{3.12.113}$$

$$w_{,yy}(x, 0) = W''(0)\sin[x - \varphi(0)] - 2\frac{\pi}{\ell}W'(0)\varphi'(0)\cos\frac{\pi}{\ell}[x - \varphi(0)]. \tag{3.12.114}$$

Because $w_{,y}(x, 0)$ is free, $W'(0) \neq 0$, so that the conditions for $w_{,yy}(x, 0) = 0$ are

$$W''(0) = 0, \quad \varphi'(0), \tag{3.12.115}$$

which means that the nodal line $x = \varphi(y)$ is perpendicular to the edge. These conditions are satisfied by choosing

$$w(x, y) = f\sin\frac{\pi}{\ell}(x - \varphi(y))\sin\frac{\pi y}{b}, \tag{3.12.116}$$

where

$$\varphi(y) = \frac{mb}{\pi}\left(1 - \cos\frac{\pi y}{b}\right). \tag{3.12.117}$$

With this choice, the result is

$$\tau_{\mathrm{cr}} = 1.352\frac{\pi^2 Eh^2}{3(1 - \upsilon^2)}, \tag{3.12.118}$$

which is already very close to the exact result.

Let us now finally consider a more exact approach. We try a solution of the form

$$w(x, y) = \sum_{k=1}^{\infty} W_k(x)\sin\frac{k\pi y}{b}, \tag{3.12.119}$$

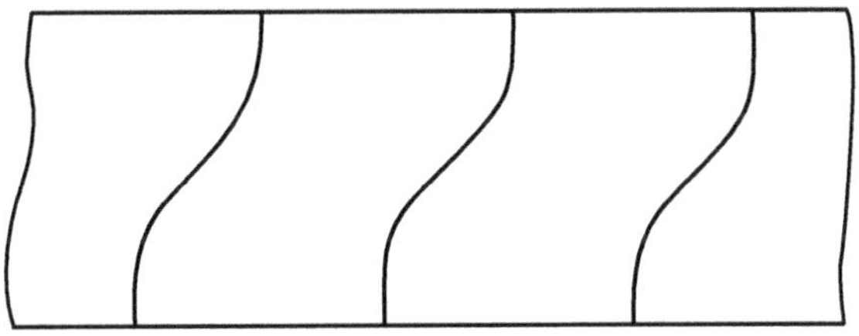

Figure 3.12.10

which satisfies the boundary conditions at $y = 0$ and $y = b$ term-wise. Furthermore, this function must be periodic in x direction,

$$W_k(x + 2\ell) = W_k(x), \tag{3.12.120}$$

where ℓ is the half wavelength. Hence, we assume

$$w(x, y) = \sum_{k=1}^{\infty} \left(a_k \cos \frac{\pi x}{\ell} + b_k \sin \frac{\pi x}{\ell} \right) \sin \frac{k\pi y}{b}. \tag{3.12.121}$$

Substitution of this expression into (3.12.100) yields

$$\begin{aligned} &\frac{Eh^3}{12(1-\upsilon^2)} \sum_{k=1}^{\infty} \left(\frac{\pi^2}{\ell^2} + \frac{k^2\pi^2}{b^2} \right)^2 \left(a_k \cos \frac{\pi x}{\ell} + b_k \sin \frac{\pi x}{\ell} \right) \sin \frac{\pi y}{\ell} \\ &\quad - 2\tau h \sum_{k=1}^{\infty} \frac{k\pi^2}{b\ell} \left(-a_k \sin \frac{\pi x}{\ell} + b_k \cos \frac{\pi x}{\ell} \right) \cos \frac{k\pi y}{b} = 0, \end{aligned} \tag{3.12.122}$$

where we have assumed that the interchanging of summation and differentiation is admissible.

Because this expression must hold for all x and y, the coefficients of $\cos \pi x/\ell$ and $\sin \pi x/\ell$ must vanish, which with $b/\ell = \mu$ yields

$$\begin{aligned} &\frac{Eh^2\pi^2}{12(1-\upsilon^2)b^2} \sum_{k=1}^{\infty} (\mu^2 + k^2)^2 a_k \sin \frac{k\pi y}{b} - 2\tau\mu \sum_{k=1}^{\infty} k b_k \cos \frac{k\pi y}{b} = 0 \\ &\frac{Eh^2\pi^2}{12(1-\upsilon^2)b^2} \sum_{k=1}^{\infty} (\mu^2 + k^2)^2 b_k \sin \frac{k\pi y}{b} + 2\tau\mu \sum_{k=1}^{\infty} k a_k \cos \frac{k\pi y}{b} = 0. \end{aligned} \tag{3.12.123}$$

Introducing the notation

$$\tau = \lambda \frac{\pi^2 Eh^2}{3(1-\upsilon^2)b^2}, \tag{3.12.124}$$

we can rewrite these equations in the form

$$\begin{aligned} &\sum_{k=1}^{\infty} (\mu^2 + k^2)^2 a_k \sin \frac{k\pi y}{b} - 8\mu\lambda \sum_{k=1}^{\infty} k b_k \cos \frac{k\pi y}{b} = 0 \\ &\sum_{k=1}^{\infty} (\mu^2 + k^2)^2 b_k \sin \frac{k\pi y}{b} + 8\mu\lambda \sum_{k=1}^{\infty} k a_k \cos \frac{k\pi y}{b} = 0. \end{aligned} \tag{3.12.125}$$

Assuming that the left-hand members can be written as a sine series, we multiply both sides of these equations by $\sin j\pi y/b$ and integrate (term-wise) from $y = 0$ to $y = b$. With

$$\int_0^b \sin \frac{j\pi y}{b} \sin \frac{k\pi y}{b}\, dy = \frac{1}{2} b\delta_{jk} \tag{3.12.126}$$

$$\int_0^b \sin\frac{j\pi y}{b}\cos\frac{k\pi y}{b}\,dy = j\frac{[1-(-1)^{j+k}]}{(j^2-k^2)\,\pi}b \quad (=0 \text{ for } j=k)\,, \tag{3.12.127}$$

we obtain

$$\begin{aligned}(\mu^2+j^2)^2 a_j - \frac{32}{\pi}\lambda\mu\sum_{k=1}^{\infty}\frac{kj}{2\,(j^2-k^2)}[1-(-1)^{k+j}]b_k = 0\\ (\mu^2+j^2)b_j + \frac{32}{\pi}\lambda\mu\sum_{k=1}^{\infty}\frac{kj}{2\,(j^2-k^2)}[1-(-1)^{k+j}]a_k = 0.\end{aligned} \tag{3.12.128}$$

There are two infinite sets of linear algebraic equations for the unknowns a_k, b_k $(k=1,2,\ldots)$. The condition for a non-trivial solution is that the determinant of the matrix of coefficients vanishes, i.e., with

$$\frac{32}{\pi}\lambda\mu = \rho \tag{3.12.129}$$

$$\begin{array}{ccccc} a_1 & b_2 & a_3 & b_4 & \cdots \end{array}$$
$$\begin{vmatrix} (\mu^2+1)^2 & \dfrac{2\rho}{3} & 0 & \dfrac{4\rho}{15} & \cdots \\ \dfrac{2\rho}{3} & (\mu^2+1)^2 & \dfrac{6\rho}{5} & 0 & \cdots \\ 0 & -\dfrac{6\rho}{5} & (\mu^2+9)^2 & \dfrac{12\rho}{7} & \cdots \\ \dfrac{4\rho}{15} & 0 & \dfrac{12\rho}{7} & (\mu^2+1)^2 & \cdots \\ \vdots & \vdots & \vdots & \vdots & \end{vmatrix}. \tag{3.12.130}$$

Taking into account two terms, we obtain

$$(\mu^2+1)^2(\mu^2+4)^2 - \frac{4\rho^2}{9} = 0, \tag{3.12.131}$$

which yields

$$(\mu^2+1)(\mu^2+4) = \frac{2\rho}{3} = \frac{64\mu\lambda}{3\pi},$$

so that

$$\lambda = \frac{3\pi}{64}\left(\mu^3 + 5\mu + \frac{4}{\mu}\right). \tag{3.12.132}$$

Minimizing with respect to μ yields

$$\mu^4 + \frac{5}{3}\mu^2 = \frac{4}{3} = 0,$$

so that

$$\mu^2 = \frac{-5+\sqrt{73}}{b} \approx 0.59 \tag{3.12.133}$$

and

$$\lambda_{\text{cr}} = 1.393 \quad (\text{“exact” } \lambda = 1.336). \tag{3.12.134}$$

Taking into account three terms, we find

$$\lambda_{\text{cr}} = 1.339, \tag{3.12.135}$$

and with four terms,

$$\lambda_{\text{cr}} = 1.336, \tag{3.12.136}$$

which shows that this process converges rapidly with the exact result.

3.13 Post-buckling behavior of plates loaded in their plane

For a square plate loaded by compressive forces σ per unit length in the x-direction, the energy functional reads

$$\begin{aligned} P[\mathbf{u}] = \iint \Bigg\{ & \frac{Eh^3}{12(1-\upsilon^2)}\left[(\Delta w)^2 - 2(1-\upsilon)\left(w_{,xx}w_{,yy} - w_{,xy}^2\right)\right] \\ & + \frac{Eh}{2(1-\upsilon^2)}\left[\left(u_{,x} + \frac{1}{2}w_{,x}^2\right)^2 + \left(v_{,y} + \frac{1}{2}w_{,y}^2\right)^2\right. \\ & + 2\upsilon\left(u_{,x} + \frac{1}{2}w_{,x}^2\right)\left(v_{,y} + \frac{1}{2}w_{,y}^2\right) \\ & \left. + \frac{1}{2}(1-\upsilon)(u_{,y} + v_{,x} + w_{,x}w_{,y})^2\right] - \frac{1}{2}\sigma h w_{,x}^2 \Bigg\}\, dx\, dy. \end{aligned} \tag{3.13.1}$$

The second variation is given by

$$\begin{aligned} P[\mathbf{u}] = \iint \Bigg\{ & \frac{Eh^3}{12(1-\upsilon^2)}[(\Delta w)^2 - 2(1-\upsilon)\left(w_{,xx}w_{,yy} - w_{,xy}^2\right)X] \\ & + \frac{Eh}{2(1-\upsilon^2)}\left[u_{,x}^2 + v_{,y}^2 + 2\upsilon u_{,x}v_{,y} + \frac{1}{2}(1-\upsilon)(u_{,y}+v_{,x})^2\right] - \frac{1}{2}\sigma h w_{,x}^2 \Bigg\}\, dx\, dy, \end{aligned} \tag{3.13.2}$$

and the third and fourth variations are given by

$$\begin{aligned} P_3[\mathbf{u}] = \frac{Eh}{2(1-\upsilon^2)} \iint & \left[u_{,x}w_{,x}^2 + v_{,v}w_{,y}^2 + \upsilon(u_{,x}w_{,y}^2 + W_{,x}^2)\right. \\ & \left. + (1-\upsilon)(u_{,y}+v_{,x})w_{,x}w_{,y}\right]\, dx\, dy \end{aligned} \tag{3.13.3}$$

and

$$P_4[\mathbf{u}] = \frac{Eh}{8(1-\upsilon^2)} \iint \left(w_{,x}^2 + w_{;y}^2\right)^2 dx\, dy, \tag{3.13.4}$$

respectively.

The displacement field for small but finite deflections from the fundamental state is written as

$$\mathbf{u} = a\mathbf{u}_1 + \bar{\mathbf{u}}, \tag{3.13.5}$$

where $\bar{\mathbf{u}}$ is orthogonal to the buckling mode. The equations for $\bar{\mathbf{u}}$ are obtained by minimizing the functional

$$P_2[\bar{\mathbf{u}}] + a(\lambda - 1)P'_{11}[\mathbf{u}_1, \bar{\mathbf{u}}] + a^2 P_{21}[\mathbf{u}_1, \bar{\mathbf{u}}], \tag{3.13.6}$$

where $\lambda = \sigma/\sigma_{\rm cr}$ (see (2.4.11). The term

$$P'_{11}[\mathbf{u}_1, \bar{\mathbf{u}}] = -\frac{Eh^2}{4(1-\upsilon^2)} \iint w_{1,x}\bar{w}_{,x}\, dx\, dy \tag{3.13.7}$$

vanishes due to the orthogonality condition

$$\begin{aligned} P_{21}[\mathbf{u}_1, \bar{\mathbf{u}}] = \frac{Eh}{2(1-\upsilon^2)} \iint &\left[(\bar{u}_{,x} + \upsilon\bar{v}_{,y})\, w_{1,x}^2 + (\bar{v}_{,y} + \upsilon\bar{u}_{,x})\, w_{1,y}^2 \right. \\ &\left. + (1-\upsilon)(\bar{u}_{,y} + \bar{v}_{,x})\, w_{1,x}w_{1,y}\right] dx\, dy, \end{aligned} \tag{3.13.8}$$

and $P_2[\bar{\mathbf{u}}]$ follows from (3.13.2) by replacing $u,\ v,\ w$ by $\bar{u},\ \bar{v},\ \bar{w}$. We now must minimize (3.13.6) with respect to $\bar{\mathbf{u}}$, i.e.,

$$\underset{\text{w.r.t.}\bar{\mathbf{u}}}{\text{Min}}\ P_2[\bar{\mathbf{u}}] + a^2 P_{21}[\mathbf{u}_1, \bar{\mathbf{u}}]\,. \tag{3.13.9}$$

Because $P_{21}[\mathbf{u}_1, \bar{\mathbf{u}}]$ does not contain $\bar{w}$, we must minimize $P_2[\bar{\mathbf{u}}]$ with respect to $\bar{w}$. $P_2[\bar{\mathbf{u}}]$ is positive-definite, and its minimum with respect to $\bar{w}$ is obtained when the first line in (3.13.2), which is positive-definite, vanishes, which yields

$$\bar{w} \equiv 0. \tag{3.13.10}$$

$P_2[\bar{\mathbf{u}}]$ is now given by

$$P_2[\bar{\mathbf{u}}] = \frac{Eh}{2(1-\upsilon^2)} \iint \left[\bar{u}_{,x}^2 + \bar{v}_{,y}^2 + 2\upsilon\bar{u}_{,x}\bar{v}_{,y} + \frac{1}{2}(1-\upsilon)(\bar{u}_{,y} + \bar{v}_{,x})^2\right] dx\, dy. \tag{3.13.11}$$

The necessary condition for (3.13.9) to be a minimum is that the variation of this functional with respect to $\bar{\mathbf{u}}$ vanishes, which yields

$$\begin{aligned} \frac{Eh}{1-\upsilon^2} \int_0^{\ell}\int_{-\frac{1}{2}b}^{\frac{1}{2}b} &\left\{\bar{u}_{,x}\xi_{,x} + \bar{v}_{,y}\eta_{,y} + \upsilon\bar{u}_{,x}\eta_{,y} + \upsilon v_{,y}\xi_{,x} + \frac{1}{2}(1-\upsilon)(\bar{u}_{,y} + \bar{v}_{,x})(\xi_{,y} + \eta_{,x}) \right. \\ &\left. + \frac{a^2}{2}\left[(\xi_{,x} + \upsilon\eta_{,y})\, w_{1,x}^2 + (\eta_{,y} + \upsilon\xi_{,x})w_{1,y}^2 + (1-\upsilon)(\xi_{,y} + \eta_{,x})\, w_{1,x}w_{1,y}\right]\right\} dx\, dy = 0. \end{aligned} \tag{3.13.12}$$

Integration by parts yields

$$\begin{aligned} &\int_{-b/2}^{b/2} \left[\bar{u}_{,x} + \upsilon\bar{v}_{,y} + \frac{a^2}{2}\left(w_{1,x}^2 + \upsilon w_{1,y}^2\right)\right]\xi\Bigg|_0^{\ell} dy + \int_0^{\ell} \left[\bar{v}_{,y} + \upsilon\bar{u}_{,x} + \frac{a^2}{2}\left(w_{1,y}^2 + \upsilon w_{1,x}^2\right)\right]\eta\Bigg|_{-b/2}^{b/2} dx \\ &+ \frac{(1-\upsilon)}{2}\int_0^{\ell} (\bar{u}_{,y} + \bar{v}_{,x} + a^2 w_{1,x}w_{1,y})\xi\Bigg|_{-b/2}^{b/2} dx + \frac{(1-\upsilon)}{2}\int_{-b/2}^{b/2} (\bar{u}_{,y} + \bar{v}_{,x} + a^2 w_{1,x}w_{1,y})\eta\Bigg|_0^{\ell} dy \end{aligned}$$

$$
\begin{aligned}
&-\int_0^{\ell}\int_{-b/2}^{b/2}\Big\{\overline{u}_{,xx}+\frac{1}{2}(1-\upsilon)\,\overline{u}_{,yy}+\frac{1}{2}(1+\upsilon)\,\overline{v}_{,xy}\\
&+\frac{a^2}{2}\Big[\left(w_{1,x}^2+w_{1,y}^2\right)_{,x}+(1-\upsilon)\left(w_{1,x}w_{1,yy}-w_{1,y}w_{1,xy}\right)\Big]\Big\}\,\xi\\
&+\Big\{\overline{v}_{,yy}+\frac{1}{2}(1-\upsilon)\,\overline{v}_{,xx}+\frac{1}{2}(1+\upsilon)\,\overline{u}_{,xy}\\
&+\frac{a^2}{2}\Big[\left(w_{1,x}^2+w_{1,y}^2\right)_{,y}+(1-\upsilon)\left(w_{1,y}w_{1,xx}-w_{1,x}w_{1,xy}\right)\Big]\,\eta\Big\}\,dx\,dy=0,
\end{aligned}
\tag{3.13.13}
$$

from which we obtain

$$
\begin{aligned}
&\overline{u}_{,xx}+\frac{1}{2}(1-\upsilon)\,\overline{u}_{,yy}+\frac{1}{2}(1+\upsilon)\overline{v}_{,xy}\\
&\quad=-\frac{a^2}{2}\Big[\left(w_{1,x}^2+w_{1,y}^2\right)_{,x}+(1-\upsilon)\left(w_{1,x}w_{1,yy}-w_{1,y}w_{1,xy}\right)\Big]\\
&\overline{v}_{,yy}+\frac{1}{2}(1-\upsilon)\overline{v}_{,xx}+\frac{1}{2}(1+\upsilon)\overline{u}_{,xy}\\
&\quad=-\frac{a^2}{2}\Big[\left(w_{1,x}^2+w_{1,y}^2\right)_{,y}+(1-\upsilon)\left(w_{1,y}w_{1,xx}-w_{1,x}w_{1,xy}\right)\Big].
\end{aligned}
\tag{3.13.14}
$$

We shall now discuss various boundary conditions. First, notice that for the determination of the buckling load we must only deal with w, whereas for the post-buckling only in-plane quantities $(\overline{u},\overline{v})$ must be dealt with. We shall now adapt this theory to the problem of a plate with many fields, loaded in compression by forces σh per unit length. The plate is simply supported at $x=\pm\,\ell/2$, and at $y=\pm((2k+1)/2)b$ $(k=0,1,2,\ldots)$ in such a way that in-plane displacements are free (see Figure 3.13.1). We shall now consider the field $\{x,y\,||x|\le\ell/2,\ |y|\le b/2\}$. The edges $y=\pm b/2$ remain straight (due to symmetry), so that

$$\overline{v}=\text{const.}\quad\text{at } y=\pm\,b/2. \tag{3.13.15}$$

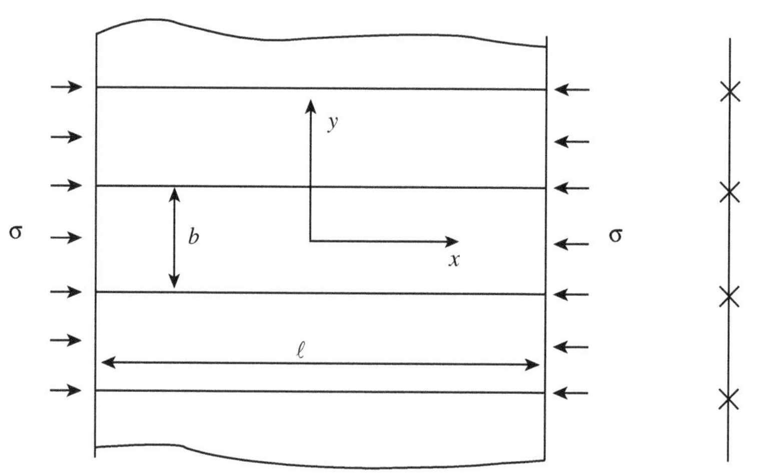

Figure 3.13.1

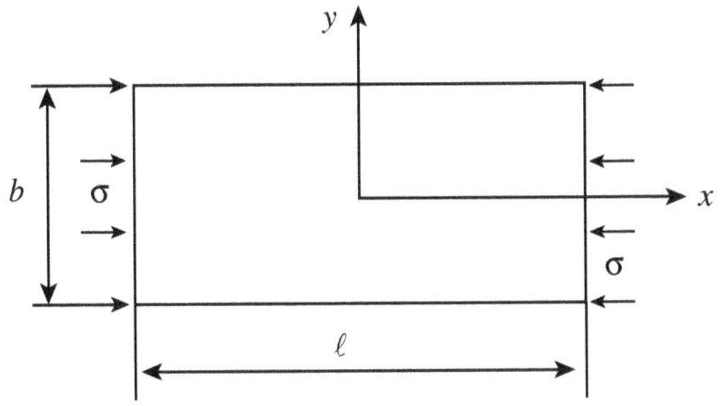

Figure 3.13.2

Further, we shall assume that at $x = \pm\, \ell/2$ the supports are such that $\overline{u} = \text{const}$.

For the rectangular plate simply supported at its edges, the buckling mode is (see (3.12.29)

$$w_1 = \cos\frac{\pi x}{\ell}\cos\frac{\pi y}{\ell} \equiv \cos\lambda x\cos\mu y. \tag{3.13.16}$$

The equations for $\overline{u}$ and $\overline{v}$ now become

$$\begin{aligned}
&\overline{u}_{,xx} + \frac{1}{2}(1-\upsilon)\overline{u}_{,yy} + \frac{1}{2}(1+\upsilon)\overline{v}_{,xy}\\
&\quad = -\frac{a^2}{4}\lambda[(\lambda^2 - \upsilon\mu^2)\sin 2\lambda x + (\lambda^2+\mu^2)\sin 2\lambda x\cos 2\mu y],\\
&\overline{v}_{,yy} + \frac{1}{2}(1-\upsilon)\overline{v}_{,xx} + \frac{1}{2}(1+\upsilon)\overline{u}_{,xy}\\
&\quad = -\frac{a^2}{4}\mu[(\mu^2 - \upsilon\lambda^2)\sin 2\mu y + (\lambda^2+\mu^2)\sin 2\mu y\cos 2\lambda x].
\end{aligned} \tag{3.13.17}$$

By inspection, we notice that particular solutions of (3.13.17) are of the form

$$\begin{aligned}
\overline{u}_{\text{part}} &= A\sin 2\lambda x + B\sin 2\lambda x\cos 2\mu y\\
\overline{v}_{\text{part}} &= C\sin 2\mu y + D\sin 2\mu y\cos 2\lambda x.
\end{aligned} \tag{3.13.18}$$

Substitution into (3.13.17) yields

$$\begin{aligned}
&4\lambda^2 A \equiv \frac{a^2}{4}\lambda(\lambda^2 - \upsilon\mu^2)\\
&[4\lambda^2 + 2(1-\upsilon)\mu^2]B + 2(1+\upsilon)\lambda\mu D \equiv \frac{a^2}{4}\lambda(\lambda^2+\mu^2)\\
&2(1+\upsilon)\lambda\mu B + [4\mu^2 + 2(1-\upsilon)\lambda^2]D \equiv \frac{a^2}{4}\mu(\lambda^2+\mu^2)\\
&4\mu^2 C = \frac{a^2}{4}\mu(\mu^2 - \upsilon\lambda^2),
\end{aligned} \tag{3.13.19}$$

from which

$$\begin{aligned}
A &= \frac{a^2}{16}\frac{\lambda^2 - \upsilon\mu^2}{\lambda}, & C &= \frac{a^2}{16}\frac{\mu^2 - \upsilon\lambda^2}{\mu},\\
B &= \frac{a^2}{16}\lambda, & D &= \frac{a^2}{16}\mu.
\end{aligned} \tag{3.13.20}$$

The particular solutions are now given by

$$\begin{aligned}\overline{u}_{\text{part}} &= \frac{a^2}{16\lambda}[(\lambda^2 - \upsilon\mu^2)\sin 2\lambda x + \lambda^2 \sin 2\lambda x \cos 2\mu y] \\ \overline{v}_{\text{part}} &= \frac{a^2}{16\mu}[(\mu^2 - \upsilon\lambda^2)\sin 2\mu y + \mu^2 \cos 2\lambda x \sin 2\mu y].\end{aligned} \tag{3.13.21}$$

To satisfy the boundary conditions, we must add solutions of the homogeneous equations. With the condition (3.13.15), the boundary conditions become

$$\begin{aligned}&\left.\begin{aligned}\overline{u} &= \text{const.} \\ \overline{u}_{,y} + \overline{v}_{,x} &= 0\end{aligned}\right\} \text{at } x = \pm\,\ell/2 \\ &\left.\begin{aligned}\overline{v} &= \text{const.} \\ \overline{u}_{,y} + \overline{v}_{,x} &= 0\end{aligned}\right\} \text{at } y = \pm\,\ell/2,\end{aligned} \tag{3.13.22}$$

which means that the shear strains vanish along the edge of the plate. Writing the full solution as the sum of the solution of the homogeneous equations and the particular solution

$$\overline{u} = \overline{u}_h + \overline{u}_{\text{part}}, \quad \overline{v} = \overline{v}_h + \overline{v}_{\text{part}}, \tag{3.13.23}$$

we find that $\overline{u}_h$ and $\overline{v}_h$ must satisfy

$$\begin{aligned}\overline{u}_{h,xx} + \frac{1}{2}(1-\upsilon)\overline{u}_{h,yy} + \frac{1}{2}(1+\upsilon)\overline{v}_{h,xy} &= 0 \\ \overline{v}_{h,yy} + \frac{1}{2}(1-\upsilon)\overline{v}_{h,xx} + \frac{1}{2}(1+\upsilon)\overline{u}_{h,xy} &= 0\end{aligned} \tag{3.13.24}$$

$$\begin{aligned}&\left.\begin{aligned}\overline{u}_h &= \text{const.} \\ \overline{u}_{h,y} + \overline{v}_{h,x} &= 0\end{aligned}\right\} \text{at } x = \pm\,\ell/2 \\ &\left.\begin{aligned}\overline{v}_h &= \text{const.} \\ \overline{u}_{h,y} + \overline{v}_{h,x} &= 0\end{aligned}\right\} \text{at } y = \pm\,\ell/2.\end{aligned} \tag{3.13.25}$$

It is easy to verify that

$$\overline{u}_h = -a^2\overline{\varepsilon}_1 x \quad \text{and} \quad \overline{v}_h = -a^2\overline{\varepsilon}_2 y \tag{3.13.26}$$

satisfy both the equations and the boundary conditions. The unknown strains $-a^2\overline{\varepsilon}_1$, $-a^2\overline{\varepsilon}_2$ follow from the conditions

$$\begin{aligned}&\int_{-b/2}^{b/2}\left(\overline{u}_{,x} + \upsilon\overline{v}_{,y} + \frac{a^2}{2}w_{1,x}^2\right)dy = 0 \quad \text{at } x = \pm\,\ell/2 \\ &\int_{-\ell/2}^{\ell/2}\left(\overline{v}_{,y} + \upsilon\overline{u}_{,x} + \frac{a^2}{2}w_{1,y}^2\right)dx = 0 \quad \text{at } y = \pm\,b/2,\end{aligned} \tag{3.13.27}$$

which yields

$$\begin{aligned}\bar{\varepsilon}_1 + \upsilon\bar{\varepsilon}_2 &= \frac{1}{8}(\lambda^2 + \upsilon\mu^2)\\ \upsilon\bar{\varepsilon}_1 + \bar{\varepsilon}_2 &= \frac{1}{8}(\upsilon\lambda^2 + \mu^2),\end{aligned} \tag{3.13.28}$$

from which we obtain the expressions for the strains $\bar{\varepsilon}_1, \bar{\varepsilon}_2$,

$$\bar{\varepsilon}_1 = \frac{1}{8}\lambda^2, \quad \bar{\varepsilon}_2 = \frac{1}{8}\mu^2. \tag{3.13.29}$$

The energy in the plate is given by (3.13.7), where $\mathbf{u} = a\mathbf{u}_1 + \bar{\mathbf{u}}$, so that with $u_1 = v_1 \equiv 0$ and $\bar{w} \equiv 0$, we obtain

$$\begin{aligned}P[a\mathbf{u}_1 + \bar{\mathbf{u}}] = \iint \Bigg\{ &\frac{Eh^3a^2}{12(1-\upsilon^2)}\left[(\Delta w_1)^2 - 2(1-\upsilon)\left(w_{1,xx}w_{1,yy} - w_{1,xy}^2\right)\right]\\ &+ \frac{Eh}{2(1-\upsilon^2)}\Bigg[\left(\bar{u}_{,x} + \frac{1}{2}a^2w_{1,x}^2\right)^2 + \left(\bar{v}_{,y} + \frac{1}{2}a^2w_{1,y}^2\right)^2\\ &+ 2\upsilon\left(\bar{u}_{,x} + \frac{a^2}{2}w_{1,x}^2\right)\left(\bar{v}_{,y} + \frac{a^2}{2}w_{1,y}^2\right)\Bigg]\\ &+ \frac{1}{2}(1-\upsilon)\left(\bar{u}_{,x} + \bar{v}_{,y} + a^2w_{1,x}w_{1,y}\right)^2 - \frac{1}{2}\sigma ha^2w_{1,x}^2\Bigg\}\,dx\,dy.\end{aligned} \tag{3.13.30}$$

Making use of the fact that

$$\begin{aligned}&\iint \frac{Eh^3}{12(1-\upsilon^2)}\left[(\Delta w_1)^2 - 2(1-\upsilon)\left(w_{1,xx}w_{1,yy} - w_{1,xy}^2\right)\right]dx\,dy\\ &= \frac{1}{2}\sigma_{\mathrm{cr}}h \iint w_{1,x}^2\,dx\,dy\end{aligned}$$

and the fact that $\bar{u}$ and $\bar{v}$ are the solutions of the minimum problem, we have

$$\begin{aligned}&\frac{Eha^2}{12(1-\upsilon^2)}\iint \left[\bar{u}_{,x}w_{1,x}^2 + \bar{v}_{,y}w_{1,y}^2 + \upsilon\left(\bar{u}_{,x}w_{1,y}^2 + \bar{v}_{,y}w_{1,x}^2\right)\right.\\ &\quad\left. + (1-\upsilon)\left(\bar{u}_{,x} + \bar{v}_{,y}\right)w_{1,x}w_{1,y}\right]dx\,dy\\ &= -\frac{Eh}{1-\upsilon^2}\iint\left[\bar{u}_{,x}^2 + \bar{v}_{,y}^2 + 2\upsilon\bar{u}_{,x}\bar{v}_{,y} + \frac{1}{2}(1-\upsilon)\left(\bar{u}_{,x} + \bar{v}_{,y}\right)^2\right]dx\,dy.\end{aligned} \tag{3.13.31}$$

Then, we may write

$$\begin{aligned}P[au_1 + \bar{\mathbf{u}}] &= \frac{Eh}{8(1-\upsilon^2)}\int_{-\ell/2}^{\ell/2}\int_{-b/2}^{b/2}\left(w_{1,x}^2 + w_{1,y}^2\right)^2dx\,dy\\ &- \frac{Eh}{2(1-\upsilon^2)}\int_{-\ell/2}^{\ell/2}\int_{-b/2}^{b/2}\left[\bar{u}_{,x}^2 + \bar{v}_{,y}^2 + 2\upsilon\bar{u}_{,x}\bar{v}_{,y} + \frac{1}{2}(1-\upsilon)\left(\bar{u}_{,y} + \bar{v}_{,x}\right)^2\right]dx\,dy\\ &+ \frac{1}{2}(\sigma_{\mathrm{cr}} - \sigma)h\int_{-\ell/2}^{\ell/2}\int_{-b/2}^{b/2} w_{1,x}^2\,dx\,dy = F(a).\end{aligned} \tag{3.13.32}$$

Evaluating these integrals, we find

$$F(a) = \frac{a^2}{8}(\sigma_{cr} + \sigma)hb\ell\lambda^2 + \frac{a^4}{256}Ehb\ell(\lambda^4 + \mu^4). \tag{3.13.33}$$

The equilibrium equation follows from

$$\frac{\partial F}{\partial a} = 0 = \frac{1}{4}ahb\ell\left[\lambda^2(\sigma_{cr} - \sigma) + \frac{E}{16}a^2(\lambda^4 + \mu^4)\right], \tag{3.13.34}$$

which yields

$$a^2 = 16\frac{(\sigma - \sigma_{cr})\lambda^2}{E(\lambda^4 + \mu^4)} = \frac{16\pi^2}{\ell^2}\frac{1}{\dfrac{\pi^4}{\ell^4} + \dfrac{\pi^4}{b^4}}\frac{\sigma - \sigma_{cr}}{E}. \tag{3.13.35}$$

For the case $\ell = b$, we have

$$a^2 = \frac{8b^2}{\pi^2}\frac{\sigma - \sigma_{cr}}{E} = \frac{8h^2}{3(1 - \upsilon^2)}\left(\frac{\sigma}{\sigma_{cr}} - 1\right), \tag{3.13.36}$$

where we have made use of (3.12.34). This means that for $\sigma = 2\sigma_{cr}$, we have

$$a = h\sqrt{\frac{8}{3(1 - \upsilon^2)}} \approx 1.7h \quad (for\ \upsilon = .25), \tag{3.13.37}$$

i.e., when the load is twice the critical load, the deflection is still only 1.7 times the plate thickness.

The displacement of the load σ after buckling is

$$\frac{\ell}{2}\frac{\sigma}{E} - \bar{u}\left(x = \frac{\ell}{2}\right) = \frac{\ell}{2}\left(\frac{\sigma}{E} + a^2\bar{\varepsilon}_1\right) \equiv \frac{\ell}{2}\varepsilon_{tot} \tag{3.13.38}$$

so that

$$\varepsilon_{tot} = \frac{\sigma}{E} + 2\frac{\sigma - \sigma_{cr}}{\text{E}}\frac{\lambda^4}{\lambda^4 + \mu^4}, \tag{3.13.39}$$

and for $\ell = b$, we have $\lambda = \mu$, and hence

$$\varepsilon_{tot} = \frac{\sigma}{\frac{1}{2}E} - \frac{\sigma_{cr}}{\text{E}} \equiv \frac{\sigma}{E_t} - \frac{\sigma_{cr}}{\text{E}}. \tag{3.13.40}$$

The so-called tangent modulus E_t is half the modulus E. This behavior is shown in Figure 3.13.3.

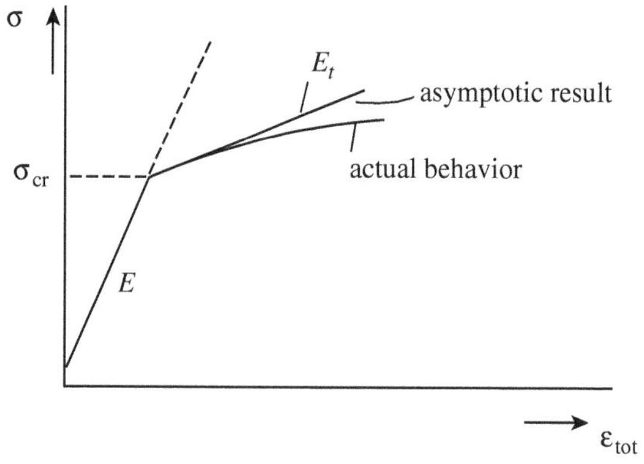

Figure 3.13.3

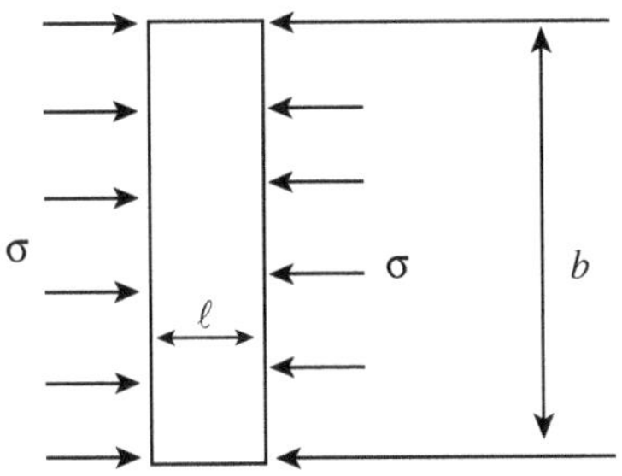

Figure 3.13.4

However, it should be noted that our solution is only valid in a small neighborhood of the critical load. Comparison of our result with exact numerical calculations show that for sufficiently long plates, the range of validity is moderately large. However, for short plates the range of validity is extremely small. Let us consider this case in some detail.

From (3.12.37), the critical load is then given by

$$\sigma_{cr} = \frac{\pi^2 E h^2}{12(1-\upsilon^2)\ell^2}, \tag{3.13.41}$$

and for $\ell/b \to 0$, we have $\mu/\lambda \to 0$ so that from (3.13.35) we obtain

$$a^2 = 16\frac{\sigma - \sigma_{cr}}{\lambda^2 E} = 16\frac{\sigma - \sigma_{cr}}{E}\frac{\ell^2}{h^2}. \tag{3.13.42}$$

The total strain ε_{tot} is now given by

$$\varepsilon_{tot} = \frac{\sigma}{E} + \frac{a^2}{8}\frac{h^2}{\ell^2} = \frac{\sigma}{\frac{1}{3}E} - \frac{2\sigma_{cr}}{E}, \tag{3.13.43}$$

so the tangent modulus now is given by $E_t = 1/3E$. This result has an extremely small range of validity because almost immediately after buckling, the plate behaves like a buckled bar.

This result is valid for the lowest bifurcation load, with only one wave.

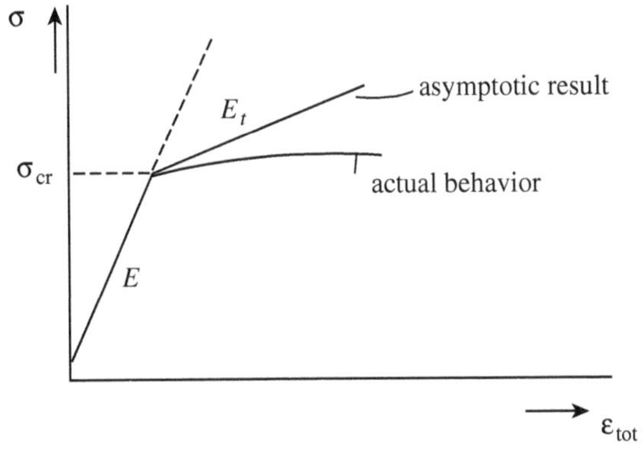

Figure 3.13.5

From (3.12.36), the exact formula for the critical load for a plate is

$$\sigma_{\mathrm{cr}} = \frac{Eh^2}{12(1-\upsilon^2)}\lambda^2\left(1+2\frac{\lambda^2}{\mu^2}+\frac{\lambda^4}{\mu^4}\right). \tag{3.13.44}$$

For fixed λ and, say, $\lambda/\mu = 10^{-2}$ for one wave, we have

$$\sigma_{\mathrm{cr}}^{(1)} \approx \frac{Eh^2}{12(1-\upsilon^2)}\lambda^2 \quad (1+2.10^{-4}). \tag{3.13.45}$$

When there are two waves, we have $\lambda/\mu = 2.10^{-2}$ and

$$\sigma_{\mathrm{cr}}^{(2)} \approx \frac{Eh^2}{12(1-\upsilon^2)}\lambda^2 \quad (1+8.10^{-4}), \tag{3.13.46}$$

which is very close to the lowest bifurcation load. This result shows that the lowest bifurcation point is a cluster point.

Let us finally make some remarks regarding the results obtained. We have assumed that the edges at $x = \pm\ell$ remain straight. This assumption holds exactly for sufficiently long plates because the nodal lines are straight lines. However, for short plates this must be accomplished with an edge beam, without constraining the strain long the edge. This condition can be done between two smooth stamps in a testing machine.

3.14 The "von Kármán-Föppl Equations"

The post-buckling behavior of plates can also be analyzed using the so-called "von Kármán-Föppl equations." For the derivation of these equations, we consider the deviation from the undeformed state. The strains in the mid-plane are

$$\begin{aligned} \gamma_{xx} &= u_{,x} + \frac{1}{2}w_{,x}^2, \quad \gamma_{yy} = v_{,y} + \frac{1}{2}w_{,y}^2, \\ 2\gamma_{xy} &= u_{,x} + v_{,y} + w_{,x}w_{,y}. \end{aligned} \tag{3.14.1}$$

The curvatures are given by

$$w_{,xx}, \quad w_{,yy}, \quad w_{,xy}. \tag{3.14.2}$$

These expressions hold for sufficiently small rotations, i.e.,

$$|w_{,x}|^2, \quad |w_{,y}|^2 \ll 1. \tag{3.14.3}$$

The elastic energy with respect to the undeformed state is now given by

$$\begin{aligned} \frac{Eh}{2(1-\upsilon^2)}\iint & \left\{\left(u_{,x}+\frac{1}{2}w_{,x}^2\right)^2 + \left(v_{,y}+\frac{1}{2}w_{,y}^2\right)^2 + 2\upsilon\left(u_{,x}+\frac{1}{2}w_{,x}^2\right)\left(v_{,y}+\frac{1}{2}w_{,y}^2\right)\right. \\ & + \frac{1}{2}(1-\upsilon)\left(u_{,y}+v_{,x}+w_{,x}w_{,y}\right)^2 \\ & \left. + \frac{h^2}{12}\left[w_{,xx}^2 + w_{,yy}^2 + 2\upsilon w_{,xx}w_{,yy} + 2(1-\upsilon)w_{,xy}^2\right]\right\} dx\, dy. \end{aligned} \tag{3.14.4}$$

This expression is in full agreement with (3.12.4) because the fundamental state is linear. Carrying out the variations with respect to u and v, we obtain a stationary value when

$$\begin{aligned}\iint \Bigg\{ & \frac{Eh}{1-\upsilon^2}\left[u_{,x}+\frac{1}{2}w_{,x}^2+\upsilon\left(v_{,y}+\frac{1}{2}w_{,y}^2\right)\right]\delta u_{,x} \\ & +\frac{Eh}{1-\upsilon^2}\left[\upsilon\left(u_{,x}+\frac{1}{2}w_{,x}^2\right)+v_{,y}+\frac{1}{2}w_{,y}^2\right]\delta v_{,y} \\ & +\frac{Eh}{2(1+\upsilon)}\left(u_{,y}+v_{,x}+w_{,x}w_{,y}\right)\left(\delta u_{,y}+\delta v_{,x}\right)\Bigg\}\,dx\,dy=0.\end{aligned} \tag{3.14.5}$$

Introducing the relations

$$\begin{aligned}&\frac{Eh}{1-\upsilon^2}\left[u_{,x}+\frac{1}{2}w_{,x}^2+\upsilon\left(v_{,y}+\frac{1}{2}w_{,y}^2\right)\right]=N_x \\ &\frac{Eh}{1-\upsilon^2}\left[\upsilon\left(u_{,x}+\frac{1}{2}w_{,x}^2\right)+\left(v_{,y}+\frac{1}{2}w_{,y}^2\right)\right]=N_y \\ &\frac{Eh}{2(1+\upsilon)}\left(u_{,y}+v_{,x}+w_{,x}w_{,y}\right)=N_{xy},\end{aligned} \tag{3.14.6}$$

we can rewrite this expression to yield

$$\iint \left[N_x\delta u_{,x}+N_y\delta v_{,y}+N_{xy}\left(\delta u_{,y}+\delta v_{,x}\right)\right]dx\,dy=0. \tag{3.14.7}$$

Applying the divergence theorem, we obtain

$$\begin{aligned}&\int_{\partial S}\left[\left(N_x\upsilon_x+N_{xy}\upsilon_y\right)\delta u+\left(N_y\upsilon_y+N_{xy}\upsilon_x\right)\delta v\right]ds \\ &-\iint_S\left[\left(N_{x,x}+N_{xy,y}\right)\delta u+\left(N_{y,y}+N_{xy,x}\right)\delta v\right]dx\,dy=0,\end{aligned} \tag{3.14.8}$$

which yields the in-plane equilibrium equations

$$N_{x,x}+N_{xy,y}=0,\quad N_{y,y}+N_{xy,x}=0, \tag{3.14.9}$$

and the in-plane boundary conditions, when either u, v are zero or free,

$$\begin{aligned}&\left(N_x\upsilon_x+N_{xy}\upsilon_y\right)\delta u\big|_{\partial S}=0 \\ &\left(N_y\upsilon_y+N_{xy}\upsilon_x\right)\delta v\big|_{\partial S}=0.\end{aligned} \tag{3.14.10}$$

The general solution to (3.14.9) is

$$N_x=F_{,yy},\quad N_y=F_{,xx},\quad N_{xy}=-F_{,xy}, \tag{3.14.11}$$

where $F(x, y)$ is Airy's stress function.

It is easily verified that (3.14.11) is a solution of (3.14.9). To show that (3.14.11) is the general solution, we write $N_{xy}=-F^*_{,xy}$, then we obtain from the first equation

$N_{x,x} = F^*_{,xyy}$, and from the second $N_{y,y} = F^*_{,xyx}$. Integrating these expressions with respect to x and y, respectively, we obtain

$$N_x = F^*_{,yy} + f(y), \quad N_y = F^*_{,xx} + g(x),$$

or defining

$$f(y) = f^*_{,yy}, \quad g(x) = g^*_{,xx},$$
$$N_x = F^*_{,yy} + f^*_{,yy}, \quad N_y = F^*_{,xx} + g^*_{,xx}.$$

We now define a function $F(x, y)$ by

$$F(x, y) = F^*(x, y) + f^*(y) + g^*(x).$$

Then

$$F_{,xy} = F^*_{,xy} = -N_{xy},$$
$$F_{,xx} = F^*_{,xx} + g^*_{,xx} = N_x, \quad F_{,yy} = F^*_{,yy} + f^*_{,yy} = N_y,$$

which concludes our proof that (3.14.11) is the general solution.

We now return to (3.14.4) and take the variation with respect to w to obtain the equilibrium equation in the direction normal to the undeformed mid-plane. Using (3.14.6), we obtain

$$\iint \Bigg\{ N_x\delta\left(\frac{1}{2}w^2_{,x}\right) + N_y\delta\left(\frac{1}{2}w^2_{,y}\right) + N_{xy}\delta(w_{,x}w_{,y}) + \frac{Eh^3}{12(1-\upsilon^2)}\left[(w_{,xx} + \upsilon w_{,yy})\,\delta w_{,xx} + (w_{,yy} + \upsilon w_{,xx})\,\delta w_{,yy} + 2(1-\upsilon)\,w_{,xy}\delta w_{,xy}\right] \Bigg\}\, dx\, dy = 0. \tag{3.14.12}$$

Equation (3.14.12) is fully identical with (3.12.9) when **S** is replaced by **N**, so we may use all the results derived from that expression. In particular, we mention the differential equation

$$\frac{Eh^3}{12(1-\upsilon^2)}\Delta\Delta w - N_x w_{,xx} - N_y w_{,yy} - 2N_{xy} w_{,xy} = 0. \tag{3.14.13}$$

Using (3.14.11), we can rewrite (3.14.13) in the form

$$\frac{Eh^3}{12(1-\upsilon^2)}\Delta\Delta w - F_{,yy} w_{,xx} + 2F_{,xy} w_{,xy} - F_{,xx} w_{,yy} = 0. \tag{3.14.14}$$

The equation for the stress function is obtained from the condition that the displacement field is compatible. To this end, we write the inverse of (3.14.6), making use of (3.14.11), which yields the following expressions:

$$\begin{aligned} u_{,x} + \frac{1}{2}w^2_{,y} &= \frac{1}{Eh}(F_{,yy} - \upsilon F_{,xx}) \\ v_{,y} + \frac{1}{2}w^2_{,y} &= \frac{1}{Eh}(F_{,xx} - \upsilon F_{,yy}) \\ u_{,y} + v_{,x} + w_{,x}w_{,y} &= -\frac{2(1-\upsilon)}{Eh}F_{,xy}. \end{aligned} \tag{3.14.15}$$

The displacements u and v are eliminated from these equations by differentiating the first equation with respect to y twice, and adding the second equation differentiated twice with respect to x, and subtracting the last equation differentiated with respect to x and y. The result is

$$\frac{1}{Eh}\Delta\Delta F = w_{,xy}^2 - w_{,xx}w_{,yy}. \tag{3.14.16}$$

This equation and (3.14.14), without the bending term, were first derived by Föppl. Later, von Kármán derived the full equations.

Notice that in this approach there are only two unknowns, Airy's stress function and the displacement w, whereas in the general theory we use the three displacements as unknowns.

3.15 Buckling and post-buckling behavior of shells using shallow shell theory

In cases where the wavelength of the buckling mode is small compared to the smallest radius of curvature, shallow shell theory may be used. Shallow shell theory can be derived as follows.

Consider a surface $Z(x, y)$, where x and y are Cartesian coordinates in a tangent plane to the surface (see Figure 3.15.1), chosen so that the projection of the lines with principal radius of curvature on the tangent plane coincide with the x and y axis, respectively. The z axis is perpendicular to the tangent plane (see Figure 3.15.1). In the undeformed state, the square of a line element on the surface $Z(x, y)$ is given by

$$\begin{aligned}(ds)^2 &= (dx)^2 + (dy)^2 + (dz)^2 \\ &= \left(1 + Z_{,x}^2\right)(dx)^2 + 2Z_{,x}Z_{,y}\, dx\, dy + \left(1 + Z_{,y}^2\right) dy^2.\end{aligned} \tag{3.15.1}$$

For the description of the surface in the deformed state, we shall make use of the fact that the in-plane displacements u, v are considerably smaller than the displacement

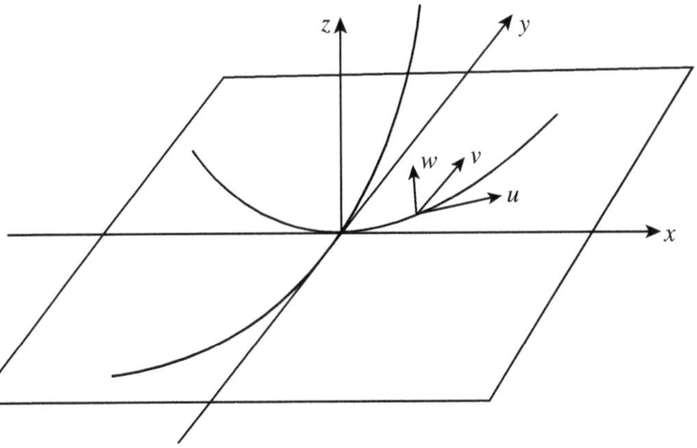

Figure 3.15.1

w, perpendicular to the mid-plane, i.e.,

$$|u|,\ |v| \ll |w|, \tag{3.15.2}$$

and that

$$w_{,x}^2,\ w_{,y}^2 \ll 1. \tag{3.15.3}$$

Under these conditions, a point (x, y, z) in the undeformed state moves to

$$(x + u - Z_{,x}w,\ y + v - Z_{,y}w,\ z + w)$$

in the deformed state. The square of the length of a line element in the deformed state is now

$$\begin{aligned}(d\bar{s})^2 &= (dx + du - Z_{,xx}w\,dx - Z_{,xy}w\,dy - Z_{,x}\,dw)^2 \\ &\quad + (dy + dv - Z_{,yy}w\,dy - Z_{,xy}w\,dx - Z_{,y}\,dw)^2 + (dz + dw)^2,\end{aligned} \tag{3.15.4}$$

where $Z_{,xy} = 0$ because x and y are the principal directions of curvature,

$$\begin{aligned}(d\bar{s})^2 - (ds)^2 &= 2(dx)^2\left[u_{,x} - Z_{,xx}w + \frac{1}{2}\,w_{,x}^2\right. \\ &\quad \left. + \frac{1}{2}(u_{,x} - Z_{,xx}w + Z_{,x}w_{,x})^2 + \frac{1}{2}(v_{,x} - Z_{,y}w_{,x})^2\right] \\ &\quad + 2dx\,dy\,[u_{,y} + v_{,x} + w_{,x}w_{,y} \\ &\quad + (u_{,x} - Z_{,xx}w - Z_{,x}w_{,x})(u_{,y} - Z_{,x}w_{,y}) \\ &\quad + (v_{,y} - Z_{,yy}w - Z_{,y}w_{,y})(v_{,x} - Z_{,y}w_{,x})] \\ &\quad + 2(dy)^2\left[v_{,y} - Z_{,yy}w + \frac{1}{2}w_{,y}^2\right. \\ &\quad \left. + \frac{1}{2}(v_{,y} - Z_{,yy}w + Z_{,y}w_{,y})^2 + \frac{1}{2}(u_{,y} - Z_{,x}w_{,y})^2\right].\end{aligned} \tag{3.15.5}$$

Neglecting higher order terms, except for $w_{,x}^2$, $w_{,y}^2$ and $w_{,x}$, $w_{,y}$, we obtain

$$\begin{aligned}(d\bar{s})^2 - (ds)^2 &= 2\left(u_{,x} - Z_{,xx}w + \frac{1}{2}w_{,x}^2\right)(dx)^2 \\ &\quad + 2(u_{,y} + v_{,x} + w_{,x}w_{,y})\,dx\,dy \\ &\quad + 2\left(u_{,y} - Z_{,yy}w + \frac{1}{2}w_{,y}^2\right)(dy)^2 \\ &\equiv 2\gamma_{xx}(dx)^2 + 4\gamma_{xy}\,dx\,dy + 2\gamma_{yy}(dx)^2.\end{aligned} \tag{3.15.6}$$

Hence, the strains are given by

$$\begin{aligned}\gamma_{xx} &= u_{,x} - \frac{w}{R_1} + \frac{1}{2}w_{,x}^2 \\ 2\gamma_{xy} &= u_{,y} + v_{,x} + w_{,x}w_{,y} \\ \gamma_{yy} &= v_{,y} - \frac{w}{R_2} + \frac{1}{2}w_{,y}^2\end{aligned} \tag{3.15.7}$$

where

$$R_1 = Z_{,xx}, \quad R_2 = Z_{,yy}, \tag{3.15.8}$$

are the principle radii of curvature. The changes of curvature are given by

$$\rho_{xx} = w_{,xx}, \quad \rho_{yy} = w_{,yy}, \quad \rho_{xy} = w_{,xy}. \tag{3.15.9}$$

Notice that all our expressions are only valid in a sufficiently small neighborhood of the tangent point. Under the assumption that the loads are dead-weight loads, which cause stresses σ_x, σ_y, and τ_{xy} in the fundamental state, the energy is given by

$$\begin{aligned} P[\mathbf{u}] = \frac{Eh}{2(1-\upsilon^2)} \iint \Bigg\{ & \left(u_{,x} - \frac{w}{R_1} + \frac{1}{2}w_{,x}^2\right)^2 + \left(v_{,y} - \frac{w}{R_2} + \frac{1}{2}w_{,y}^2\right)^2 \\ & + 2\upsilon\left(u_{,x} - \frac{w}{R_1} + \frac{1}{2}w_{,x}^2\right)\left(v_{,y} - \frac{w}{R_2} + \frac{1}{2}w_{,y}^2\right) + \frac{1}{2}(1-\upsilon)(u_{,y}+v_{,x}+w_{,x}w_{,y})^2 \\ & + \frac{h^2}{12}\left[w_{,xx}^2 + w_{,yy}^2 + 2\upsilon w_{,xx}w_{,yy} + 2(1-\upsilon)w_{,xy}^2\right] \\ & + \frac{1-\upsilon^2}{E}\left(\sigma_x w_{,x}^2 + \sigma_y w_{,y}^2 + 2\tau_{xy}w_{,x}w_{,y}\right)\Bigg\}\, dx\, dy, \end{aligned} \tag{3.15.10}$$

where the radii of the curvature can be assumed to be constants. Later we shall show that this expression may also be used for a shallow shell under uniform pressure (which is not a dead-weight load).

The second and the third variations are given by

$$\begin{aligned} P_2[\mathbf{u}] = \frac{Eh}{2(1-\upsilon^2)} \iint \Bigg\{ & \left(u_{,x} - \frac{w}{R_1}\right)^2 + \left(v_{,y} - \frac{w}{R_2}\right)^2 \\ & + 2\upsilon\left(u_{,x} - \frac{w}{R_1}\right)\left(v_{,y} - \frac{w}{R_2}\right) + \frac{1}{2}(1-\upsilon)(u_{,y}+v_{,x})^2 \\ & + \frac{h^2}{12}\left[w_{,xx}^2 + w_{,yy}^2 + 2\upsilon w_{,xx}w_{,yy} + 2(1-\upsilon)w_{,xy}^2\right] \\ & + \frac{1-\upsilon^2}{E}\left(\sigma_x w_{,x}^2 + \sigma_y w_{,y}^2 + 2\tau_{xy}w_{,x}w_{,y}\right)\Bigg\}\, dx\, dy \end{aligned} \tag{3.15.11}$$

$$\begin{aligned} P_3[\mathbf{u}] = \frac{Eh}{2(1-\upsilon^2)} \iint \Bigg\{ & \left[u_{,x} - \frac{w}{R_1} + \upsilon\left(v_{,y} - \frac{w}{R_2}\right)\right] w_{,x}^2 \\ & + \left[v_{,y} - \frac{w}{R_2} + \upsilon\left(u_{,x} - \frac{w}{R_1}\right)\right] w_{,y}^2 + (1-\upsilon)(u_{,y}+v_{,x})w_{,x}w_{,y}\Bigg\}\, dx\, dy. \end{aligned} \tag{3.15.12}$$

The variational equation for neutral equilibrium is $P_{11}[\mathbf{u}, \boldsymbol{\zeta}] = 0$, where $\boldsymbol{\zeta} = (\xi, \eta, \zeta)$ is a kinematically admissible displacement field,

$$\begin{aligned} P_{11}[\mathbf{u}, \boldsymbol{\zeta}] = \frac{Eh}{1-\upsilon^2} \iint \Bigg[& \left(u_{,x} - \frac{w}{R_1}\right)\left(\xi_{,x} - \frac{\zeta}{R_1}\right) + \left(v_{,y} - \frac{w}{R_2}\right)\left(\eta_{,y} - \frac{\zeta}{R_2}\right) \\ & + \upsilon\left(u_{,x} - \frac{w}{R_1}\right)\left(\eta_{,y} - \frac{\zeta}{R_2}\right) + \upsilon\left(v_{,y} - \frac{w}{R_2}\right)\left(\xi_{,x} - \frac{\zeta}{R_1}\right) \end{aligned}$$

$$+\frac{1}{2}(1-\upsilon)(u_{,y}+v_{,x})(\xi_{,y}+\eta_{,x}) \tag{3.15.13}$$
$$+\frac{h^2}{12}\left\{w_{,xx}\zeta_{,xx}+w_{,yy}\zeta_{,yy}+\upsilon w_{,xx}\zeta_{,yy}+\upsilon w_{,yy}\zeta_{,xx}+2(1-\upsilon)w_{,xy}\zeta_{,xy}\right\}$$
$$\left.+\frac{1-\upsilon^2}{E}(\sigma_x w_{,x}\zeta_{,x}+\sigma_y w_{,y}\zeta_{,y}+\tau_{xy}w_{,x}\zeta_{,y}+\tau_{xy}w_{,y}\zeta_{,x})\right]dx\,dy.$$

Repeated application of the divergence theorem yields

$$\frac{Eh}{1-\upsilon^2}\oint_{\partial S}\left[\left[\left\{\left(u_{,x}-\frac{w}{R_1}\right)+\upsilon\left(v_{,y}-\frac{w}{R_2}\right)\right\}\upsilon_x+\frac{1}{2}(1-\upsilon)(u_{,y}+v_{,x})\upsilon_y\right]\xi\right.$$
$$+\left[\left\{v_{,y}-\frac{w}{R_2}+\upsilon\left(u_{,x}-\frac{w}{R_1}\right)\right\}\upsilon_y+\frac{1}{2}(1-\upsilon)(u_{,y}+v_{,x})\upsilon_x\right]\eta$$
$$+\frac{h^2}{12}\left[\left\{(w_{,xx}+\upsilon w_{,yy})\,\upsilon_x+(1-\upsilon)w_{,xy}\upsilon_y\right\}\zeta_{,x}\right.$$
$$\left.+\left\{(w_{,yy}+\upsilon w_{,xx})\upsilon_y+(1-\upsilon)w_{,xy}\upsilon_x\right\}\zeta_{,y}\right]$$
$$+\left\{\frac{1-\upsilon^2}{E}\left[\sigma_x w_{,x}\upsilon_x+\sigma_y w_{,y}\upsilon_y+\tau_{xy}(w_{,x}\upsilon_{,y}+w_{,y}\upsilon_x)\right]\right.$$
$$\left.\left.-\frac{h^2}{12}\left[(w_{,xx}+w_{,yy})_{,x}\upsilon_x+(w_{,xx}+w_{,yy})_{,y}\upsilon_y\right]\right\}\zeta\right]ds \tag{3.15.14}$$
$$-\frac{Eh}{1-\upsilon^2}\iint_S\left\{u_{,xx}+\frac{1}{2}(1-\upsilon)u_{,yy}+\frac{1}{2}(1+\upsilon)v_{,xy}-\left(\frac{1}{R_1}+\frac{\upsilon}{R_2}\right)w_{,x}\right\}\xi$$
$$+\left\{v_{,yy}+\frac{1}{2}(1-\upsilon)v_{,xx}+\frac{1}{2}(1+\upsilon)u_{,xy}-\left(\frac{1}{R_2}+\frac{\upsilon}{R_1}\right)w_{,y}\right\}\eta$$
$$\left\{-\left(\frac{1}{R_1}+\frac{\upsilon}{R_2}\right)u_{,x}-\left(\frac{1}{R_2}+\frac{\upsilon}{R_1}\right)v_{,y}+\left(\frac{1}{R_1^2}+\frac{2\upsilon}{R_1R_2}+\frac{1}{R_2^2}\right)w\right.$$
$$\left.\left.+\frac{h^2}{12}\Delta\Delta w-\frac{1-\upsilon^2}{E}\left\{(\sigma_x w_{,x})_{,x}+(\sigma_y w_{,y})_{,y}+(\tau_{xy}w_{,x})_{,y}+(\tau_{xy}w_{,y})_{,x}\right\}\right\}\zeta\right]dx\,dy=0,$$

where υ is the unit normal vector at the edge. This yields the differential equations

$$\begin{aligned}&u_{,xx}+\frac{1}{2}(1-\upsilon)u_{,yy}+\frac{1}{2}(1+\upsilon)v_{,xy}-\left(\frac{1}{R_1}+\frac{\upsilon}{R_2}\right)w_{,x}=0\\&v_{,yy}+\frac{1}{2}(1-\upsilon)v_{,xx}+\frac{1}{2}(1+\upsilon)u_{,xy}-\left(\frac{1}{R_2}+\frac{\upsilon}{R_1}\right)w_{,y}=0\end{aligned} \tag{3.15.15}$$

$$\begin{aligned}&\frac{h^2}{12}\Delta\Delta w-\left(\frac{1}{R_1}+\frac{\upsilon}{R_2}\right)u_{,x}-\left(\frac{1}{R_2}+\frac{\upsilon}{R_1}\right)v_{,y}+\left(\frac{1}{R_1^2}+\frac{2\upsilon}{R_1R_2}+\frac{1}{R_2^2}\right)w\\&\quad-\frac{1-\upsilon^2}{E}(\sigma_x w_{,xx}+\sigma_y w_{,yy}+2\tau_{xy}w_{,xy})=0,\end{aligned} \tag{3.15.16}$$

where we have used the fact that σ_x, σ_y, and τ_{xy} (approximately) satisfy the equations

$$\sigma_{x,x}+\tau_{yx,y}=0,\quad \tau_{xy,x}+\sigma_{y,y}=0. \tag{3.15.17}$$

From the line integral, we obtain the conditions in the mid-plane,

$$\left[\left\{\left(u_{,x}-\frac{w}{R_1}\right)+\upsilon\left(v_{,y}-\frac{w}{R_2}\right)\right\}\upsilon_x+\frac{1}{2}(1-\upsilon)(u_{,y}+v_{,x})\upsilon_y\right]\xi\bigg|_{\partial S}=0$$
$$\left[\left\{\left(v_{,y}-\frac{w}{R_2}\right)+\upsilon\left(u_{,x}-\frac{w}{R_1}\right)\right\}\upsilon_y+\frac{1}{2}(1-\upsilon)(u_{,y}+v_{,x})\upsilon_x\right]\eta\bigg|_{\partial S}=0. \qquad (3.15.18)$$

For the reduction of the remaining terms, we make use of the relations

$$\zeta_{,x}=-\upsilon_y\zeta_{,s}+\upsilon_x\zeta_{,\nu},\qquad \zeta_{,y}=\upsilon_x\zeta_{,s}+\upsilon_y\zeta_{,\nu}, \qquad (3.15.19)$$

and (3.12.17) to obtain

$$\left[(w_{,xx}+\upsilon w_{,yy})\upsilon_x^2+2(1-\upsilon)w_{,xy}\upsilon_x\upsilon_y+(w_{,yy}+\upsilon w_{,xx})\upsilon_y^2\right]\zeta_{,\nu}\,|_{\partial S}=0 \qquad (3.15.20)$$

$$\Big\{(1-\upsilon)\left[(w_{,yy}-w_{,xx})\upsilon_x\upsilon_y+w_{,xy}\left(\upsilon_x^2-\upsilon_y^2\right)\right]_{,s}$$
$$+(w_{,xx}+w_{,yy})_{,x}\upsilon_x+(w_{,xx}+w_{,yy})_{,y}\upsilon_y \qquad (3.15.21)$$
$$-\frac{12(1-\upsilon^2)}{Eh^3}\left[(\sigma_x w_{,x}+\tau_{xy}w_{,y})\upsilon_x+(\sigma_y w_{,y}+\tau_{xy}w_{,x})\upsilon_y\right]\Big\}\zeta\bigg|_{\partial S}=0.$$

Notice that only the in-plane boundary conditions are affected by the curvature of the surface.

We shall now apply this theory to a spherical shell under a uniform external pressure p per unit area of the middle surface.† The stresses in the fundamental state are given by

$$\sigma_x=\sigma_y=-\sigma=-\frac{pR}{2h},\qquad \tau_{xy}=0. \qquad (3.15.22)$$

The equations now read

$$u_{,xx}+\frac{1}{2}(1-\upsilon)u_{,yy}+\frac{1}{2}(1+\upsilon)v_{,xy}-\frac{1+\upsilon}{R}w_{,x}=0$$
$$v_{,yy}+\frac{1}{2}(1-\upsilon)v_{,xx}+\frac{1}{2}(1+\upsilon)u_{,xy}-\frac{1+\upsilon}{R}w_{,y}=0 \qquad (3.15.23)$$

$$\frac{Eh^2}{12(1-\upsilon^2)}\Delta\Delta w-\frac{E}{(1-\upsilon)R}\left(u_{,x}+v_{,y}-2\frac{w}{R}\right)+\sigma(w_{,xx}+w_{,yy})=0. \qquad (3.15.24)$$

We try a solution of the form

$$u=A\sin px/R\cos qy/R$$
$$v=B\cos px/R\sin qy/R \qquad (3.15.25)$$
$$w=C\cos px/R\cos qy/R,\qquad (p,q\in\mathbb{R}).$$

Substitution into (3.15.23) yields

$$-\left[p^2+\frac{1}{2}(1-\upsilon)q^2\right]A-\frac{1}{2}(1+\upsilon)pqB+(1+\upsilon)pC=0$$
$$-\frac{1}{2}(1+\upsilon)pqA-\left[\frac{1}{2}(1-\upsilon)p^2+q^2\right]B+(1+\upsilon)qC=0. \qquad (3.15.26)$$

† See the remark below (3.15.10).

We can now solve A/C and B/C from these equations, which yield

$$\frac{A}{C} = (1+\upsilon)\frac{p}{p^2+q^2}, \quad \frac{B}{C} = (1+\upsilon)\frac{q}{p^2+q^2}. \tag{3.15.27}$$

In our derivation of the shallow shell equations, we have use the assumption $|w| \gg |u|, |v|$, which means $A/C,\ B/C \ll 1$, so that for p or $q \gg 1$, our assumptions are satisfied.

Substitution of the displacement field (3.15.25) into the third equilibrium equation and using (3.15.27), we obtain

$$\frac{\sigma}{E} = \frac{1}{p^2+q^2} + \frac{h^2}{4c^2R^2}(p^2+q^2), \quad c = \sqrt{3(1-\upsilon^2)}. \tag{3.15.28}$$

We still must minimize this expression with respect to p and q, but because the expression only contains the combination p^2+q^2, we can minimize with respect to this parameter, which yields

$$p^2+q^2 = \frac{2cR}{h}. \tag{3.15.29}$$

This means that our assumptions are satisfied as $R/h \gg 1$. The critical load is now given by

$$\sigma_{\mathrm{cr}} = \frac{Eh}{cR}. \tag{3.15.30}$$

Notice that the critical load is proportional to h/R, whereas for plates the critical load was proportional to h^2/b^2.

Because p and q must only satisfy the condition (3.15.29), there is an infinite number of buckling modes. In the $p-q$ plane, (3.15.29) represents a circle with radius $p_0 = \sqrt{2cR/h}$ (see Figure 3.15.2).

Because we have a continuous spectrum, the displacement field is given by integrals instead of a series. However, for our purpose we may consider some discrete values, $p_i,\ q_i$. Because for the evaluation of $P_3[\mathbf{u}_1]$ we must deal with cubic terms,

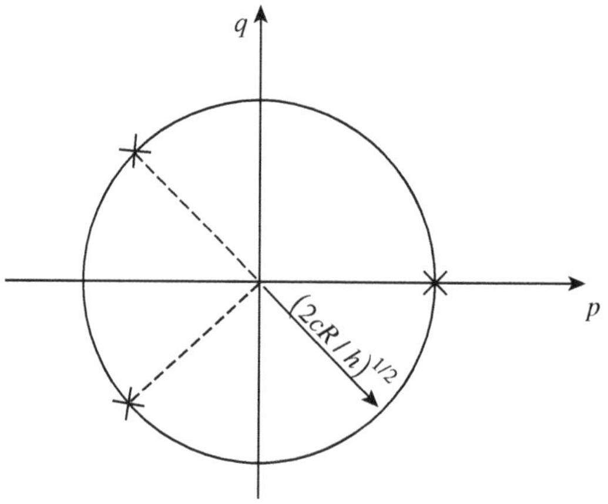

Figure 3.15.2

we can consider the discrete pairs

$$(p_1, q_1), \quad (p_2, q_2), \quad (p_3, q_3), \tag{3.15.31}$$

which in the evaluation of $P_3\,[\mathbf{u}_1]$ lead to integrals of the form

$$\int \left(\begin{matrix}\sin\\ \cos\end{matrix} p_1 \frac{x}{R} \quad \begin{matrix}\sin\\ \cos\end{matrix} p_2 \frac{x}{R} \quad \sin p_3 \frac{x}{R} \right) dx, \tag{3.15.32}$$

which can be expressed in integrals of the form

$$\int \begin{matrix}\sin\\ \cos\end{matrix} (p_1 \pm p_2 \pm p_3) \frac{x}{R}\, dx. \tag{3.15.33}$$

For large values of x, these integrals vanish† except when the argument of the cosine is equal to zero, i.e.,

$$p_1 \pm p_2 \pm p_3 = 0, \tag{3.15.34}$$

and similarly for the terms in y,

$$q_1 \pm q_2 \pm q_3 = 0. \tag{3.15.35}$$

This means that the points (p_1, q_1), (p_2, q_2), (p_3, q_3) divide the circle in three equal parts. We shall now consider the special case of (see Figure 3.15.2)

$$\begin{aligned} (p_1, q_1) = (p_0, 0), \quad (p_2, q_2) &= \left(-\frac{1}{2} p_0, \frac{1}{2} p_0 \sqrt{3} \right), \\ (p_3, q_3) &= \left(-\frac{1}{2} p_0, -\frac{1}{2} p_0 \sqrt{3} \right). \end{aligned} \tag{3.15.36}$$

Let us introduce the notation

$$p = \frac{1}{2} p_0, \quad q = \frac{1}{2} \sqrt{3}\, p_0; \tag{3.15.37}$$

then our displacement field is given by

$$\begin{aligned} u &= (1+\upsilon) \frac{h^2}{2cR} \left(2p a_0 \sin 2p \frac{x}{R} + p a_1 \sin p \frac{x}{R} \cos q \frac{y}{R} \right) \\ v &= (1+\upsilon) \frac{h^2}{2cR} \left(\qquad\qquad q a_1 \cos p \frac{x}{R} \sin q \frac{y}{R} \right) \\ w &= h \left(a_0 \cos 2p \frac{x}{R} + a_1 \cos p \frac{x}{R} \cos q \frac{y}{R} \right), \end{aligned} \tag{3.15.38}$$

where we have used (3.15.27).

For the evaluation of $P_3[\mathbf{u}_1]$, we shall need the following expressions,

$$\begin{aligned} u_{,x} - \frac{w}{R} &= \frac{h}{R} \left[\upsilon a_0 \cos 2p \frac{x}{R} - \frac{1}{4} (3 - \upsilon) a_1 \cos p \frac{x}{R} \cos q \frac{y}{R} \right] \\ v_{,y} - \frac{w}{R} &= \frac{h}{R} \left[-a_0 \cos 2p \frac{x}{R} - \frac{1}{4} (1 - 3\upsilon) a_1 \cos p \frac{x}{R} \cos q \frac{y}{R} \right] \end{aligned}$$

† Let $L \ll x \ll R$, e.g., $x = O(R^{3/4} h^{1/4})$, so that our shallow shell assumptions still hold approximately.

$$u_{,x} - \frac{w}{R} + \upsilon\left(v_{,y} - \frac{w}{R}\right) = \frac{h}{R}\left[-\frac{3}{4}a_1(1-\upsilon^2)\cos p\frac{x}{R}\cos q\frac{y}{R}\right]$$
$$v_{,y} - \frac{w}{R} + \upsilon\left(u_{,x} - \frac{w}{R}\right) = \frac{h}{R}\left[-(1-\upsilon^2)a_0\cos 2p\frac{x}{R} - \frac{1}{4}(1-\upsilon^2)a_1\cos p\frac{x}{R}\cos q\frac{y}{R}\right]$$
$$u_{,y} + v_{,x} = (1+\upsilon)\frac{h}{R}\left[-\frac{1}{2}\sqrt{3}\,a_1\sin p\frac{x}{R}\sin q\frac{y}{R}\right] \tag{3.15.39}$$
$$w_{,x} = 2\frac{ph}{R}\left(-a_0\sin 2p\frac{x}{R} - \frac{1}{2}a_1\sin p\frac{x}{R}\cos q\frac{y}{R}\right)$$
$$w_{,y} = 2\frac{ph}{R}\left(-\frac{1}{2}\sqrt{3}\,a_1\cos p\frac{x}{R}\sin q\frac{y}{R}\right).$$

Introducing the notation

$$\overline{(\quad)} = \frac{1}{4R^2}\int_{-R}^{R}\int_{-R}^{R}(\quad)dx\,dy, \tag{3.15.40}$$

which is the average value of (), we obtain

$$\overline{\left[u_{,x} - \frac{w}{R} + \upsilon\left(v_{,y} - \frac{w}{R}\right)\right]w_{,x}^2 + \left[v_{,y} - \frac{w}{R} + \upsilon\left(u_{,x} - \frac{w}{R}\right)\right]w_{,y}^2}$$
$$\overline{+(1-\upsilon)(u_{,y}+v_{,x})w_{,x}w_{,y}}$$
$$= p^2\frac{h^3}{R^3}\left[-\frac{3}{8}(1-\upsilon^2)a_0a_1^2 - \frac{3}{8}(1-\upsilon^2)a_0a_1^2 - \frac{3}{8}(1-\upsilon^2)a_0a_1^2\right] \tag{3.15.41}$$
$$= -\frac{9}{16}(1-\upsilon^2)\frac{ch^2}{R^2}a_0a_1^2,$$

so the average value of $P_2[\mathbf{u}_1]$ is

$$\overline{P_3[\mathbf{u}_1]} = -\frac{9}{32}\frac{Ech^3}{R^2}a_0a_1^2. \tag{3.15.42}$$

For the evaluation of $\overline{P_2[\mathbf{u};\lambda]}$, we need

$$\overline{w_{,x}^2 + w_{,y}^2} = 4p^2\frac{h^2}{R^2}\left(\frac{1}{2}a_0^2 + \frac{1}{4}a_1^2\right), \tag{3.15.43}$$

so that

$$\overline{P_2[\mathbf{u}_1;\lambda]} = (1-\lambda)\frac{Eh^3}{R^2}\left(\frac{1}{2}a_0^2 + \frac{1}{4}a_1^2\right). \tag{3.15.44}$$

In a sufficiently small neighborhood of the critical load, the average energy $\overline{P[\mathbf{u},\lambda]}$ is given by

$$F(a_i,\lambda) = \overline{P[\mathbf{u},\lambda]} = \frac{Eh^3}{R^2}\left[\frac{1}{2}(1-\lambda)\left(a_0^2 + \frac{1}{2}a_1^2\right) - \frac{9c}{32}a_0a_1^2\right]. \tag{3.15.45}$$

The equilibrium equations are given by

$$\frac{\partial F}{\partial a_0} = \frac{Eh^3}{R^2}\left[(1-\lambda)a_0 - \frac{9c}{32}a_1^2\right] = 0$$
$$\frac{\partial F}{\partial a_1} = \frac{Eh^3}{R^2}\left[\frac{1}{2}(1-\lambda)a_1 - \frac{9c}{16}a_0 a_1\right] = 0. \tag{3.15.46}$$

From the second of these equations, we obtain the non-trivial solution

$$a_0 = \frac{8}{9c}(1-\lambda), \tag{3.15.47}$$

and from the first equation,

$$a_1 = \pm\frac{16}{9c}(1-\lambda) = \pm 2a_0. \tag{3.15.48}$$

It is sufficient to consider only the positive solution because the negative sign is obtained by shifting the origin of our coordinate system.

As

$$\frac{\partial^2 F}{\partial a_0^2}\frac{\partial^2 F}{\partial a_1^2} - \left(\frac{\partial^2 F}{\partial a_0 \partial a_1}\right)^2 = -\left(\frac{Eh^3}{R^2}\right)^2(1-\lambda)^2 < 0, \tag{3.15.49}$$

the equilibrium states are unstable for both $\lambda < 1$ and $\lambda > 1$. Notice that for $\upsilon = 0.272$ and $c = 5/3$, we have

$$a_0 = \frac{8}{15}(1-\lambda), \quad a_1 = \frac{16}{15}(1-\lambda), \tag{3.15.50}$$

so that for $\lambda = 1/2$, $a_0 = 4/15$, $a_1 = 8/15$, $w = 0.8h$. This means that for deflections of the order of the plate thickness, the load is reduced to about 50% of the critical load.

Assuming that our average energy expression holds on the whole surface of the sphere, the total energy is given by

$$4\pi R^2 F(a_i, \lambda). \tag{3.15.51}$$

The work of the uniform pressure p is $-p\Delta V$, where ΔV is the volume increment passing from the fundamental path to the branched path. It follows that

$$4\pi R^2 \frac{\partial F}{\partial p} = -\Delta V, \tag{3.15.52}$$

or with

$$p = 2\lambda\frac{Eh^2}{cR^2} \tag{3.15.53}$$

$$\frac{2\pi R^4 c}{Eh^2}\frac{\partial F}{\partial \lambda} = -\Delta V. \tag{3.15.54}$$

Using (3.15.45), we obtain

$$\Delta V = -\frac{64\pi}{27c}hR^2(1-\lambda)^2. \tag{3.15.55}$$

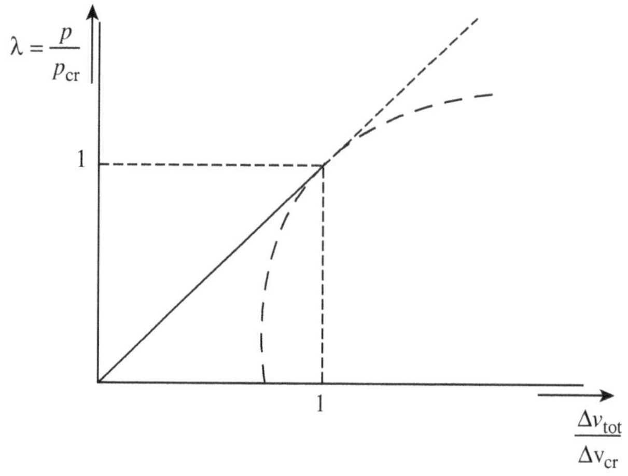

Figure 3.15.3

The volume change at the critical load is

$$\Delta V_{cr} = -4\pi R^2 w_{cr} = -\frac{\sigma_{cr}}{E}(1-\upsilon)R \cdot 4\pi R = -\frac{4\pi}{c}(1-\upsilon)hR^2, \tag{3.15.56}$$

so that

$$\frac{\Delta V_{tot}}{\Delta V_{cr}} = \frac{16}{27(1-\upsilon)}(1-\lambda)^2 + \lambda. \tag{3.15.57}$$

This relation is illustrated in Figure 3.15.3. Notice that there is no stable equilibrium state in the vicinity of the bifurcation point, and that the load drops very rapidly after bifurcation, which means that the shell is very imperfection-sensitive. We will come back to this later.

Let us now first have a closer look at the buckling mode,

$$w = ha_0\left(\cos 2p\frac{x}{R} + 2\cos p\frac{x}{R}\cos p\sqrt{3}\frac{y}{R}\right). \tag{3.15.58}$$

Introducing the notation

$$2p\frac{x}{R} = \alpha, \quad 2p\frac{y}{R} = \beta, \tag{2.15.59}$$

we can rewrite this expression as

$$w = ha_0 f(\alpha, \beta), \quad f(\alpha, \beta) = \cos\alpha + 2\cos\alpha/2\cos\sqrt{3}\beta/2. \tag{3.15.60}$$

To recognize the buckling pattern, we investigate $f(\alpha, \beta)$ more closely. First, we notice that

$$\begin{aligned} f_{\max} = 3 \quad \text{for } \alpha = 4j\pi, \quad \beta = \frac{4k\pi}{\sqrt{3}}, \\ \alpha = (2j+2)\pi, \quad \beta = (2k+2)\frac{\pi}{\sqrt{3}}. \end{aligned} \tag{3.15.61}$$

Necessary conditions for a minimum value of $f(\alpha, \beta)$ are

$$\begin{aligned}\frac{\partial f}{\partial \alpha} &= -\sin\alpha - \sin\alpha/2\cos\sqrt{3}\,\beta/2 = 0\\ \frac{\partial f}{\partial \beta} &= -\sqrt{3}\cos\alpha/2\sin\sqrt{3}\beta/2 = 0.\end{aligned} \tag{3.15.62}$$

From the second equation, we find

$$\begin{aligned}&\cos\alpha/2 = 0, \qquad \alpha = (2k+1)\pi,\\ &\sin\sqrt{3}\,\beta/2 = 0, \quad \beta = \frac{2j\pi}{\sqrt{3}},\end{aligned} \tag{3.15.63}$$

and from the first equation,

$$\begin{aligned}&\sin\alpha/2 = 0, \quad \alpha = 2k\pi,\\ &2\cos\alpha/2 + \cos\sqrt{3}\,\beta/2 = 0.\end{aligned} \tag{3.15.64}$$

Both equations are satisfied when

$$\alpha = (2k+1)\pi, \quad \beta = \frac{2j+1}{\sqrt{3}}\pi \tag{3.15.65}$$

$$\beta = 2j\frac{\pi}{\sqrt{3}}, \qquad \alpha = \begin{cases}\pm\dfrac{2\pi}{3} + 4k\pi \ \ (j\,\text{odd})\\ \pm\dfrac{4\pi}{3} + 4k\pi \ (j\,\text{even})\end{cases} \tag{3.15.66}$$

$$\alpha = 2k\pi, \quad \beta = \frac{2j\pi}{\sqrt{3}}. \tag{3.15.67}$$

We now obtain the following values for $f(\alpha, \beta)$,

$$\begin{aligned}&f\left((2k+1)\pi, \ \ \frac{2j+1}{\sqrt{3}}\pi\right) = 1\\ &f\left(\pm\frac{2\pi}{3} + 4k\pi, \ \ \frac{2j\pi}{\sqrt{3}}\right) = \begin{cases}+\dfrac{1}{2} & j\,\text{even}\\ -\dfrac{3}{2} & j\,\text{odd}\end{cases}\\ &f\left(\pm\frac{4\pi}{3} + 4k\pi, \ \ \frac{2j\pi}{\sqrt{3}}\right) = -\frac{3}{2}\\ &f\left(2k\pi, \ \ \frac{2j\pi}{\sqrt{3}}\right) = \begin{cases}-1 & j+k\,\text{odd}\\ 3 & j+k\,\text{even.}\end{cases}\end{aligned} \tag{3.15.68}$$

We may now draw the graph shown in Figure 3.15.4.

We have a pattern of regular hexagons. Ordinarily, this pattern is not observed because this buckling mode is unstable. It can be observed when the buckling mode is artificially fixed, which may be done by fitting a solid sphere with a radius slightly

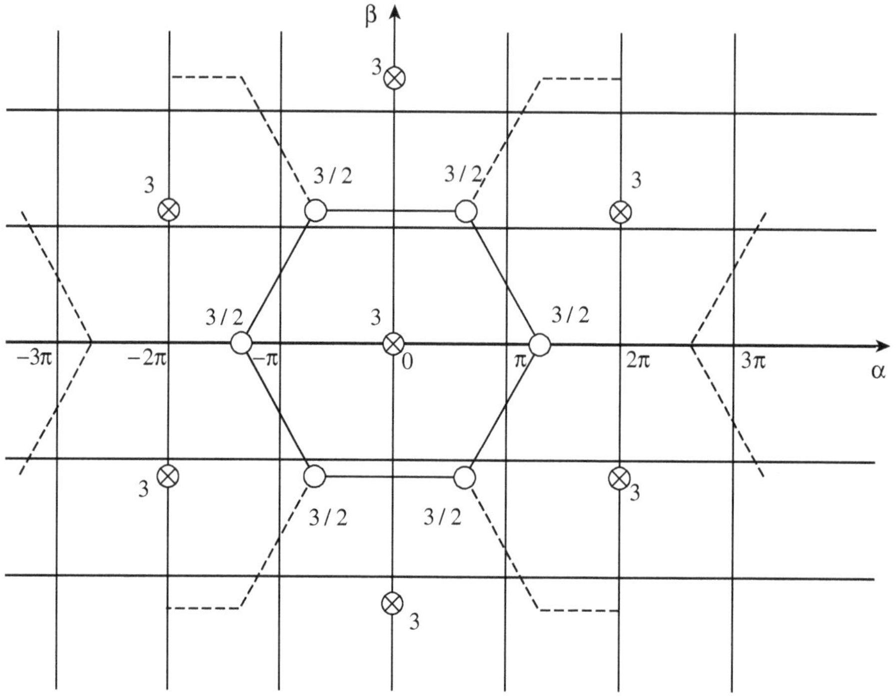

Figure 3.15.4

smaller than that of the shell inside the shell.† The experimental results are in agreement with our buckling pattern.

As we have seen previously, the load drops rapidly after bifurcation, which implies that the shell is very imperfection-sensitive. To get quantitative information about the imperfection sensitivity, we assume imperfections corresponding to the buckling mode, which according to the general theory are the worst type of imperfections.

Let the imperfections be given by

$$w_0 = h\left[\bar{a}_0 \cos 2p\frac{x}{R} + \bar{a}_1 \cos p\frac{x}{R} \cos\sqrt{3}\,p\frac{y}{R}\right], \tag{3.15.69}$$

where $\bar{a}_1 = 2\bar{a}_0$. In the presence of imperfections, the function $F(a_i;\lambda)$ (see 3.15.45) must be replaced by (see 2.6.39)

$$F^*(a_i;\lambda) = \frac{Eh^3}{R^2}\left[\frac{1}{2}(1-\lambda)\left(a_0^2 + \frac{1}{2}a_1^2\right) - \frac{9c}{32}a_0 a_1^2 - \lambda\left(a_0\bar{a}_0 + \frac{1}{2}a_0\bar{a}_1\right)\right]. \tag{3.15.70}$$

With $\bar{a}_1 = 2\bar{a}_0$ and a_0 and a_1 given by (3.15.47) and (3.15.48), we can rewrite this expression in the form

$$F^*(a_i;\lambda) = \frac{Eh^3}{2R^2}\left[3(1-\lambda)a_0^2 - \frac{9c}{4}a_1^3 - 6\lambda\bar{a}_0 a_0\right]. \tag{3.15.71}$$

† Cf. R. L. Carlson, R. L. Sendelbeck, and N. J. Hoff, Experimental studies of the buckling of complete spherical shells. *Experimental Mechanics*, **7**, no. 7, 281–288 (1967).

The equilibrium equation then reads

$$\frac{\partial F^*}{\partial a_0} = \frac{Eh^3}{2R^2}\left[6(1-\lambda)a_0 - \frac{27c}{4}a_0^2 - 6\lambda\bar{a}_0\right] = 0. \tag{3.15.72}$$

The minimum value of a_0 is obtained when the discriminant of this quadratic equation vanishes, which yields

$$(1-\lambda^*)^2 = \frac{9c}{2}\lambda^*\bar{a}_0. \tag{3.15.73}$$

The corresponding amplitude is given by

$$a_0^* = \frac{4(1-\lambda^*)}{9c}. \tag{3.15.74}$$

For $\lambda^* = 0.5$, we have $\bar{a}_0 = 1/15$ (for $\upsilon = 0.28$), so the maximum imperfection amplitude is $3\bar{a}_0 = 1/5$, i.e., imperfections of about 20% of the shell thickness reduce the critical load by 50%. This result shows that the shell is extremely imperfection-sensitive.

The present approach is due to HUTCHINSON.† Let us now reconsider the approach. The starting point was shallow shell theory, so the results are only valid in a sufficiently small neighborhood of the tangent point, i.e., $x, y \ll R$. However, for our evaluation of the average value of the energy we have carried out the integration over the domain $-R \leq x, y \leq R$. Furthermore, we have ignored the fact that the hexagonal buckling pattern cannot be extended over the complete spherical shell, which implies a relaxation of the geometric conditions, which should result in decreased stability in the post-buckling range. This means that Hutchinson's analysis yields a lower bound for the post-buckling load. The difficulties with the domain of validity for x and y can be overcome by using a displacement field (u^*, v^*, w^*), defined by

$$(u^*, v^*, w^*) = (u, v, w)\exp\left[-\frac{1}{2}\mu^2(x^2+y^2)/R^2\right], \tag{3.15.75}$$

where $1 \ll \mu^2 \ll p^2 + q^2$. Under this restriction, the exponential factor can be considered to be a constant factor when differentiation is carried out, which means that this modified displacement field satisfies the same equations as our original field. Furthermore, $(u^*, v^*, w^*) \sim (u, v, w)$ for sufficiently small values of x and y, and $(u^*, v^*, w^*) \to 0$ for sufficiently large values of x and y.

Such a type of displacement field may also be used to describe localized imperfections.‡ In the next section, we shall treat the problem of the spherical shell again using the general theory of shells, and verify the validity of results obtained in this section.

† J. W. HUTCHINSON, Imperfections sensitivity of externally pressurized spherical shells. *Journal Appl. Mech.* **34**, 49–55 (1967).

‡ Cf. W. T. KOITER, The influence of more or less localized short-wave imperfections.... Rep. Lab. For *Appl. Mech. Delft*, 534. V. Z. GRISTCHAK, Asymptotic formula for... WTHD rep. No. 88.

3.16 Buckling behavior of a spherical shell under uniform external pressure using the general theory of shells

In this section, we shall again consider the behavior of a spherical shell under uniform external pressure, but now we shall employ the general theory of shells. For the derivation of the equations of the theory of shells, we refer to the literature.† However, we shall give a short survey of the most important quantities.

Let $\mathbf{r}(x^\alpha)$ $(\alpha \in 1, 2)$ be the position vector from a fixed origin in space to a generic point on the middle surface of the undeformed shell. The tangential base vectors are $\mathbf{a}_\alpha = \mathbf{r}_{,\alpha} = \partial\mathbf{r}/\partial x^\alpha$. The reciprocal base is defined by $\mathbf{a}_\alpha \cdot \mathbf{a}^\beta = \delta_\alpha^\beta$. The covariant and contravariant metric tensors are given by $\mathbf{a}_{\alpha\beta} = \mathbf{a}_\alpha \cdot \mathbf{a}_\beta$ and $\mathbf{a}^{\alpha\beta} = \mathbf{a}^\alpha \cdot \mathbf{a}^\beta$, respectively. The unit normal to the middle surface is $\mathbf{n} = \frac{1}{2}\varepsilon^{\alpha\beta}\mathbf{a}_\alpha \times \mathbf{a}_\beta$ where $\boldsymbol{\varepsilon}^{\alpha\beta}$ is the contravariant alternating tensor. The second fundamental tensor is specified by $b_{\alpha\beta} = \mathbf{n} \cdot \mathbf{r}_{,\alpha\beta}$.

A point in shell space is identified by its distance z to the middle surface and by the surface coordinates of its projection on to the middle surface. The coordinate z is orthonormal to the middle surface. The shell faces $z = \pm\frac{1}{2}h$, where h is the constant shell thickness, are surfaces parallel to the mid-surface.

The covariant components of the spatial metric tensor g_{ij} are specified by

$$\begin{aligned} g_{\alpha\beta} &= a_{\alpha\beta} - 2zb_{\alpha\beta} + z^2 c_{\alpha\beta} \\ g_{13} &= g_{23} = 0, \quad g_{33} = 1, \end{aligned} \tag{3.16.1}$$

where $c_{\alpha\beta} = b_\alpha^\kappa b_{\kappa\beta}$ is the third fundamental tensor. On the mid-surface, we have $g_{\alpha\beta} = a_{\alpha\beta}$. Let g be the determinant of g_{ij} and a be the determinant of $\alpha_{\alpha\beta}$; then

$$\sqrt{\frac{g}{a}} = 1 - 2zH + z^2 K, \tag{3.16.2}$$

where $H = \frac{1}{2}b_\alpha^\alpha$ is the mean curvature in a point (x^1, x^2) of the middle surface and $K = b_1^1 b_2^2 - b_2^1 b_1^2$ is the Gaussian curvature.

The edge of the shell is assumed to be a ruled surface formed by normals to the middle surface along an edge curve on this surface. Let $\boldsymbol{v}$ be the unit vector in the tangent plane, normal to the edge curve and positive outward. The positive sense on the edge curve is defined by the tangential unit vector $t = \mathbf{n} \times \boldsymbol{v}$.

A deformation of the middle surface is described by the two-dimensional displacement field,

$$\mathbf{u}(x^\kappa) = u_\alpha \mathbf{a}^\alpha + w\mathbf{n}. \tag{3.16.3}$$

The strain tensor in the mid-plane is given by

$$\gamma_{\alpha\beta} = \theta_{\alpha\beta} + \frac{1}{2}a^{\kappa\lambda}(\theta_{\kappa\alpha} - \omega_{\kappa\alpha})(\theta_{\lambda\beta} - \omega_{\lambda\beta}) + \frac{1}{2}\varphi_\alpha\varphi_\beta, \tag{3.16.4}$$

† Cf. W. T. Koiter and J. G. Simmonds, Foundations of shell theory. *Proc. 13th IUTAM Congress*, (Springer Verlag, 1972), 150–176 (also WTHD Rep. No. 40).

where $\theta_{\alpha\beta}$ is the linearized strain tensor,

$$\theta_{\alpha\beta} = \frac{1}{2}(u_{\alpha|\beta} + u_{\beta|\alpha}) - b_{\alpha\beta} w \tag{3.16.5}$$

$$\omega_{\alpha\beta} = \frac{1}{2}\left(u_{\beta|\alpha} - u_{\alpha|\beta}\right) \tag{3.16.6}$$

is a rotation in the tangent plane to the shell, and

$$\varphi_\alpha = w_{,\alpha} + b^\kappa_\alpha u_\kappa \tag{3.16.7}$$

is a rotation of the normal vector **n**. The linearized tensor of changes of curvature is

$$\rho_{\alpha\beta} = \frac{1}{2}\left(\varphi_{\alpha|\beta} + \varphi_{\beta|\alpha}\right) - \frac{1}{2}\left(b^\kappa_\alpha \omega_{\kappa\beta} + b^\kappa_\beta \omega_{\kappa\alpha}\right). \tag{3.16.8}$$

After this survey of expressions from shell theory, we return to our stability problem.

According to the general theory of elastic stability (2.6.18), the energy functional for a three-dimensional body is given by

$$P[\mathbf{u}] = \int\limits_V \left(\frac{1}{2} S_{ij} u_{h,i} u_{h,j} + \frac{1}{2} E_{ijk\ell} \gamma_{ij} \gamma_{k\ell}\right) dV, \tag{3.16.9}$$

where **u** denotes the three-dimensional increment of the displacement field, passing from the fundamental state to the adjacent state, and γ_{ij} is the strain tensor

$$\gamma_{ij} = \frac{1}{2}(u_{i,j} + u_{j,i} + u_{h,i} u_{h,j}). \tag{3.16.10}$$

The second variation of this energy is given by

$$P_2[\mathbf{u}] = \int\limits_V \left(\frac{1}{2} S_{ij} u_{h,i} u_{h,j} + \frac{1}{2} E_{ijk\ell} \theta_{ij} \theta_{k\ell}\right) dV, \tag{3.16.11}$$

where θ_{ij} is the linearized strain tensor,

$$\theta_{ij} = \frac{1}{2}(u_{i,j} + u_{j,i}). \tag{3.16.12}$$

In shell theory, the state of stress is assumed to be approximately plain and parallel to the mid-surface. This results in a decoupling of the membrane energy and the bending energy. The energy functional can then be written as

$$P[\mathbf{u}] = \int\limits_A \left[\frac{1}{2} S_{ij} u_{h,i} u_{h,j} + \frac{h}{2} E^{\alpha\beta\lambda\mu}\left(\gamma_{\alpha\beta}\gamma_{\lambda\mu} + \frac{h^2}{12}\rho_{\alpha\beta}\rho_{\lambda\mu}\right)\right] dA, \tag{3.16.13}$$

where the integration is to be carried out over the mid-surface of the shell. Here $E^{\alpha\beta\lambda\mu}$ is the tensor of elastic moduli corresponding to a plate state of stress

$$E^{\alpha\beta\lambda\mu} = \mathcal{G}\left[a^{\alpha\lambda} a^{\beta\mu} + a^{\alpha\mu} a^{\beta\lambda} + \frac{2\upsilon}{1-\upsilon} a^{\alpha\beta} a^{\lambda\mu}\right], \tag{3.16.14}$$

where $\mathcal{G}$ is the shear modulus,

$$\mathcal{G} = \frac{E}{2(1+\upsilon)}. \qquad (3.16.15)$$

The second term in (3.16.13) can now be rewritten to yield

$$\begin{aligned}&\frac{1}{2}hE^{\alpha\beta\lambda\mu}\left(\gamma_{\alpha\beta}\gamma_{\lambda\mu} + \frac{h^2}{12}\rho_{\alpha\beta}\rho_{\lambda\mu}\right) \qquad (3.16.16)\\ &= \frac{Eh}{2(1-\upsilon^2)}\left\{(1-\upsilon)\gamma^{\alpha\beta}\gamma_{\alpha\beta} + \upsilon\,(\gamma^{\kappa}_{\kappa})^2 + \frac{h^2}{12}\left[(1-\upsilon)\rho^{\alpha\beta}\rho_{\alpha\beta} + \upsilon(\rho^{\kappa}_{\kappa})^2\right]\right\}.\end{aligned}$$

Notice that the elastic energy density is expressed in terms of invariants of the strain tensor and the tensor of changes of curvature.

Let $N^{\alpha\beta}$ be the membrane stresses in the fundamental state, due to dead-weight loads. The energy functional can then be written as

$$\begin{aligned}P[\mathbf{u}] = \int_A \Bigg\{&\frac{Eh}{2(1-\upsilon^2)}\left[(1-\upsilon)\gamma^{\alpha\beta}\gamma_{\alpha\beta} + \upsilon(\gamma^{\kappa}_{\kappa})^2\right] + \frac{h^2}{12}\left[(1-\upsilon)\rho^{\alpha\beta}\rho_{\alpha\beta} + \upsilon(\rho^{\kappa}_{\kappa})^2\right]\\ &+ \frac{1}{2}N^{\alpha\beta}\left[a^{\kappa\lambda}\left(\theta_{\kappa\alpha} - \omega_{\kappa\alpha}\right)\left(\theta_{\lambda\beta} - \omega_{\lambda\beta}\right) + \varphi_\alpha\varphi_\beta\right]\Bigg\}\, dA. \qquad (3.16.17)\end{aligned}$$

In the case of uniform pressure p, the fundamental state is nonlinear, because the potential energy $p\Delta V$ is proportional to the volume V.

Let us now consider this functional more closely. Let V_0 be the volume in the undeformed state. The volume in the deformed state is then

$$V = \int_{V_0} \frac{1}{6}\,\varepsilon_{ijk}\,\varepsilon_{pqr}\,y_{i,p}\,y_{j,q}\,y_{k,r}\,dV_0, \qquad (3.16.18)$$

where y_i are the coordinates in the deformed state of a point with coordinates x_i in the undeformed state. Let $\mathbf{u}$ be the displacement field, then

$$y_i = x_i + u_i. \qquad (3.16.19)$$

Expression (3.16.18) can then be written as

$$\begin{aligned}V &= \int_{V_0} \frac{1}{6}\,\varepsilon_{ijk}\,\varepsilon_{pqr}\,(\delta_{ip} + u_{i,p})(\delta_{jq} + u_{j,q})(\delta_{kr} + u_{k,r})\,dV_0\\ &= V_0 + \int_{V_0} \frac{1}{2}\,\varepsilon_{ijk}\,\varepsilon_{pqr}\delta_{ip}\delta_{jq}u_{k,r}\,dV_0 \qquad (3.16.20)\\ &\quad + \int_{V} \frac{1}{2}\,\varepsilon_{ijk}\,\varepsilon_{pqr}\delta_{ip}u_{j,q}u_{k,r}\,dV_0 + \int_{V_0} \frac{1}{6}\,\varepsilon_{ijk}\,\varepsilon_{pqr}u_{i,p}u_{j,q}u_{k,r}\,dV_0.\end{aligned}$$

For the term linear in $\mathbf{u}$, we obtain

$$\begin{aligned}\int_{V_0} \frac{1}{2}\,\varepsilon_{ijk}\,\varepsilon_{pqr}\,\delta_{ip}\,\delta_{jq}u_{k,r}\,dV_0 &= \int_{V_0} \frac{1}{2}\left(\delta_{jq}\delta_{kr} - \delta_{jr}\delta_{kq}\right)\delta_{jq}u_{k,r}\,dV_0\\ &= \int_{V_0} \delta_{kr}u_{k,r}\,dV_0 = \int u_{k,k}\,dV_0 = \int_A u_k n_k\,dA = \int_A w\,dA,\end{aligned}$$

(3.16.21)

where we have used the divergence theorem on a closed surface so that there are no stock terms. The reduction of the term quadratic in **u** is considerably more difficult.

$$\begin{aligned}\int_{V_0} \frac{1}{2}\varepsilon_{ijk}\varepsilon_{pqr}\delta_{ip}u_{j,q}u_{k,r}\,dV_0 &= \int_{V_0}\frac{1}{2}(\delta_{jq}\delta_{kr}-\delta_{jr}\delta_{kq})\,u_{j,q}u_{k,r}\,dV_0 \\ &= \int_{V_0}\frac{1}{2}(u_{j,j}u_{k,k}-u_{j,k}u_{k,j})\,dV_0 \\ &= \int_A \left(\frac{1}{2}u_{j,j}u_k n_k\right)dA - \int_{V_0}\frac{1}{2}u_{j,jk}u_k\,dV_0 \\ &\quad - \int_A \frac{1}{2}u_{j,k}u_k n_j\,dA + \int_{\Gamma_0}\frac{1}{2}u_{j,kj}u_k\,d\Gamma_0 \\ &= \int_A \frac{1}{2}(u_{j,j}n_k - u_{j,k}n_j)\,u_k\,dA.\end{aligned} \tag{3.16.22}$$

Because the integrand is an invariant, it can readily be rewritten in general coordinates, which yields

$$\int_A \frac{1}{2}\left(u^j\,\|_j u^k n_k - u^j\,\|_k u^k n_j\right)dA, \tag{3.16.23}$$

where a subscript preceded by a double vertical bar denotes three-dimensional covariant differentiation with respect to the undeformed metric. In our coordinate system on the shell surface, **n** is perpendicular to the coordinates x^α, so that

$$n_1 = n_2 = 0, \qquad n_3 = -1, \tag{3.16.24}$$

which means that (3.16.23) can be written as

$$\begin{aligned}-\int_A \frac{1}{2}\left(u^j\,\|_j w - w\,\|_k u^k\right)dA &= -\int_A \frac{1}{2}\left(u^\alpha\,\|_\alpha w + w\,\|_3 w - w\,\|_\alpha u^\alpha - w\,\|_3 w\right)dA \\ &= -\int_A \frac{1}{2}\left(u^\alpha\,\|_\alpha w - w\,\|_\alpha u^\alpha\right)dA.\end{aligned} \tag{3.16.25}$$

Now

$$u^i\,\|_\alpha = u^i_{,\alpha} + \Gamma^i_{k\alpha}u^k, \tag{3.16.26}$$

where $\Gamma^i_{k\alpha}$ is a Christoffel symbol of the second kind, so that

$$\begin{aligned}u^\alpha\,\|_\alpha &= u^\alpha_{,\alpha} + \Gamma^\alpha_{\beta\alpha}u^\beta + \Gamma^\alpha_{3\alpha}w \\ w\,\|_\alpha = u^3\,\|_\alpha &= w_{,\alpha} + \Gamma^3_{\beta\alpha}u^\beta + \Gamma^3_{3\alpha}w.\end{aligned} \tag{3.16.27}$$

The Christoffel symbols can be expressed as

$$\Gamma^{\alpha}_{3\alpha} = g^{\alpha k}\Gamma_{k3\alpha} = g^{\alpha\beta}\Gamma_{\beta 3\alpha} + g^{\alpha 3}\Gamma_{33\alpha}$$
$$= g^{\alpha\beta}\frac{1}{2}\left(g_{\beta 3,\alpha} + g_{\beta\alpha,3} - g_{3\alpha,\beta}\right) = -g^{\alpha\beta}b_{\alpha\beta} = -b^{\alpha}_{\alpha}, \tag{3.16.28}$$

where we have made use of (3.16.1),

$$\Gamma^{3}_{\beta\alpha} = g^{3k}\Gamma_{k\beta\alpha} = g^{33}\Gamma_{3\beta\alpha} = \frac{1}{2}\left(g_{3\beta,\alpha} + g_{3\alpha,\beta} - g_{\beta\alpha,3}\right) = b_{\alpha\beta} \tag{3.16.29}$$

$$\Gamma^{3}_{3\alpha} = g^{3k}\Gamma_{k3\alpha} = g^{33}\Gamma_{33\alpha} = \frac{1}{2}\left(g_{33,\alpha} + g_{3\alpha,3} - g_{3\alpha,3}\right) = 0. \tag{3.16.30}$$

Here, Γ_{ijk} are Christoffel symbols of the first kind.

Using the relation

$$u^{\alpha}\left.\right|_{\alpha} = u^{\alpha}_{,\alpha} + \Gamma^{\alpha}_{\beta\alpha}u^{\beta}\,, \tag{3.16.31}$$

which is the covariant derivative of u^{α} with respect to the surface coordinates, we can now rewrite (3.16.25) to yield

$$-\int_A \frac{1}{2}\left[\left(u^{\alpha}\left.\right|_{\alpha} - b^{\alpha}_{\alpha}w\right)w - \left(w_{,\alpha} + b_{\alpha\beta}u^{\beta}\right)u^{\alpha}\right]dA = -\int_A \frac{1}{2}\left(w\theta^{\alpha}_{\alpha} - \varphi_{\alpha}u^{\alpha}\right)\,dA. \tag{3.16.32}$$

The energy of the uniform pressure load is now

$$-p\int_A wA - p\int\left(\frac{1}{2}w\theta^{\alpha}_{\alpha} - \frac{1}{2}\varphi_{\alpha}u^{\alpha}\right)dA - p\int_{V_0}\frac{1}{6}\,\varepsilon_{ijk}\,\varepsilon_{pqr}\,u_{i,p}\,u_{j,q}\,u_{k,r}\,dV_0. \tag{3.16.33}$$

Because $p = O(N/R)$, where N denotes the maximum of the membrane forces and R the minimum radius of curvature, the second term is of order $O(Nw\theta^{\alpha}_{\alpha}/R) = O(Nw^2/R^2)$, whereas the first term is of order $O(Nw/R)$. The potential energy of the membrane stresses in the fundamental state is $O(Nw^2/L^2)$, so that for $L/R \ll 1$ the contribution to the energy of the second term in (3.16.33) may be neglected. The third term, which is even smaller, may then certainly be neglected. It follows that for $L/R \ll 1$ (shallow shell theory), only the first (linear) term must be taken into account, which means that to a first approximation, uniform external pressure may be considered as a dead-weight load. This implies that the energy functional is given by (3.16.17).

We shall now apply these results to a spherical shell under uniform external pressure. However, we shall also take into account the second term in (3.16.33) and show that this term may indeed be neglected. The basic property of a spherical shell is that the first and second fundamental tensors are proportional to each other,

$$b_{\alpha\beta} = \frac{1}{R}a_{\alpha\beta}, \quad b^{\alpha}_{\beta} = \frac{1}{R}a^{\alpha}_{\beta}, \quad b^{\alpha\beta} = \frac{1}{R}a^{\alpha\beta}, \tag{3.16.34}$$

where R is the radius of the middle surface. The sign convention in (3.16.34) implies that the positive direction of $\mathbf{n}$ points inward. The linearized strain tensor (3.16.5) can now be written as

$$\theta_{\alpha\beta} = \frac{1}{2}(u_{\alpha|\beta} + u_{\beta|\alpha}) - \frac{1}{R}a_{\alpha\beta}w. \tag{3.16.35}$$

The rotation of **n** (3.16.7) now reads

$$\varphi_\alpha = w_{,\alpha} + \frac{1}{R}a_\alpha, \tag{3.16.36}$$

and the linearized tensor of changes of curvature (3.16.8) is now given by

$$\rho_{\alpha\beta} = w\,|_{\alpha\beta} + \frac{1}{2R}(u_{\alpha|\beta} + u_{\beta|\alpha}) = w\,|_{\alpha\beta} + a_{\alpha\beta}\frac{w}{R^2} + \frac{1}{R}\theta_{\alpha\beta}. \tag{3.16.37}$$

In the following, we shall make use of the property that the tangential displacement field can always be expressed in terms of the two invariants

$$u_\alpha = \varphi_{,\alpha} + \varepsilon_{\alpha\lambda}\,\psi|^{\lambda}\,.^{\dagger} \tag{3.16.38}$$

To prove this property, we notice that

$$u^\alpha\,|_\alpha = \varphi\,|^\alpha_\alpha + \varepsilon_{\cdot\lambda}^{\alpha}\psi|^{\lambda}_{\cdot\alpha} = \Delta\varphi + \varepsilon^{\alpha\lambda}\psi|_{\lambda\alpha} = \Delta\varphi, \tag{3.16.39}$$

where the term $\varepsilon^{\alpha\lambda}\psi|_{\lambda\alpha}$ vanishes because $\psi|_{\lambda\alpha}$ is symmetric, and Δ is the Laplacian. This is for a known displacement field for an equation with φ, which has a unique solution[‡] for sufficiently smooth displacement fields. To get an equation for ψ, we multiply (3.16.38) by $\varepsilon^{\alpha\mu}$ to obtain

$$u_\alpha\varepsilon^{\alpha\mu} = \varepsilon^{\alpha\mu}\varphi_{,\alpha} + \varepsilon^{\alpha\mu}\varepsilon_{\alpha\lambda}\psi|^{\lambda} = \varepsilon^{\alpha\mu}\varphi_{,\alpha} + \psi|^{\mu}\,,$$

so that

$$\psi_{,\mu} = \varepsilon_{\cdot\mu}^{\alpha}(u_\alpha - \varphi_{,\alpha}). \tag{3.16.40}$$

This equation is integrable when the right-hand side of

$$\psi|_{\mu\upsilon} = \varepsilon_{\cdot\mu}^{\alpha}\,(u_{\alpha|\mu} - \varphi\,|_{\alpha\upsilon}) \tag{3.16.41}$$

is symmetric in μ and υ, so that the left-hand term must also be symmetric, which means

$$\varepsilon^{\mu\upsilon}\,\varepsilon_{\cdot\mu}^{\alpha}\,(u_{\alpha|\upsilon} - \varphi\,|_{\alpha\upsilon}) = u^\alpha\,|_\alpha - \Delta\varphi = 0. \tag{3.16.42}$$

This is exactly the equation for $\Delta\varphi$, and this completes our proof.

As we shall show, the advantage of expressing u_α in terms of φ and ψ is that we obtain a separate equation for ψ in which φ and w do not appear. Using (3.16.38), we can rewrite (3.16.35) to (3.16.37) to yield

$$\theta_{\alpha\beta} = \varphi\,|_{\alpha\beta} + \frac{1}{2}\varepsilon_{\alpha\lambda}\psi|^{\lambda}_{\cdot\beta} + \frac{1}{2}\varepsilon_{\beta\lambda}\psi|^{\lambda}_{\cdot\alpha} - \frac{1}{R}a_{\alpha\beta}w \tag{3.16.43}$$

† Cf. A. van der Neut, *De elastische stabiliteit vaan den dunwandigen bol* (Thesis, Delft; H. J. Paris, Amsterdam, 1932).

‡ Apart from an arbitrary constant, which is unimportant as only derivatives are needed.

$$\varphi_\alpha = w_{,\alpha} + \frac{1}{R}\varphi_{,\alpha} + \frac{1}{R}\varepsilon_{\alpha\lambda}\psi|^{\lambda} \tag{3.16.44}$$

$$\rho_{\alpha\beta} = w\,|_{\alpha\beta} + \frac{1}{R}\varphi\,|_{\alpha\beta} + \frac{1}{2R}\varepsilon_{\alpha\lambda}\psi|_{\cdot\beta}^{\lambda} + \frac{1}{2R}\varepsilon_{\cdot\beta\lambda}\psi|_{\cdot\alpha}^{\lambda} \tag{3.16.45}$$

We now return to our energy functional (3.16.17). The fundamental state of a spherical shell under uniform external pressure p is a membrane state of stress, and the contravariant tensor of stress resultants is

$$N^{\alpha\beta} = \sigma h a^{\alpha\beta} = \frac{1}{2}pRa^{\alpha\beta}. \tag{3.16.46}$$

It will be convenient to introduce a dimensionless load parameter λ, defined by

$$\sigma = \frac{Eh}{cR}\lambda, \qquad c = \sqrt{3(1-\upsilon^2)}. \tag{3.16.47}$$

Also taking into account the second term of (3.16.33), the second variation of our energy functional can now be written as

$$P_2[\mathbf{u};\lambda] = \int_A \left\{\frac{Eh}{2(1-\upsilon^2)}\left[(1-\upsilon)\theta^{\alpha\beta}\theta_{\alpha\beta} + \upsilon(\theta^\kappa_\kappa)^2 + \frac{h^2}{12}\left\{(1-\upsilon)\rho_{\alpha\beta}\rho^{\alpha\beta} + \upsilon(\theta^\kappa_\kappa)^2\right\}\right]\right.$$
$$\left. - \frac{Eh^2}{2cR}\lambda\left[(\theta_{\alpha\beta}-\omega_{\alpha\beta})(\theta^{\alpha\beta}-\omega^{\alpha\beta}) + \varphi_\alpha\varphi^\alpha\right] + \frac{Eh^2}{cR^2}\lambda\left(w\theta^\alpha_\alpha - u^\alpha\varphi_\alpha\right)\right\}dA. \tag{3.16.48}$$

Let us now consider the term

$$\int_A \theta_{\alpha\beta}\theta^{\alpha\beta}\,dA = \int\left(\varphi\,|_{\alpha\beta} + \frac{1}{2}\varepsilon_{\alpha\lambda}\psi|_{\cdot\beta}^{\lambda} + \frac{1}{2}\varepsilon_{\beta\lambda}\psi|_{\cdot\alpha}^{\lambda} - \frac{1}{R}a_{\alpha\beta}w\right)\left(\varphi\,|^{\alpha\beta} + \frac{1}{2}\varepsilon_{\cdot\lambda}^{\alpha}\,w\,|^{\lambda\beta}\right.$$
$$\left. + \frac{1}{2}\varepsilon_{\cdot\lambda}^{\beta}\psi|^{\lambda\alpha} - \frac{1}{R}a^{\alpha\beta}w\right)dA. \tag{3.16.49}$$

We shall now show that the coupling terms between ψ and φ and between ψ and w vanish. The term

$$\int_A a_{\alpha\beta}w\varepsilon_{\cdot\lambda}^{\alpha}\psi|^{\lambda\beta}\,dA = \int_A w\varepsilon_{\beta\lambda}\psi|^{\lambda\beta}\,dA = 0 \tag{3.16.50}$$

because $\psi|^{\lambda\beta}$ is symmetric. Thus,

$$\int \varphi\,|_{\alpha\beta}\varepsilon_{\cdot\lambda}^{\alpha}\psi|^{\lambda\beta}\,dA = \int \varphi\,|_{\alpha\beta}\varepsilon^{\alpha\lambda}\psi|_{\lambda}^{\cdot\beta}\,dA$$
$$= \int \left(\varphi\,|_{\alpha\beta}\varepsilon^{\alpha\lambda}\psi|^{\beta}\right)\Big|_{\lambda}dA - \int \varphi\,|_{\alpha\beta\lambda}\,\varepsilon^{\alpha\lambda}\psi|^{\beta}\,dA.$$

The first of these integrals vanishes after the application of the divergence theorem on the closed surface. Changing the order of covariant differentiation in the remaining integral, we can write

$$-\int_A \varphi\,|_{\alpha\beta\lambda}\varepsilon^{\alpha\lambda}\psi|^{\beta}\,dA = -\int_A \varphi\,|_{\alpha\lambda\beta}\,\varepsilon^{\alpha\lambda}\psi|^{\beta}\,dA - \int_A R^{\kappa}_{\cdot\alpha\beta\lambda}\varphi_{,\kappa}\varepsilon^{\alpha\lambda}\psi|^{\beta}\,dA.$$

Here the first integral on the right-hand side vanishes because $\varphi\,|_{\alpha\lambda\beta}$ is symmetric in α and λ ; $R^{\kappa}_{\cdot\alpha\beta\lambda}$ is the Riemann-Christoffel tensor of the surface, which is related to the Gaussian curvature by

$$R^{\kappa}_{\cdot\alpha\beta\lambda} = K\varepsilon^{\kappa}_{\cdot\alpha}\varepsilon_{\cdot\beta\lambda}. \tag{3.16.51}$$

For a sphere $K = R^{-2}$, we can write

$$\begin{aligned}
-\int_A R^{\kappa}_{\cdot\alpha\beta\lambda}\varphi_{,\kappa}\varepsilon^{\alpha\lambda}\psi|^{\beta}\, dA &= -\frac{1}{R^2}\int_A \varepsilon^{\kappa}_{\cdot\alpha}\varepsilon_{\beta\lambda}\varepsilon^{\alpha\lambda}\varphi_{,\kappa}\psi|^{\alpha}\, dA = -\frac{1}{R^2}\int_A \varepsilon_{\kappa\alpha}\varphi\,|^{\kappa}\,\psi|^{\alpha}\, dA \\
&= -\frac{1}{R^2}\int_A (\varepsilon_{\kappa\alpha}\varphi\,|^{\kappa}\,\psi)|^{\alpha}\, dA + \frac{1}{R^2}\int_A \varepsilon_{\kappa\alpha}\varphi\,|^{\kappa\alpha}\psi dA = 0.
\end{aligned}$$

The first of these integrals vanishes after the application of the divergence theorem on a closed surface, and the second integral vanishes because $\varphi\,|^{\kappa\alpha}$ is symmetric in κ and α.

We may now rewrite (3.16.49) in the form

$$\begin{aligned}
\int_A \theta_{\alpha\beta}\theta^{\alpha\beta}\, dA &= \int_A \Big[\Big(\varphi\,|_{\alpha\beta} - \frac{1}{R}a_{\alpha\beta}w\Big)\Big(\varphi\,|^{\alpha\beta} - \frac{1}{R}a^{\alpha\beta}w\Big) \\
&\quad + \frac{1}{4}\left(\varepsilon_{\alpha\lambda}\psi|^{\lambda}_{\cdot\beta} + \varepsilon_{\beta\lambda}\psi|^{\lambda}_{\cdot\alpha}\right)\left(\varepsilon^{\alpha}_{\cdot\kappa}\psi|^{\kappa\beta} + \varepsilon^{\beta}_{\cdot\kappa}\psi|^{\kappa\alpha}\right)\Big]\, dA \\
&= \int_A \Big(\varphi\,|_{\alpha\beta}\varphi\,|^{\alpha\beta} - \frac{2}{R}w\Delta\varphi + \frac{2}{R^2}w^2\Big)\, dA \\
&\quad + \frac{1}{2}\int_A \left(\psi|^{\lambda}_{\cdot\beta}\,\psi|^{\cdot\beta}_{\lambda} + \varepsilon_{\alpha\lambda}\varepsilon^{\beta}_{\cdot\kappa}\psi|^{\lambda}_{\cdot\beta}\,\psi|^{\kappa\alpha}\right)\, dA.
\end{aligned} \tag{3.16.52}$$

Let us now consider various terms separately,

$$\int_A \varphi\,|_{\alpha\beta}\,\varphi\,|^{\alpha\beta}\, dA = \int_A \left(\varphi\,|^{\alpha\beta}\varphi\,|_{\beta}\right)\Big|_{\alpha}\, dA - \int_A \varphi\,|^{\alpha\beta}_{\cdot\cdot\alpha}\varphi\,|_{\beta}\, dA.$$

Applying the divergence theorem on the first term, this term vanishes for a closed surface. Changing the order of differentiation, in the second term we obtain

$$\begin{aligned}
\int_A \varphi\,|_{\alpha\beta}\,\varphi\,|^{\alpha\beta}\, dA &= \int_A \varphi\,|^{\alpha\beta}_{\cdot\alpha}\varphi\,|_{\beta}\, dA - \int_A \frac{1}{R^2}\varepsilon^{\kappa\alpha}\varepsilon^{\beta}_{\cdot\alpha}\varphi_{,\kappa}\varphi_{,\beta}\, dA \\
&= -\int_A (\Delta\varphi)\,|^{\beta}\varphi\,|_{\beta}\, dA - \frac{1}{R^2}\int_A \varphi\,|^{\kappa}\varphi_{,\kappa}\, dA \\
&= -\int \left((\Delta\varphi)\,|^{\beta}\,\varphi\right)\big|_{\beta} dA + \int \varphi\Delta\Delta\varphi\, dA - \frac{1}{R^2}\int \left(\varphi\,|^{\kappa}\varphi\right)\big|_{\kappa}\, dA \\
&\quad + \frac{1}{R^2}\int \varphi\Delta\varphi\, dA.
\end{aligned}$$

Applying the divergence theorem to the first and the third terms, these terms vanish because we are dealing with a closed surface. Our final result can now be written as

$$\int_A \varphi\,|_{\alpha\beta}\,\varphi\,|^{\alpha\beta}\,dA = \int_A \left(\varphi\Delta\Delta\varphi + \frac{1}{R^2}\varphi\Delta\varphi\right)\,dA, \tag{3.16.53}$$

or in the equivalent form

$$\int_A \varphi\,|_{\alpha\beta}\,\varphi\,|^{\alpha\beta}\,dA = \int_A \left\{(\Delta\varphi)^2 + \frac{1}{R^2}\varphi\Delta\varphi\right\}\,dA. \tag{3.16.54}$$

Similarly,

$$\begin{aligned}
\int_A \psi\big|^{\lambda}_{\cdot\beta}\,\psi\big|^{\cdot\beta}_{\lambda}\,dA &= \int_A \left(\psi\big|^{\lambda}_{\cdot\beta}\,\psi\big|^{\beta}\right)\Big|_{\lambda}\,dA - \int_A \psi\big|^{\lambda}_{\cdot\beta\lambda}\psi\big|^{\beta}\,dA \\
&= -\int_A \psi\big|^{\lambda}_{\cdot\lambda\beta}\psi\big|^{\beta}\,dA - \int_A \frac{1}{R^2}\varepsilon^{\kappa\lambda}\varepsilon_{\beta\lambda}\psi_{,\kappa}\psi\big|^{\beta}\,dA \\
&= -\int_A (\Delta\psi)\big|_{\beta}\psi\big|^{\beta}\,dA - \frac{1}{R^2}\int_A \psi\big|^{\lambda}\,\psi\big|_{\lambda}\,dA \\
&= -\int_A \left((\Delta\psi)\,\psi\big|^{\beta}\right)\Big|_{\beta}\,dA + \int_A (\Delta\psi)^2\,dA \\
&\quad -\frac{1}{R^2}\int_A \left(\psi\big|^{\lambda}\,\psi\right)\Big|_{\lambda}\,dA + \frac{1}{R^2}\int_A \psi\Delta\psi dA \\
&= \int_A \left[(\Delta\psi)^2 + \frac{1}{R^2}\psi\Delta\psi\right]dA \equiv \int_A \left(\psi\Delta\Delta\psi + \frac{1}{R^2}\psi\Delta\psi\right)\,dA.
\end{aligned} \tag{3.16.55}$$

Finally,

$$\begin{aligned}
\int_A \varepsilon_{\alpha\lambda}\varepsilon^{\beta}_{\cdot\kappa}\psi\big|^{\lambda}_{\cdot\beta}\,\psi\big|^{\kappa\alpha}\,dA &= \int_A \varepsilon_{\alpha\lambda}\varepsilon^{\beta\kappa}\psi\big|^{\lambda}_{\cdot\beta}\,\psi\big|^{\alpha}_{\kappa}\,dA \\
&= \int_A \left(\varepsilon_{\alpha\lambda}\varepsilon^{\beta\kappa}\psi\big|^{\lambda}_{\cdot\beta}\psi\big|^{\alpha}\right)\Big|_{\kappa}\,dA - \int_A \varepsilon_{\alpha\lambda}\varepsilon^{\beta\kappa}\psi\big|^{\lambda}_{\beta\kappa}\psi\big|^{\alpha}\,dA \\
&= -\frac{1}{2}\varepsilon_{\alpha\lambda}\varepsilon^{\beta\kappa}\left(\psi\big|^{\lambda}_{\cdot\beta\kappa} - \psi\big|^{\lambda}_{\cdot\kappa\beta}\right)\psi\big|^{\alpha}\,dA.
\end{aligned}$$

Interchanging the order of covariant differentiation in the second term, we obtain

$$\begin{aligned}
\int_A \varepsilon_{\alpha\lambda}\varepsilon^{\beta}_{\cdot\kappa}\psi\big|^{\lambda}_{\cdot\beta}\,\psi\big|^{\kappa\alpha}\,dA &= \frac{1}{2R^2}\int_A \varepsilon_{\alpha\lambda}\varepsilon^{\beta\kappa}\varepsilon^{\mu\lambda}\varepsilon_{\kappa\beta}\psi_{,\mu}\psi\big|^{\alpha}\,dA \\
&= -\frac{1}{2R^2}\int_A 2\delta^{\mu}_{\alpha}\psi\big|^{\alpha}\psi_{,\mu}\,dA = -\frac{1}{R^2}\int_A \psi\big|^{\mu}\psi\big|_{\mu}\,dA \\
&= -\frac{1}{R^2}\int_A \left(\psi\big|^{\mu}\psi\right)\Big|_{\mu}\,dA + \frac{1}{R^2}\int_A \psi\Delta\psi dA = \frac{1}{R^2}\int_A \psi\Delta\psi dA.
\end{aligned} \tag{3.16.56}$$

Using these results, we may rewrite (3.16.52) to yield

$$\int_A \theta_{\alpha\beta}\theta^{\alpha\beta}\, dA = \int_A \left[(\Delta\varphi)^2 + \frac{1}{R^2}\,\varphi\Delta\varphi - \frac{2}{R} w\Delta\varphi + \frac{2}{R^2} w^2 \right.$$

$$\left. + \frac{1}{2}(\Delta\psi)^2 + \frac{1}{2R^2}\psi\Delta\psi + \frac{1}{2R^2}\psi\Delta\psi \right] dA \tag{3.16.57}$$

$$= \int_A \left[(\Delta\varphi)^2 + \frac{1}{R^2}\varphi\Delta\varphi - \frac{2}{R} w\Delta\varphi + \frac{2}{R^2} w^2 + \frac{1}{2}(\Delta\psi)^2 + \frac{1}{R^2}\psi\Delta\psi \right] dA.$$

We now return to (3.16.48) and consider the term

$$\int_A (\theta^{\kappa}_{\kappa})^2\, dA = \int_A \left(\Delta\varphi - \frac{2}{R} w \right)^2 dA, \tag{3.16.58}$$

so that the membrane energy is given by

$$P_2^{(m)}[\mathbf{u}, \lambda] = \frac{Eh}{2(1-\upsilon^2)} \int_A \left[(\Delta\varphi)^2 + \frac{1-\upsilon}{R^2}\,\varphi\Delta\varphi - \frac{2(1+\upsilon)}{R} w\Delta\varphi \right.$$

$$\left. + \frac{2(1+\upsilon)}{R^2} w^2 + \frac{1}{2}(1-\upsilon)\,\Delta\psi\left(\Delta\psi + \frac{2}{R^2}\psi \right) \right] dA. \tag{3.16.59}$$

We shall now proceed with the evaluation of the bending terms in (3.16.48). First, consider

$$\int_A \rho_{\alpha\beta}\rho^{\alpha\beta}\, dA = \int_A \left[w\,|_{\alpha\beta} + \frac{1}{R}\varphi\,|_{\alpha\beta} + \frac{1}{2R}\left(\varepsilon_{\alpha\lambda}\psi|^{\lambda}_{\cdot\beta} + \varepsilon_{\beta\lambda}\psi|^{\lambda}_{\alpha} \right) \right]$$

$$\times \left[w\,|^{\alpha\beta} + \frac{1}{R}\varphi\,|^{\alpha\beta} + \frac{1}{2R}\left(\varepsilon^{\alpha}_{\cdot\kappa}\psi|^{\kappa\beta} + \varepsilon^{\beta}_{\cdot\kappa}\psi|^{\kappa\alpha} \right) \right] dA. \tag{3.16.60}$$

We shall show that here again the coupling terms between ψ and w and ψ and φ vanish. Because w and φ appear in this expression equivalently, it is sufficient to show that the coupling terms between ψ and w vanish. To show this, consider the term

$$\int_A w\,|_{\alpha\beta}\varepsilon^{\alpha}_{\cdot\kappa}\psi|^{\kappa\beta}\, dA = \int_A w\,|_{\alpha\beta}\varepsilon^{\alpha\kappa}\,\psi|^{\beta}_{\cdot\kappa}\, dA = \int_A \left(\varepsilon^{\alpha\kappa}\psi|^{\beta}_{\cdot\kappa} w_{,\alpha} \right)\Big|_{\beta} dA - \int_A \varepsilon^{\alpha\kappa}\psi\Big|^{\cdot\beta}_{\kappa\beta} w_{,\alpha}\, dA.$$

Applying the divergence theorem to the first integral, this term vanishes on a closed surface. Changing the order of covariant differentiation in the second term, we obtain

$$\int_A w\,|_{\alpha\beta}\varepsilon^{\alpha}_{\cdot\kappa}\psi|^{\kappa\beta}\, dA = -\int_A \varepsilon^{\alpha\kappa}\psi\Big|^{\cdot\beta}_{\beta\cdot\kappa} w_{,\alpha}\, dA - \frac{1}{R^2}\int_A \varepsilon^{\alpha\kappa}\varepsilon^{\lambda\beta}\varepsilon_{\kappa\beta}\psi_{,\lambda} w_{,\alpha}\, dA$$

$$= -\int_A \left(\varepsilon^{\alpha\kappa}\Delta\psi w_{,\alpha} \right)\Big|_{\kappa} dA + \int_A \varepsilon^{\alpha\kappa}\Delta\psi w\,|_{\alpha\kappa}\, dA - \frac{1}{R^2}\int_A \varepsilon^{\alpha\kappa}\psi_{,\kappa} w_{,\alpha}\, dA.$$

Applying the divergence theorem to the first of these integrals, this term vanishes because we are dealing with a closed surface. The second integral vanishes because $w\,|_{\alpha\kappa}$ is symmetric in α and κ. Hence, we are left only with the third term, which can be rewritten as

$$-\frac{1}{R^2}\int_A \left(\varepsilon^{\alpha\kappa}\psi w_{,\alpha}\right)\Big|_\kappa dA + \frac{1}{R^2}\int_A \varepsilon^{\alpha\kappa}\psi w\,|_{\alpha\kappa}\; dA = 0,$$

where the first term vanishes on a closed surface and the second term vanishes due to symmetry in α and κ of $w\,|_{\alpha\kappa}$. Our result is thus

$$\int_A w\,|_{\alpha\beta}\,\varepsilon^{\alpha}_{\cdot\kappa}\psi\,|^{\kappa\beta}\; dA = 0. \tag{3.16.61}$$

This means that (3.16.60) can be rewritten in the form

$$\begin{aligned}\int_A \rho_{\alpha\beta}\rho^{\alpha\beta}\,dA &= \int_A \Bigg[\left(w\,|_{\alpha\beta} + \frac{1}{R}\varphi\,|_{\alpha\beta}\right)\left(w\,|^{\alpha\beta} + \frac{1}{R}\varphi\,|^{\alpha\beta}\right)\\ &\quad + \frac{1}{4R^2}\left(\varepsilon_{\alpha\lambda}\psi|^{\lambda}_{\cdot\beta} + \varepsilon_{\beta\lambda}\psi|^{\lambda}_{\cdot\alpha}\right)\left(\varepsilon^{\alpha}_{\cdot\kappa}\psi|^{\kappa\beta} + \varepsilon^{\beta}_{\kappa}\psi|^{\kappa\alpha}\right)\Bigg]\; dA\\ &= \int_A \Big[w\,|_{\alpha\beta}\, w\,|^{\alpha\beta} + \frac{2}{R}w\,|_{\alpha\beta}\,\varphi\,|^{\alpha\beta} + \frac{1}{R^2}\varphi\,|_{\alpha\beta}\,\varphi\,|^{\alpha\beta}\\ &\quad + \frac{1}{2R^2}\left(\psi|^{\lambda}_{\cdot\beta}\,\psi|^{\cdot\beta}_{\lambda} + \varepsilon_{\alpha\lambda}\varepsilon^{\beta}_{\cdot\kappa}\psi|^{\lambda}_{\cdot\beta}\,\psi|^{\kappa\alpha}\right)\Big]\; dA,\end{aligned} \tag{3.16.62}$$

where the terms involving ψ have been evaluated in (3.16.55) and (3.16.56). The term containing only φ is given in (3.16.53) or (3.16.54), which are obviously also valid when φ is replaced by w.

Completely analogously, we obtain

$$\int_A w\,|_{\alpha\beta} w\,|^{\alpha\beta}\; dA = \int_A \Delta\varphi\left(\Delta w + \frac{w}{R^2}\right) dA = \int_A \Delta w\left(\Delta\varphi + \frac{\varphi}{R^2}\right) dA, \tag{3.16.63}$$

so that (3.16.62) can be rewritten to yield

$$\begin{aligned}\int_A \rho_{\alpha\beta}\rho^{\alpha\beta}\,dA &= \int_A \Bigg[\Delta w\left(\Delta w + \frac{w}{R^2}\right) + \frac{1}{R}\Delta w\left(\Delta\varphi + \frac{\varphi}{R^2}\right)\\ &\quad + \frac{1}{R}\Delta\varphi\left(\Delta w + \frac{w}{R^2}\right) + \frac{1}{R^2}\Delta\varphi\left(\Delta\varphi + \frac{\varphi}{R^2}\right)\\ &\quad + \frac{1}{2R^2}\Delta\psi\left(\Delta\psi + \frac{\psi}{R^2}\right) + \frac{1}{2R^4}\psi\Delta\psi\Bigg]\; dA.\end{aligned} \tag{3.16.64}$$

We now return to (3.16.48) and consider the term

$$\int_A (\rho^{\kappa}_{\kappa})^2\, dA = \int_A \left(\Delta w + \frac{1}{R}\Delta\varphi\right)^2 dA, \tag{3.16.65}$$

so that the bending energy is given by

$$P_2^{(b)}[\mathbf{u},\lambda] = \frac{Eh^2}{2(1-\upsilon^2)}\frac{h^2}{12}\int_A \left[\frac{(1-\upsilon)}{2R^2}\Delta\psi\left(\Delta\psi+\frac{2\psi}{R^2}\right)\right.$$
$$\left. + \left(\Delta w + \frac{1}{R}\Delta\varphi\right)^2 + (1-\upsilon)\left(\Delta w + \frac{1}{R}\Delta\varphi\right)\left(\frac{w}{R^2}+\frac{\varphi}{R^3}\right)\right] dA. \tag{3.16.66}$$

Next, we must determine the contribution $P_2[\mathbf{u},\lambda]$ of the stresses in the fundamental state. From (3.16.48), consider the integral

$$\int_A \left[(\theta_{\alpha\beta}-\omega_{\alpha\beta})(\theta^{\alpha\beta}-\omega^{\alpha\beta}) + \varphi_\alpha\varphi^\alpha\right] dA. \tag{3.16.67}$$

From (3.16.5) and (3.16.6), we obtain

$$\theta_{\alpha\beta}-\omega_{\alpha\beta} = u_{\alpha|\beta} - b_{\alpha\beta}w = u_{\alpha|\beta} - \frac{1}{R}a_{\alpha\beta}w, \tag{3.16.68}$$

and using (3.16.38), we can rewrite this expression in the form

$$\theta_{\alpha\beta}-\omega_{\alpha\beta} = \varphi\,|_{\alpha\beta} + \varepsilon_{\alpha\lambda}\psi|^{\lambda}_{\cdot\beta} - \frac{1}{R}a_{\alpha\beta}w. \tag{3.16.69}$$

The integral (3.16.67) can now be written as

$$\int_A \left[\left(\varphi\,|_{\alpha\beta} + \varepsilon_{\alpha\lambda}\psi|^{\lambda}_{\cdot\beta} - \frac{1}{R}a_{\alpha\beta}w\right)\left(\varphi\,|^{\alpha\beta} + \varepsilon^{\alpha}_{\cdot\kappa}\psi|^{\kappa\beta} - \frac{1}{R}a^{\alpha\beta}w\right)\right.$$
$$\left. + \left(w_{,\alpha} + \frac{1}{R}\varphi_{,\alpha} + \frac{1}{R}\varepsilon_{\alpha\lambda}\psi|^{\lambda}\right)\left(w\left|^{\alpha} + \frac{1}{R}\varphi\right|^{\alpha} + \frac{1}{R}\varepsilon^{\alpha}_{\cdot\kappa}\psi|^{\kappa}\right)\right] dA, \tag{3.16.70}$$

where we have used the relation (3.16.44).

By arguments similar to those used for the evaluation of the membrane terms and the bending terms, it follows that the coupling terms between ψ and φ and ψ and w vanish, so we can rewrite this integral in the form

$$\int_A \left[\left(\varphi\,|_{\alpha\beta} - \frac{1}{R}a_{\alpha\beta}w\right)\left(\varphi\,|^{\alpha\beta} - \frac{1}{R}a^{\alpha\beta}w\right) + \left(w\,|_\alpha + \frac{1}{R}\varphi\,|_\alpha\right)\left(w\,|^\alpha + \frac{1}{R}\varphi\,|^\alpha\right)\right.$$
$$\left. + \psi|^{\kappa}_{\cdot\beta}\psi|^{\cdot\beta}_{\kappa} + \frac{1}{R^2}\psi|^{\lambda}\,\psi|_{\lambda}\right] dA, \tag{3.16.71}$$

Consider

$$\int_A \left(\varphi\,|_{\alpha\beta}\,\varphi\,|^{\alpha\beta} - \frac{2w}{R}\varphi\,|^{\alpha}_{\alpha} + \frac{2}{R^2}w^2\right) dA = \int_A \left[(\Delta\varphi)\left(\Delta\varphi + \frac{1}{R^2}\varphi\right) - \frac{2w}{R}\Delta\varphi + \frac{2}{R^2}w^2\right] dA,$$

where we have made use of (3.16.54). For the second term in (3.16.71), we obtain

$$\int_A \left(w\,|_\alpha w\,|^\alpha + \frac{2}{R}w\,|_\alpha\varphi\,|^\alpha + \frac{1}{R^2}\varphi\,|^\alpha\,\varphi\,|_\alpha\right) dA$$
$$= \int_A \left[(w\,|_\alpha\, w)\big|_\alpha - w\Delta w + \frac{2}{R}(\varphi w\,|^\alpha)\big|_\alpha - \frac{2}{R}w\Delta\varphi + \frac{1}{R^2}(\varphi\,|^\alpha\,\varphi)\big|_\alpha - \frac{1}{R^2}\varphi\Delta\varphi\right] dA$$

$$= -\int_A \left[w\Delta w + \frac{2}{R} w\Delta\varphi + \frac{1}{R^2}\varphi\Delta\varphi \right] dA.$$

For the last two terms in (3.16.71), we obtain

$$\int_A \left(\psi|^{\kappa}_{\cdot\beta}\, \psi|^{\cdot\beta}_{\kappa} + \frac{1}{R^2}\psi|^{\lambda}\psi|_{\lambda} \right) dA = \int_A \left[(\Delta\psi)\left(\Delta\psi + \frac{1}{R^2}\psi \right) - \frac{1}{R^2}\psi\Delta\psi \right] dA$$
$$= \int_A (\Delta\psi)^2 \, dA,$$

where we have made use of (3.16.55). The contribution of the stresses in the fundamental state to the energy functional is thus given by

$$\begin{aligned} P_2^{(s)}[\mathbf{u};\lambda] &= \frac{Eh^2}{2cR}\lambda \int_A \Big[(\Delta\varphi)^2 + \frac{1}{R^2}\,\varphi\Delta\varphi - \frac{2w}{R}\Delta\varphi + \frac{2}{R^2}w^2 \\ &\quad - w\Delta w - \frac{2}{R}w\Delta\varphi - \frac{1}{R^2}\varphi\Delta\varphi + (\Delta\psi)^2 \Big]\, dA \\ &= -\frac{Eh^2}{2cR}\lambda^2 \int_A \left[(\Delta\varphi)^2 - \frac{4w}{R}\Delta\varphi - w\Delta w + \frac{2}{R^2}w^2 + (\Delta\psi)^2 \right] dA. \end{aligned} \tag{3.16.72}$$

Finally, we shall consider the last term in (3.16.48),

$$\begin{aligned} \int_A \left[w\theta^{\alpha}_{\alpha} - u^{\alpha}\varphi_{\alpha} \right] dA &= \int_A \Big[w\left(\Delta\varphi - \frac{2}{R}w \right) \\ &\quad - \left(w\Big|^{\alpha} + \frac{1}{R}\varphi\Big|^{\alpha} + \frac{1}{R}\varepsilon^{\alpha}_{\cdot\lambda}\psi\Big|^{\lambda} \right)\left(\varphi_{,\alpha} + \varepsilon_{\alpha\kappa}\psi|^{\kappa} \right) \Big]\, dA \\ &= \int_A \left[w\Delta\varphi - \frac{2}{R}w^2 - \left(w\,|^{\alpha} + \frac{1}{R}\varphi\,|^{\alpha} \right)\varphi\,|_{\alpha} - \frac{1}{R}\psi|_{\kappa}\psi|^{\kappa} \right] dA \\ &= \int_A \left[w\Delta\varphi - \frac{2}{R}w^2 + w\Delta\varphi + \frac{1}{R}\varphi\Delta\varphi + \frac{1}{R}\psi\Delta\psi \right] dA. \end{aligned} \tag{3.16.73}$$

The second variation of the potential energy is now given by

$$\begin{aligned} P_2[\mathbf{u};\lambda] &= \frac{Eh}{2(1-\upsilon^2)} \int_A \Big[(\Delta\varphi)^2 + \frac{1-\upsilon}{R^2}\,\varphi\Delta\varphi - \frac{2(1+\upsilon)}{R}w\Delta\varphi \\ &\quad + \frac{2(1+\upsilon)}{R^2}w^2 + \frac{1}{2}(1-\upsilon)\,\Delta\psi\left(\Delta\psi + \frac{2}{R^2}\varphi \right) \\ &\quad + \frac{h^2}{12}\Big\{ \frac{1-\upsilon}{2R^2}\,\Delta\psi\left(\Delta\psi + \frac{2\psi}{R^2} \right) + \left(\Delta w + \frac{1}{R}\Delta\varphi \right)^2 \\ &\quad + (1-\upsilon)\left(\Delta w + \frac{1}{R}\Delta\varphi \right)\left(\frac{w}{R^2} + \frac{\varphi}{R^3} \right) \Big\} \\ &\quad - (1-\upsilon^2)\,\frac{\sigma}{E}\left\{ \underline{\underline{(\Delta\varphi)^2} + \frac{2}{R^2}\varphi\Delta\varphi - w\Delta w - \frac{2}{R^2}w^2 + \underline{\underline{\Delta\psi\left(\Delta\psi + \frac{2}{R}\psi \right)}}} \right\} \Big]\, dA. \end{aligned} \tag{3.16.74}$$

To be in the elastic range, we must have $\sigma/E <<< 1\ (\sigma/E = O(10^{-3}))$ for most engineering materials, which means that the underlined terms in (3.16.74) may be neglected, implying a relative error of $O(\sigma/E)$ compared to unity. Further, we may neglect the term with ψ in the third line compared to that in the second line, which introduces a relative error of $O(h^2/R^2)$, which is smaller than the error in shell theory. Similarly, we may neglect the term with $(\Delta\varphi)^2$ in the third line compared to the corresponding term in the first line, again with a relative error of $O(h^2/R^2)$. To show that the term $2R^{-1}h^2\Delta w\Delta\varphi$ in the third line may also be neglected, we employ the inequality

$$\frac{2h}{R}\int_A h\Delta w\Delta\varphi\, dA \leq \frac{h}{R}\int_A \left\{(h\Delta w)^2 + (\Delta\varphi)^2\right\} dA, \tag{3.16.75}$$

which shows that this term may be neglected in comparison with the sum of the terms $h^2(\Delta w)^2$ in the third line and $(\Delta\varphi)^2$ in the first line, introducing a relative error of $O(h/R)$, which is consistent with the error inherent to the shell equations. However, we shall accept this error. To discuss the term $h^2 w\Delta w/R^2$ in the fourth line, we consider the inequality

$$2\int_A \frac{h^2}{R^2} w\Delta w\, dA \leq \frac{h}{R}\int_A \left\{\frac{w^2}{R^2} + h^2(\Delta w)^2\right\} dA, \tag{3.16.76}$$

which shows that this term may be neglected compared to the sum of the terms with w^2/R^2 in the second line and $h^2(\Delta w)^2$ in the third line, again with a relative error of $O(h/R)$. The terms $(h^2/R^2)\, w\Delta\varphi/R$ and $(h^2/R^2)\,\varphi\Delta\varphi/R^2$ in the fourth line may be neglected compared to the terms $w\Delta\varphi/R$ and $\varphi\Delta\varphi/R^2$ in the first line with a relative error of $O(h^2/R^2)$. Using the relation

$$\int_A \varphi\Delta w\, dA = \int_A w\Delta\varphi\, dA, \qquad \forall\varphi, w|w, \varphi \in C^2, \tag{3.16.77}$$

it follows that the term $h^2\varphi\Delta w/R^3$ in the fourth line may also be neglected with a relative error of $O(h^2/R^2)$. This means that the second variation (3.16.74) can be simplified to yield

$$\begin{aligned} P_2[\mathbf{u};\lambda] = \frac{Eh}{2(1-\upsilon^2)}\int_A &\Big[(\Delta\varphi)^2 + \frac{1-\upsilon}{R^2}\varphi\Delta\varphi - \frac{2(1+\upsilon)}{R}w\Delta\varphi \\ &+ \frac{2(1+\upsilon)}{R^2}w^2 + \frac{1}{2}(1-\upsilon)\Delta\psi\Big(\Delta\psi + \frac{2}{R^2}\psi\Big) \\ &+ \frac{h^2}{12}(\Delta w)^2 + (1-\upsilon^2)\frac{\sigma}{E}w\Delta w\Big]\, dA\left[1 + O\left(\frac{h}{R}\right)\right]. \end{aligned} \tag{3.16.78}$$

A necessary condition for stability is that $P_2[\mathbf{u};\lambda]$ is positive-definite. Because ψ appears uncoupled from φ and w, we shall first consider the contribution of the terms with ψ, i.e., we consider

$$\int_A \Delta\psi\left(\Delta\psi + \frac{2}{R^2}\psi\right)\, dA. \tag{3.16.79}$$

Any continuously differentiable function on a spherical surface can be expanded in a series of spherical surface harmonics. Hence, we can write

$$\psi = \sum_{n=0}^{\infty} C_n S_n\left(x^\alpha\right), \tag{3.16.80}$$

where $S_n\left(x^\alpha\right)$ is a spherical surface harmonic of integral degree n, characterized by the differential equation

$$\Delta S_n\left(x^\alpha\right) = -\frac{n\left(n+1\right)}{R^2} S_n\left(x^\alpha\right). \tag{3.16.81}$$

$S_n\left(x^\alpha\right)$ can be represented by

$$S_n\left(x^\alpha\right) = a_0 P_n\left(\cos\theta\right) + \sum_{m=1}^{n}\left(a_m \cos m\beta + b_m \sin m\beta\right) P_n^m\left(\cos\theta\right), \tag{3.16.82}$$

where θ and β are surface coordinates (see Figure 3.16.1).

$P_n\left(\cos\theta\right)$ is a Legendre polynomial of degree n, and $P_n^m\left(\cos\theta\right)$ is a Legendre function of the first kind, of degree n and order m. Further, $S_0\left(x^\alpha\right) = \text{constant}$.

Let n_1 and n_2 be two distinct integral numbers, then spherical surface harmonics of degree n_1 and n_2, respectively, satisfy the orthogonality relation

$$\int_A S_{n_1}\left(x^\alpha\right) S_{n_2}\left(x^\alpha\right)\, dA = 0. \tag{3.16.83}$$

Using this relation and (3.16.82), we can rewrite (3.16.79) to yield

$$\frac{1}{R^4}\int_A \sum_{n=0}^{\infty}\left\{\left[n\left(n+1\right)-1\right]^2 - 1\right\} C_n^2 S_n^2\, dA. \tag{3.16.84}$$

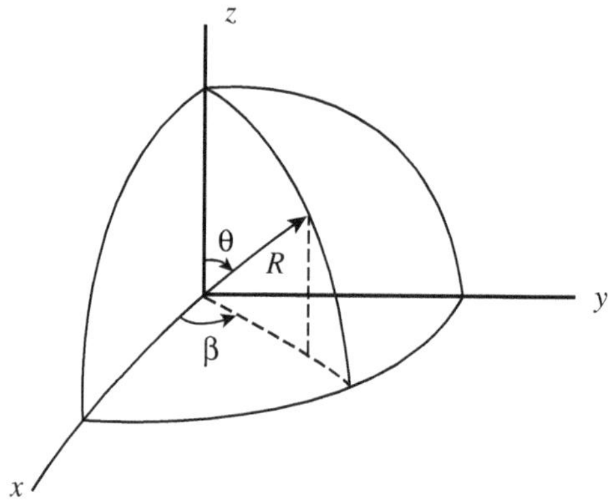

Figure 3.16.1

This expression vanishes for $n = 0$ and $n = 1$ and is positive for $n > 1$. Because $n = 1$ implies a rigid body motion, we are only interested in values of $n > 1$. This means that the terms with ψ are positive-definite, so a minimum of $P_2\,[\mathbf{u};\lambda]$ can only be obtained when the terms with ψ vanish, i.e.,

$$\Delta\psi = 0 \quad \text{or} \quad \Delta\psi + \frac{2}{R^2}\psi = 0. \qquad (3.16.85)$$

Later, we shall see that only the first of these equations holds. From (3.16.6), the rotation in the tangent plane is

$$\Omega = \frac{1}{2}\varepsilon^{\alpha\beta}\omega_{\alpha\beta} = \frac{1}{2}\varepsilon^{\alpha\beta}\left(\varphi\,|_{\beta\alpha} + \varepsilon_{\beta\lambda}\psi|_{\alpha}^{\lambda} - \varphi\,|_{\alpha\beta} - \varepsilon_{\alpha\lambda}\psi|_{\beta}^{\lambda}\right) = -\frac{1}{2}\psi|_{\lambda}^{\lambda} = -\frac{1}{2}\Delta\psi, \qquad (3.16.86)$$

so that the first of the equations of (3.16.86) means that the rotation Ω around the normal vanishes identically in all buckling modes.

We may now restrict our discussion to the functional

$$P_2\,[\mathbf{u},\omega;\lambda] = \frac{Eh}{2\left(1-\upsilon^2\right)}\int_A \left[(\Delta\varphi)^2 + \frac{1-\upsilon}{R^2}\,\varphi\Delta\varphi - \frac{2\,(1+\upsilon)}{R} w\Delta\varphi \right.$$
$$\left. + \frac{2\,(1+\upsilon)}{R^2} w^2 + \frac{h^2}{12}\,(\Delta w)^2 + \left(1-\upsilon^2\right)\frac{\sigma}{E} w\Delta w \right] dA. \qquad (3.16.87)$$

A necessary condition for a minimum value of this functional is that its variation with respect to φ and w vanishes. For the reduction of the various terms, we need the following results:

$$\int_A \Delta\varphi\Delta\delta\varphi\, dA = \int_A \Delta\varphi\delta\varphi\,|_{\alpha}^{\alpha}\; dA = -\int_A \Delta\varphi\,|_{\alpha}\; \delta\varphi\,|^{\alpha}\; dA$$
$$= -\int_A \Delta\varphi\,|^{\alpha}\; \delta\varphi\,|_{\alpha}\; dA = \int_A \Delta\varphi\,|_{\alpha}^{\alpha}\; \delta\varphi\, dA = \int_A \Delta\Delta\varphi\delta\varphi\, dA \qquad (3.16.88)$$
$$\textit{delta} \int_A w\Delta\varphi\, dA = \int_A \Delta\varphi\delta w\; dA + \int_A w\delta\varphi\,|_{\alpha}^{\alpha}\; dA = \int_A \Delta\varphi\delta w\; dA + \int_A \Delta w\delta\varphi\, dA.$$

Our result can now be written as

$$\int_A \left[2\Delta\Delta\varphi\,\delta\varphi + \frac{2\,(1-\upsilon)}{R^2}\,\Delta\varphi\delta\varphi - \frac{2\,(1+\upsilon)}{R}\Delta\varphi\delta w \right.$$
$$- \frac{2\,(1+\upsilon)}{R}\Delta w\delta\varphi + \frac{2\,(1+\upsilon)}{R^2} w\delta w + \frac{2h^2}{12}\Delta\Delta w\delta w \qquad (3.16.89)$$
$$\left. + 2\left(1-\upsilon^2\right)\frac{\sigma}{E}\Delta w\delta w \right] dA = 0, \quad \forall\delta\varphi,\ \delta w,$$

which yields the following equations of neutral equilibrium,

$$\Delta\Delta\varphi + \frac{1-\upsilon}{R^2}\,\Delta\varphi - \frac{1+\upsilon}{R}\,\Delta w = 0$$
$$\frac{h^2}{12}\,\Delta\Delta w + \left(1-\upsilon^2\right)\frac{\sigma}{E}\,\Delta w + \frac{(1+\upsilon)}{R^2}\,w - \frac{(1+\upsilon)}{R}\,\Delta\varphi = 0. \qquad (3.16.90)$$

Expand φ and w in a series of spherical surface harmonics,

$$\varphi = R\sum_{n=0}^{\infty} D_n S_n, \quad w = \sum_{n=0}^{\infty} C_n S_n. \tag{3.16.91}$$

Assuming that these series can be differentiated term-wise four times and using the relation (3.16.81), we obtain from (3.16.90) the following equations:

$$\begin{aligned} &\left\{[n(n+1)]^2 - (1-\upsilon)\,n(n+1)\right\} D_n + (1-\upsilon)\,n(n+1)\,C_n = 0 \\ &(1-\upsilon)\,n(n+1)\,D_n + \left\{\frac{h^2}{12R^2}[n(n+1)]^2 - \left(1-\upsilon^2\right)\frac{\sigma}{E}n(n+1)\right\} C_n = 0. \end{aligned} \tag{3.16.92}$$

The condition for a non-trivial solution is

$$\begin{aligned} &\left(1-\upsilon^2\right)[n(n+1)-2] + [n(n+1)-(1-\upsilon)] \\ &\times\left[\frac{h^2}{12R^2}n^2(n+1)^2 - \left(1-\upsilon^2\right)\frac{\sigma}{E}n(n+1)\right] = 0, \end{aligned} \tag{3.16.93}$$

which yields

$$\frac{\sigma}{E} = \frac{n(n+1)-2}{n(n+1)[n(n+1)-(1-\upsilon)]} + \frac{h^2}{12R^2}\frac{n(n+1)}{1-\upsilon^2}. \tag{3.16.94}$$

For values of n of $O(1)$, the first term on the right-hand side is of order unity, which means $\sigma/E = O(1)$. Because we must stay in the elastic range, we require $\sigma/E \ll 1$, which is only possible for $n \gg 1$. This means that (3.16.94) can be simplified to yield

$$\frac{\sigma}{E} = \frac{1}{n(n+1)} + \frac{h^2}{12R^2}\frac{n(n+1)}{1-\upsilon^2}. \tag{3.16.95}$$

It follows that because $n \gg 1$, the wavelength L of the deformation pattern is small, i.e., $L/R \ll 1$, which means that we can restrict ourselves to shallow shell theory. The functional (3.16.78) can now further be simplified to yield

$$\begin{aligned} P_2[\mathbf{u};\lambda] = \frac{Eh}{2\left(1-\upsilon^2\right)}\int_A &\left[(\Delta\varphi)^2 - 2(1+\upsilon)\frac{w}{R}\Delta\varphi + \frac{1}{2}(1-\upsilon)(\Delta\psi)^2\right. \\ &\left.+\frac{h^2}{12}(\Delta w)^2 + \left(1-\upsilon^2\right)\frac{\sigma}{E}w\Delta w\right] dA\left[1+O\left(\frac{h}{R}+\frac{L^2}{R^2}\right)\right]. \end{aligned} \tag{3.16.96}$$

Omitting the error term, we rewrite this functional in the form

$$\begin{aligned} P_2[\mathbf{u};\lambda] = \frac{Eh}{2\left(1-\upsilon^2\right)}\int_A &\left\{\left[\Delta\varphi - (1+\upsilon)\frac{w}{R}\right]^2 + \frac{1}{2}(1-\upsilon)(\Delta\psi)^2\right. \\ &\left.-\left(1+\upsilon^2\right)\frac{w^2}{R^2} + \frac{h^2}{12}(\Delta w)^2 + \left(1-\upsilon^2\right)\frac{\sigma}{E}w\Delta w\right\} dA. \end{aligned} \tag{3.16.97}$$

Notice that for the present simplified functional, we did not need any *a priori* assumptions with respect to the wavelength of the deformation pattern. The condition that $L/R \ll 1$ follows from the condition that $\sigma/E \ll 1$ to stay in the elastic range.

In our present functional, we only have quadratic terms with positive constants, except for the load term. For the functional to become semi-positive-definite, the term with ψ must vanish, which leads to the first of the equations of (3.16.85), and the term with φ must also vanish, which yields the equation

$$\Delta\varphi = (1 + \upsilon)\,\frac{w}{R}. \tag{3.16.98}$$

This admits of a single-valued and non-singular solution for the potential φ if the integral of the normal deflection w over the entire surface of the shell vanishes for all buckling modes. To show this, apply the operator $\int_A ()\, dA$ to both sides of this equation. The left-hand member then vanishes (divergence theorem), so the right-hand side must also vanish. This condition is satisfied because $w = \sum_n^\infty C_n S_n\left(x^\alpha\right)$ and $\int_A S_n dA = 0$ $(n \geq 1)$. Apart from irrelevant arbitrary constants, the solution to (3.16.98) is

$$\varphi\left(x^\alpha\right) = -\sum_n^\infty \frac{1+\upsilon}{n\,(n+1)} R C_n S_n\left(x^\alpha\right). \tag{3.16.99}$$

Let us now return to the functional (3.16.97). The remaining terms contain only w, so that a stationary value of $P_2\left[\mathbf{u};\lambda\right]$ is obtained by putting the variation with respect to w equal to zero. Using (3.16.91), the condition for a non-trivial solution yields again the expression for σ/E given in (3.16.96). To obtain the critical value of σ/E, we minimize this expression with respect to $n\,(n+1)$, which yields (for a continuous variable n^*)

$$n^*\left(n^*+1\right) = \frac{2cR}{h}, \tag{3.16.100}$$

so that

$$\sigma_{cr} = \frac{Eh}{cR}. \tag{3.16.101}$$

Notice that this is in complete agreement with our earlier result of (3.15.30).

The positive root n^* of (3.16.100) is in general non-integral. However, the nearest integers do not differ more from n^* than $O\left(n^{*-1}\right)$, which means a relative error of $O\,(h/R)$, and which is negligible within the framework of shell theory. The critical load factor λ_1 is therefore always equal to unity within the limits of accuracy of shell theory. The associated buckling mode is any spherical surface harmonic of integral degree $[n^*]$ or $[n^*]+1$ associated with the integer that yields the smallest value of the right-hand member of (3.16.95). The corresponding buckling mode is

$$w_n = ah\, S_n\left(x^\alpha\right), \tag{3.16.102}$$

where A is the amplitude factor and $S_n\left(x^\alpha\right)$ is given in (3.16.82). This representation shows that at the same value of the load factor λ_1, there are $2n+1$ independent buckling modes.

In the special case that R/h is such that (3.16.100) yields two integer values for n^*, the right-hand side of (3.16.95) is the same for both values. In that case, the corresponding buckling mode is given by

$$w_n = ah\, S_n(x^\alpha) + a^* h\, S_{n-1}(x^\alpha), \tag{3.16.103}$$

so that in this case, we have $4n$ independent buckling modes. Further, we have

$$\psi = 0, \quad \forall n. \tag{3.16.104}$$

A necessary condition for stability at the critical bifurcation load is that the third variation vanishes identically for any linear combination of buckling modes. Returning to (3.16.17) and remembering that $\gamma_{\alpha\beta} = \theta_{\alpha\beta} + \frac{1}{2} w_{,\alpha} w_{,\beta}$, we see that the only contribution to $P_3[\mathbf{u}]$ originates from the membrane strains,

$$\begin{aligned} P_3[\mathbf{u}] &= \frac{Eh}{2(1-\upsilon^2)} \int_A \left[(1-\upsilon)\theta_{\alpha\beta} w|^\alpha w|^\beta + \upsilon \theta^\alpha_\alpha w_{,\kappa} w|^\kappa \right] dA \\ &= \frac{Eh}{2(1-\upsilon^2)} \int_A \left\{ (1-\upsilon)\left[\frac{1}{2}(u_{\alpha|\beta} + u_{\beta|\alpha}) - a_{\alpha\beta}\frac{w}{R} \right] w|^\alpha w|^\beta \right. \\ &\quad \left. + \upsilon (u^\alpha|_\alpha - 2\frac{w}{R}) w_{,\kappa} w|^\kappa \right\} dA. \end{aligned} \tag{3.16.105}$$

For the evaluation of this functional, we recall that

$$\varphi = -\frac{(1-\upsilon)R}{n(n+1)} w, \quad \psi = 0, \quad u_\alpha = \varphi_{,\alpha} = -\frac{(1-\upsilon)R}{n(n+1)} w_{,\alpha}, \tag{3.16.106}$$

so the functional can be rewritten to yield

$$\begin{aligned} P_3[a_h \mathbf{u}_h] &= \frac{Eh}{2(1-\upsilon^2)} \int_A \left\{ (1-\upsilon)\left[-\frac{(1-\upsilon)R}{n(n+1)} w|_{\alpha\beta} - a_{\alpha\beta}\frac{w}{R} \right] w|^\alpha w|^\beta \right. \\ &\quad \left. + \upsilon \left[-\frac{(1-\upsilon)R}{n(n+1)} \Delta w - 2\frac{w}{R} \right] w_{,\kappa} w|^\kappa \right\} dA \\ &= \frac{Eh}{2(1-\upsilon^2)} \int_A \left[-\frac{(1-\upsilon^2)R}{n(n+1)} w|_{\alpha\beta} w|^\alpha w|^\beta - (1-\upsilon)\frac{w}{R} w|_\beta w|^\beta \right. \\ &\quad \left. - \upsilon(1-\upsilon)\frac{w}{R} w_{,\kappa} w|^\kappa \right] dA, \end{aligned} \tag{3.16.107}$$

where we have used (3.16.81).

Let us now consider the various terms

$$\begin{aligned} \int_A w|_{\alpha\beta} w|^\alpha w|^\beta \, dA &= \int_A (w|_\alpha w|^\alpha)|_\beta w|^\beta \, dA - \int_A w|_\alpha w|^\alpha_\beta w|^\beta \, dA \\ &= \int_A \left[\left(w|_\alpha w|^\alpha w|^\beta \right)|_\beta - w|_\alpha w|^\alpha w|^\beta_\beta \right] dA - \int_A w|_{\alpha\beta} w|^\alpha w|^\beta \, dA. \end{aligned}$$

so that (because the first term vanishes on a closed surface)

$$\int_A w|_{\alpha\beta} w|^{\alpha} w|^{\beta} = -\frac{1}{2}\int_A w|_{\alpha} w|^{\alpha} \Delta w \, dA = \frac{n(n+1)}{2R^2}\int_A w|_{\alpha} w|^{\alpha} w \, dA$$
$$= \frac{n(n+1)}{4R^2}\int_A \left[\left(w^2 w|^{\alpha}\right)|_{\alpha} - w^2 \Delta w\right] dA = \frac{n^2(n+1)}{4R^4}\int_A w^3 \, dA \tag{3.16.108}$$

$$\int_A w|_{\beta} w|^{\beta} w \, dA = \frac{n(n+1)}{2R^2}\int_A w^3 \, dA. \tag{3.16.109}$$

The final result for $P_3[a_h\mathbf{u}_h]$ is now

$$P_3[a_h\mathbf{u}_h] = -\frac{3}{8}\frac{Eh}{R^3}n(n+1)\int_A w^3 \, dA = -\frac{3}{4}\frac{cE}{R^2}\int_A w^3 \, dA, \tag{3.16.110}$$

where we have made use of (3.16.100). For the evaluation of this term and a further discussion of the problem we refer to KOITER's paper.†

The result is that $P_3[a_h\mathbf{u}_h]$ vanishes if the degree of the buckling mode is odd, but it does not vanish when the degree of the buckling mode is even. This means that in the latter case, the critical bifurcation point is unstable. In the case of an even degree of the buckling mode, stability is governed by the sign of A_4; cf. (3.4.38) and (3.4.53). It turns out that A_4 is negative for both even and odd degrees of buckling modes, so the critical bifurcation point is unstable.

3.17 Buckling of circular cylindrical shells

In the following discussion, we shall assume that the fundamental state is a membrane state of stress. (This assumption is usually violated in experiments.) The second variation of the energy functional is then given by

$$P_2[\mathbf{u}] = \int_A \left\{ \frac{Eh}{2(1-\upsilon^2)}\left[(1-\upsilon)\,\theta_{\alpha\beta}\theta^{\alpha\beta} + \upsilon(\theta^{\kappa}_{\kappa})^2 + \frac{h^2}{12}\left\{(1-\upsilon)\,\rho_{\alpha\beta}\rho^{\alpha\beta} + \upsilon(\rho^{\kappa}_{\kappa})^2\right\}\right]\right.$$
$$+\frac{1}{2}N^{\alpha\beta}\left[(\theta^{\kappa}_{\cdot\alpha} - \omega^{\kappa}_{\cdot\alpha})(\theta_{\kappa\beta} - \omega_{\kappa\beta}) + \varphi_{\alpha}\varphi_{\beta}\right] \tag{3.17.1}$$
$$\left. +\frac{1}{2}P\left(w\theta^{\alpha}_{\alpha} - u^{\alpha}\varphi_{\alpha}\right)\right\} dA.$$

It is convenient to employ Cartesian coordinates (x, y) on the surface, and let $\mathbf{n}$ be the unit vector normal to this surface and positive in the outward direction. Let R be the radius of the shell, and then we can introduce dimensionless coordinates so

† W.T. KOITER, The nonlinear buckling problem of a complete spherical shell under uniform external pressure. *Proc. Kon Ned. Ak Wit., Series B*, **72**, 40–123 (1969).

that

$$\frac{\partial}{\partial x}(\,) \equiv \frac{1}{R}(\,)', \qquad \frac{\partial}{\partial y}(\,) \equiv \frac{1}{R}(\,)^{\cdot}. \tag{3.17.2}$$

In the following, we shall use the KOITER-SANDERS strain-displacement relations for cylindrical shells.[†]

$$\begin{aligned} \theta_{xx} &= \frac{1}{R}u' & \theta_{yy} &= \frac{1}{R}(\dot{v} + w) & \theta_{xy} &= \frac{1}{2R}(\dot{u} + v') \\ \omega_{xy} &= \frac{1}{2}(v' - \dot{u}) & \varphi_x &= \frac{1}{R}w' & \varphi_y &= \frac{1}{R}(\dot{w} - v) \\ \rho_{xx} &= \frac{1}{R^2}w'' & \rho_{yy} &= \frac{1}{R^2}(\dot{w} - v)^{\cdot} & \rho_{xy} &= \frac{1}{R^2}\left(\dot{w}' - \frac{3}{4}v' + \frac{1}{4}\dot{u}\right) \end{aligned}\quad. \tag{3.17.3}$$

Because we are dealing with Cartesian coordinates, the functional (3.17.1) can readily be rewritten to yield

$$\begin{aligned} P_2[\mathbf{u}] = &\int_A \frac{Eh}{2(1-\upsilon^2)}\Big\{\theta_{xx}^2 + \theta_{yy}^2 + 2\upsilon\theta_{xx}\theta_{yy} + 2(1-\upsilon)\theta_{xy}^2 \\ &+ \frac{h^2}{12}\left[\rho_{xx}^2 + \rho_{yy}^2 + 2\upsilon\rho_{xx}\rho_{yy} + 2(1-\upsilon)\rho_{xy}^2\right]\Big\}\, dA \\ &+ \int_A \Big\{\frac{1}{2}N_{xx}\left[\theta_{xx}^2 + (\theta_{yx} - \omega_{yx})^2 + \varphi_x^2\right] + \frac{1}{2}N_{yy}\left[\theta_{yy}^2 + (\theta_{xy} - \omega_{xy})^2 + \varphi_y^2\right] \\ &+ \frac{1}{2}P\left[w(\theta_{xx} + \theta_{yy}) - u\varphi_x - v\varphi_y\right]\Big\}\, dA. \end{aligned} \tag{3.17.4}$$

The term with N_{xy} is absent because there are no shear forces. The terms $N_{xx}\theta_{xx}^2$ and $N_{yy}\theta_{yy}^2$ can be neglected compared to the terms $Eh\left(\theta_{xx}^2 + \theta_{yy}^2 + 2\upsilon\theta_{xx}\theta_{yy}\right)$.

KOITER[‡] has shown that within the context of the classical theory of shells, the changes of curvature can be altered by adding terms of the order of the strain in the middle surface divided by the minimum radius of curvature without altering the accuracy of the theory. We now consider the following modified energy functional,

$$\begin{aligned} P_2^*[\mathbf{u}] = P_2[\mathbf{u}] + &\int_A \frac{Eh^3}{24(1-\upsilon^2)}\Big[-\frac{2(1-\upsilon)}{R}\theta_{xx}\rho_{yy} + \frac{2}{R}\theta_{yy}(\rho_{xx} + \rho_{yy}) \\ &+ \frac{1}{R^2}\theta_{yy}^2 + \frac{2(1-\upsilon)}{R}\theta_{xy}\rho_{xy} - \frac{3(1-\upsilon)}{2R^2}\theta_{xy}^2\Big]\, dA. \end{aligned} \tag{3.17.5}$$

KOITER has shown that an upper bound for the change in strain energy is

$$\left|P_2^*[\mathbf{u}] - P_2[\mathbf{u}]\right| < 0.467\frac{h}{R}P_2[\mathbf{u}]\left[1 + O\left(\frac{h}{R}\right)\right]. \tag{3.17.6}$$

[†] Cf. W. T. KOITER, Summary of equations for modified simplistic possible accurate linear theory of thin circular cylindrical shells. Rep. Lab. For *Appl. Mech. Delft*, 442.

[‡] W. T. KOITER, A consistent first approximation in the general theory of thin elastic shells. Proc. IUTAM Symp. *On Theory of Thin Elastic Shells* (Delft, Aug. 1959; North-Holland Publ. Co., 1960), pp. 12–33.

Substitution of the relations (3.17.3) into this modified energy functional yields

$$
\begin{aligned}
P_2^*[\mathbf{u}] = \int_A \frac{Eh}{2(1-\upsilon^2)R^2} \Big\{ & u'^2 + (v^{\cdot} + w)^2 + 2\upsilon u'(v^{\cdot} + w) + \frac{1}{2}(1-\upsilon)(u^{\cdot} + v')^2 \\
& + \frac{h^2}{12R^2}\left[w''^2 + w^{\cdot\cdot 2} + 2\upsilon w'' w^{\cdot\cdot} + 2(1-\upsilon) w'^{\cdot 2} + 2w(w'' + w^{\cdot\cdot}) + w^2\right] \\
& + 2(1-\upsilon)\frac{h^2}{12R^2}\left[u' v^{\cdot} - u^{\cdot} v' + u^{\cdot} w'^{\cdot} - u' w^{\cdot\cdot} + v^{\cdot} w'' - v' w'^{\cdot}\right]\Big\}\, dA \qquad (3.17.7) \\
& + \int_A \Big\{ \frac{N_{xx}}{2R^2}(v'^2 + w'^2) + \frac{N_{yy}}{2R^2}\left\{u^{\cdot 2} + (w^{\cdot} - v)^2\right] \\
& + \frac{P}{2R}\left(u' w - u w' + v^{\cdot} w - v w^{\cdot} + v^2 + w^2\right)\Big\}\, dA
\end{aligned}
$$

This modified functional has the advantage that the first term with h^2/R^2 contains only the displacement w, whereas the second term with h^2/R^2 contains only expressions of the divergence type, which do not alter the differential equation. Consider, e.g.,

$$
\begin{aligned}
R^{-1}\int_A (u' v^{\cdot} - u^{\cdot} v')\, dA &= \int_{\partial S} (u' v v_y - u^{\cdot} v v_x)\, ds \\
-\int_A (u'^{\cdot} - u^{\cdot\prime}) v\, dA &= -\int_{\partial S} u^{\cdot} v v_x\, ds,
\end{aligned}
$$

because at $x = 0$ and $x = \ell$, $v_y = 0$. Alternatively, we can also write

$$
\begin{aligned}
R^{-1}\int (u' v^{\cdot} - u^{\cdot} v')\, dA &= \int_{\partial S} (u v^{\cdot} v_x - u v' v_y)\, ds \\
-\int_A (v^{\cdot\prime} - v'^{\cdot})\, u\, dA &= \int_{\partial S} u v^{\cdot} v_x\, ds,
\end{aligned}
$$

i.e., these terms contribute to the boundary conditions.

Let us now introduce the following load parameters,

$$
\lambda_x = (1-\upsilon^2)\frac{N_{xx}}{Eh}, \quad \lambda_y = (1-\upsilon^2)\frac{N_{yy}}{Eh}, \quad N_{yy} = -PR; \qquad (3.17.8)
$$

then the last integral in (3.17.7) can be rewritten in the form

$$
\begin{aligned}
\frac{Eh}{2(1-\upsilon^2)R^2}\int_A & \left[\lambda_x (v'^2 + w'^2)\right. \\
& \left. + \lambda_y \left(u^{\cdot 2} + w^{\cdot 2} - w^{\cdot} v - u' w + u w' - v^{\cdot} w - w^2\right)\right] dA. \qquad (3.17.9)
\end{aligned}
$$

A necessary condition for neutral equilibrium is that

$$
\begin{aligned}
P_{11}[\mathbf{u}, \zeta] = 0 = \frac{Eh}{2(1-\upsilon^2)R^2}\int_A & \Big\{ 2u'\xi' + 2(v^{\cdot} + w)(\eta^{\cdot} + \zeta) \\
& + 2\upsilon u'(\eta^{\cdot} + \zeta) + 2\upsilon \xi'(v^{\cdot} + w) + (1-\upsilon)(u^{\cdot} + v')(\xi^{\cdot} + \eta') \\
& + \frac{h^2}{12R^2}\left[2w''\zeta'' + 2w^{\cdot\cdot}\zeta^{\cdot\cdot} + 2\upsilon w''\zeta^{\cdot\cdot} + 2\upsilon\zeta'' w^{\cdot\cdot}\right.
\end{aligned}
$$

$$
\begin{aligned}
&+4\,(1-\upsilon)\,w'^{\cdot}\zeta'^{\cdot} + 2w\,(\zeta'' + \zeta^{\cdot\cdot}) + 2\zeta(w'' + w^{\cdot\cdot}) + 2w\zeta] \\
&+\frac{2\,(1-\upsilon)}{12}\frac{h^2}{R^2}\,[u'\eta^{\cdot} + \xi'v^{\cdot} - u^{\cdot}\eta' - \xi^{\cdot}v' \\
&+ u^{\cdot}\zeta'^{\cdot} + \xi^{\cdot}w'^{\cdot} - u'\zeta^{\cdot\cdot} - \xi'w^{\cdot\cdot} + v^{\cdot}\zeta'' + \eta^{\cdot}w'' - v'\zeta'^{\cdot} - \eta'w'^{\cdot}] \\
&+2\lambda_x\,(v'\eta' + w'\zeta') + \lambda_y\,(2u^{\cdot}\xi^{\cdot} + 2w^{\cdot}\zeta^{\cdot} - w^{\cdot}\eta - \zeta^{\cdot}v \\
&-\,u'\zeta - \xi'w + u\zeta' + \xi w' - v^{\cdot}\zeta - \eta^{\cdot}w - 2w\zeta)\Big\}\;dA
\end{aligned}
\tag{3.17.10}
$$

for all kinematically admissible displacement fields (ξ, η, ζ).

Let us first consider all terms with ξ,

$$
\begin{aligned}
R^2\int_0^{2\pi}\!\!\int_0^{\ell}\Big\{&\left[u' + \upsilon(v^{\cdot} + w) + \frac{(1-\upsilon)}{12}\frac{h^2}{R^2}\{v^{\cdot} - w^{\cdot\cdot}\} - \frac{1}{2}\lambda_y w\right]\xi' \\
&+\left[\frac{1}{2}(1-\upsilon)\,(u^{\cdot} + v') + \frac{(1-\upsilon)}{12}\frac{h^2}{R^2}\,(w'^{\cdot} - v') + \lambda_y u^{\cdot}\right]\xi^{\cdot} \\
&\frac{1}{2}\lambda_y w'\xi\Big\}\;d\theta\,dx/R = 0.
\end{aligned}
\tag{3.17.11}
$$

By integration by parts, we obtain

$$
\begin{aligned}
-R^2\int_0^{2\pi}\!\!\int_0^{\ell}&\Big[u'' + \upsilon(v^{\cdot\prime} + w') + \frac{(1-\upsilon)}{12}\frac{h^2}{R^2}\,(v^{\cdot\prime} - w^{\cdot\cdot\prime}) - \frac{1}{2}\lambda_y w' \\
&+\frac{1}{2}(1-\upsilon)\,(u^{\cdot\cdot} + v'^{\cdot}) + \frac{1-\upsilon}{12}\frac{h^2}{R^2}\,(w'^{\cdot\cdot} - v'^{\cdot}) + \lambda_y\left(u^{\cdot\cdot} - \frac{1}{2}w'\right)\Big]\;d\theta\,dx/R \\
&+R^2\int_0^{2\pi}\left[u' + \upsilon(v^{\cdot} + w) + \frac{(1-\upsilon)}{12}\frac{h^2}{R^2}\,(v^{\cdot} - w^{\cdot\cdot}) - \frac{1}{2}\lambda_y w\right]\xi\Big|_0^{\ell}\,d\theta \\
&+R^2\int_0^{\ell}\left[\frac{1}{2}(1-\upsilon)\,(u^{\cdot} + v') + \frac{(1-\upsilon)}{12}\frac{h^2}{R^2}w'^{\cdot} - v' + \lambda_y\dot{u}\right]\xi\Big|_0^{2\pi}\frac{dx}{R} = 0,
\end{aligned}
\tag{3.17.12}
$$

which yields the equilibrium equation

$$
u'' + \frac{1}{2}(1-\upsilon)\,u^{\cdot\cdot} + \frac{1}{2}(1+\upsilon)\,v^{\cdot\prime} + \upsilon w' + \lambda_y\,(u^{\cdot\cdot} - w') = 0. \tag{3.17.13}
$$

In the following, we shall consider the case of a simply supported edge at $x = 0$ and $x = \ell$, i.e., we have the kinematic conditions

$$
v = w = 0 \quad \text{at } x = 0,\; x = \ell. \tag{3.17.14}
$$

We then obtain from (3.17.12) the dynamic boundary conditions

$$
u' + \upsilon(v^{\cdot} + w) + \frac{1-\upsilon}{12}\frac{h^2}{R^2}\,(v^{\cdot} - w^{\cdot\cdot}) - \frac{1}{2}\lambda_y w = 0 \quad \text{at } x = 0,\; x = \ell \tag{3.17.15}
$$

which with (3.17.14) reduce to

$$
u' = 0 \quad \text{at} \quad x = 0, \quad x = \ell. \tag{3.17.16}
$$

The second line integral in (3.17.12) vanishes identically because all the functions must be single-valued at $\theta = 0$, and $\theta = 2\pi$.

Let us now consider the terms with η,

$$R^2 \int_0^{2\pi} \int_0^{\ell} \left\{ \left[v^{\cdot} + w + \upsilon u' + \frac{(1-\upsilon)}{12} \frac{h^2}{R^2} (u' + w'') - \frac{1}{2} \lambda_y w \right] \eta^{\cdot} \right.$$
$$\times \left[+ \frac{1}{2} (1-\upsilon)(u^{\cdot} + v') + \frac{(1-\upsilon)}{12} \frac{h^2}{R^2} (-u^{\cdot} - w'^{\cdot}) + \lambda_x v' \right] \eta' \tag{3.17.17}$$
$$\left. - \frac{1}{2} \lambda_y w^{\cdot} \eta \right\} d\theta \frac{dx}{R} = 0.$$

By integration by parts, we obtain

$$-R^2 \int_0^{2\pi} \int_0^{\ell} \left[v^{\cdot\cdot} + w^{\cdot} + \upsilon u'^{\cdot} + \frac{(1-\upsilon)}{12} \frac{h^2}{R^2} (u'^{\cdot} + w''^{\cdot}) - \frac{1}{2} \lambda_y w^{\cdot} \right.$$
$$\left. + \frac{1}{2} (1-\upsilon)(u'^{\cdot} + v'') - \frac{(1-\upsilon)}{12} \frac{h^2}{R^2} (u'^{\cdot} + w'^{\cdot\cdot}) + \lambda_x v'' + \frac{1}{2} \lambda_y w^{\cdot} \right] \eta \, d\theta \frac{dx}{R}$$
$$+ R^2 \int_0^{\ell} \left[v^{\cdot} + w + \upsilon u' + \frac{(1-\upsilon)}{12} \frac{h^2}{R^2} (u' + w'') - \frac{1}{2} \lambda_y w \right] \eta \Bigg|_0^{2\pi} \frac{dx}{R} \tag{3.17.18}$$
$$+ R^2 \int_0^{2\pi} \left[\frac{1}{2} (1-\upsilon)(u^{\cdot} + v') - \frac{(1-\upsilon)}{12} \frac{h^2}{R^2} (u^{\cdot} + w'^{\cdot}) + \lambda_x v' \right] \eta \Bigg|_0^{\ell} d\theta = 0,$$

which yields the equilibrium equation

$$v^{\cdot\cdot} + \frac{1}{2} (1-\upsilon) v'' + \frac{1}{2} (1+\upsilon) u'^{\cdot} + w^{\cdot} + \lambda_x v'' = 0, \tag{3.17.19}$$

and in the case of a simply supported edge, both line integrals vanish.

Finally, we consider the terms with ζ,

$$\int_0^{2\pi} \int_0^{\ell} \left\{ \left[v^{\cdot} + w + \upsilon u' + \frac{h^2}{12R^2} (w'' + w^{\cdot\cdot} + w) + \frac{1}{2} \lambda_y (-u' - v^{\cdot} - 2w) \right] \zeta \right.$$
$$+ \left[\lambda_x w' + \frac{1}{2} \lambda_y u \right] \zeta' + \lambda_y \left(w^{\cdot} - \frac{1}{2} v \right) \zeta^{\cdot}$$
$$+ \frac{h^2}{12R^2} [w'' + \upsilon w^{\cdot\cdot} + w + (1-\upsilon) v^{\cdot}] \, \zeta'' \tag{3.17.20}$$
$$+ \frac{h^2}{12R^2} [w^{\cdot\cdot} + \upsilon w'' - (1-\upsilon) u' + w] \, \zeta^{\cdot\cdot}$$
$$\left. + \frac{h^2}{12R^2} [2(1-\upsilon) w'^{\cdot} + (1-\upsilon) u^{\cdot} - (1-\upsilon) v'] \zeta'^{\cdot} \right\} d\theta \frac{dx}{R} = 0.$$

By integration by parts once, we obtain

$$\begin{aligned}
&\int_0^{2\pi}\int_0^{\ell}\left\{\left[v^{\cdot}+w+\upsilon u'+\frac{h^2}{12R^2}\left(w''+w^{\cdot\cdot}+w\right)-\lambda_x w''-\lambda_y\left(u'+w^{\cdot\cdot}+w\right)\right]\zeta\right.\\
&\quad-\frac{h^2}{12R^2}\left[w'''+\upsilon w^{\cdot\cdot\prime}+w'+(1-\upsilon)\,v^{\cdot\prime}\right]\zeta'\\
&\quad-\frac{h^2}{12R^2}\left[w^{\cdot\cdot\cdot}+\upsilon w''^{\cdot}-(1-\upsilon)\,u^{\cdot\prime}+w^{\cdot}+2\,(1-\upsilon)\,w'^{\cdot\prime}+(1-\upsilon)\,u^{\cdot\prime}\right.\\
&\quad\left.\left.-(1-\upsilon)\,v''\right]\zeta^{\cdot}\right\}d\theta\frac{dx}{R}\\
&\quad+\int_0^{2\pi}\left(\lambda_x w'+\frac{1}{2}\lambda_y u^{\cdot}\right)\zeta\Big|_0^{\ell}d\theta+\int_0^{\ell}\lambda_y\left(w^{\cdot}-\frac{1}{2}v'\right)\zeta\Big|_0^{2\pi}\frac{dx}{R}\\
&\quad+\frac{h^2}{12R^2}\int_0^{2\pi}\left\{\left[w''+\upsilon w^{\cdot\cdot}+w+(1-\upsilon)\,v^{\cdot}\right]\zeta'\right.\\
&\quad\left.+\left[2\,(1-\upsilon)\,w'^{\cdot}+(1-\upsilon)\,u^{\cdot}-(1-\upsilon)\,v'\right]\zeta'\right\}\Big|_0^{\ell}d\theta\\
&\quad+\frac{h^2}{12R^2}\int_0^{\ell}\left(w^{\cdot\cdot}+\upsilon w''-(1-\upsilon)\,u'+w\right)\zeta^{\cdot}\Big|_0^{2\pi}\frac{dx}{R}=0,
\end{aligned}\tag{3.17.21}$$

where the second and the fourth line integrals vanish because the displacements and their derivatives are periodic in the y-direction. For simply supported edges, the first line integral and the second term in the third line integral vanish. Integrating by parts once again, we obtain

$$\begin{aligned}
&\int_0^{2\pi}\int_0^{\ell}\left\{v^{\cdot}+w+\upsilon u'+\frac{h^2}{12R^2}\left[w''+w^{\cdot\cdot}+w+w''''+\upsilon w^{\cdot\cdot\prime\prime}+w''+(1-\upsilon)\,v^{\cdot\prime\prime}\right.\right.\\
&\quad\left.+w^{\cdot\cdot\cdot\cdot}+\upsilon w''^{\cdot\cdot}-(1-\upsilon)\,u'^{\cdot\cdot}+w^{\cdot\cdot}+2\,(1-\upsilon)\,w''^{\cdot\cdot}+(1-\upsilon)\,u'^{\cdot\cdot}-(1-\upsilon)\,v''^{\cdot}\right]\\
&\quad\left.-\lambda_x w''-\lambda_y\left(u'+w^{\cdot\cdot}+w\right)\right\}\zeta\,d\theta\frac{dx}{R}\\
&\quad+\frac{h^2}{12R^2}\int_0^{2\pi}\left[w''+\upsilon w^{\cdot\cdot}+w+(1-\upsilon)\,v^{\cdot}\right]\zeta'\Big|_0^{\ell}d\theta\\
&\quad-\frac{h^2}{12R^2}\int_0^{2\pi}\left[w'''+\upsilon w^{\cdot\cdot\prime}+w'+(1-\upsilon)\,v^{\cdot\prime}\right]\zeta\Big|_0^{\ell}d\theta\\
&\quad-\frac{h^2}{12R^2}\int_0^{\ell}\left[w^{\cdot\cdot\cdot}+\upsilon w''^{\cdot}-(1-\upsilon)\,u'^{\cdot}+w^{\cdot}+2\,(1-\upsilon)\,w''^{\cdot}\right.\\
&\quad\left.+(1-\upsilon)\,u'^{\cdot}-(1-\upsilon)\,v''\right]\zeta\Big|_0^{2\pi}\frac{dx}{R}=0.
\end{aligned}\tag{3.17.22}$$

Here, the last line integral vanishes because the displacements and their derivatives are periodic in the y-direction. For simply supported edges, the second of the line integrals vanishes. We now obtain the third equilibrium equation,

$$\begin{aligned} & v^{\cdot} + w + \upsilon u' + \frac{h^2}{12R^2}\left[w'''' + 2w''^{\cdot\cdot} + w^{\cdot\cdot\cdot\cdot} + 2\left(w'' + w^{\cdot\cdot}\right) + w\right] \\ & - \lambda_x w'' - \lambda_y \left(u' + w^{\cdot\cdot} + w\right) = 0, \end{aligned} \tag{3.17.23}$$

and the dynamic boundary conditions

$$w'' + \upsilon w^{\cdot\cdot} + w + (1-\upsilon)\, v^{\cdot} = 0 \quad \text{at}\, x = 0, \quad x = \ell, \tag{3.17.24}$$

which with (3.17.14) can be simplified to yield

$$w'' = 0 \quad \text{at}\, x = 0,\ x = \ell. \tag{3.17.25}$$

Notice that (3.17.23) can be written as

$$v^{\cdot} + \upsilon u' + w + \frac{h^2}{12R^2}\left(\Delta + 1\right)^2 w - \lambda_x w'' - \lambda_y\left(u' + w^{\cdot\cdot} + w\right) = 0,$$

where

$$\Delta = R^2\left(\partial^2/\partial x^2 + \partial^2/\partial y^2\right). \tag{3.17.26}$$

The equations (3.17.13), (3.17.15), and (3.17.26) are known as the KOITER-MORLEY EQUATIONS.

We now try to find a solution of these equations of the form

$$\begin{aligned} u &= A\cos(px/R)\sin(qy/R) \\ v &= B\sin(px/R)\cos(qy/R) \\ w &= C\sin(px/R)\sin(qy/R), \end{aligned} \tag{3.17.27}$$

where $p = k\pi R/\ell$, $k \in \mathbb{N}^+$, and $q \in \mathbb{N}^+$. These expressions satisfy all the boundary conditions of (3.17.14), (3.17.16), and (3.17.24). Substitution of these expressions into the KOITER-MORLEY equations yields three homogeneous equations for the unknown constants A, B, and C. The condition for a non-trivial solution is then

$$\begin{vmatrix} p^2 + \frac{1}{2}(1-\upsilon)q^2 + \lambda_y q^2 & \frac{1}{2}(1+\upsilon)pq & -\upsilon p + \lambda_y p \\ \frac{1}{2}(1+\upsilon)pq & \frac{1}{2}(1-\upsilon)p^2 + \lambda_x p^2 & -q \\ -\upsilon p + \lambda_y p & -q & 1 + \frac{h^2}{2R^2}\left(p^2+q^2-1\right)^2 + \lambda_x p^2 + \lambda_y\left(q^2-1\right) \end{vmatrix} = 0. \tag{3.17.28}$$

After some algebra, we find the following result:

$$\begin{aligned} & \frac{1}{2}(1-\upsilon)\left[\left(1-\upsilon^2\right)p^4 + \frac{h^2}{12R^2}\left(p^2+q^2\right)^2\left(p^2+q^2-1\right)^2\right] \\ & + p^2\left\{\frac{1}{2}(1-\upsilon)\left(p^2+q^2\right)^2 \underline{-\upsilon^2 p^2} + \left[\underline{p^2} + \frac{1}{2}(1-\upsilon)q^2\right]\left[1 + \underline{\frac{h^2}{12R^2}\left(p^2+q^2-1\right)^2}\right]\right\}\lambda_x \end{aligned}$$

$$+\left\{\frac{1}{2}(1-\upsilon)q^2\left[\left(p^2+q^2\right)^2-3p^2-q^2\right]-\underline{\frac{1}{2}(1-\upsilon)(1-2\upsilon)p^4}\right. \tag{3.17.29}$$

$$\left.+\underline{\frac{h^2}{12R^2}\left(p^2+q^2-1\right)^2\left[\frac{1}{2}(1-\upsilon)p^2q^2+q^4\right]}\right\}\lambda_y$$

$$+\left[q^2\left(q^2-1\right)-p^2\right]\left[\frac{1}{2}(1-\upsilon)p^2+q^2\right]\lambda_y^2+\cdots=0.$$

Because $\lambda_x, \lambda_y \ll 1$, the underlined terms may be neglected compared to the principal terms. Further, we may neglect the higher order terms in λ_x and λ_y, e.g., the coefficient of λ_y^2 is of the same order of magnitude as the coefficient of λ_y, so that the term with λ_y^2 may be neglected compared to the term with λ_y. The resulting equation now becomes

$$\left(1-\upsilon^2\right)p^4+\frac{h^2}{12R^2}\left(p^2+q^2\right)^2\left(p^2+q^2-1\right)^2+\lambda_x p^2\left[\left(p^2+q^2\right)^2+q^2\right]$$
$$+\lambda_y q^2\left[\left(p^2+q^2\right)^2-\left(3p^2+q^2\right)\right]=0. \tag{3.17.30}$$

We shall now discus some special cases:

i) $q=0$. In this case, the term with λ_y vanishes, i.e., there is no influence of the external pressure P. The resulting equation now reads

$$\left(1-\upsilon^2\right)p^4+\frac{h^2}{12R^2}p^4\left(p^2-1\right)^2+\lambda_x p^6=0, \tag{3.17.31}$$

from which (for $p\neq 0$)

$$\lambda_x=-\frac{1-\upsilon^2}{p^2}-\frac{h^2}{12R^2}\frac{\left(p^2-1\right)^2}{p^2}. \tag{3.17.32}$$

To stay in the elastic range, we must have large values of p, which means that we can neglect unity with respect to p^2, so that the expression for λ_x can be simplified to yield

$$\lambda_x=-\frac{1-\upsilon^2}{p^2}-\frac{h^2}{12R^2}p^2. \tag{3.17.33}$$

Notice that λ_x is negative, i.e., the cylinder is loaded in compression. The critical value for λ_x follows from (3.17.33) by minimizing the right-hand side with respect to p^2, which yields

$$\left(p^2\right)_{\min}=\frac{2cR}{h},\qquad \left(\lambda_x\right)_{\min}=-\left(1-\upsilon^2\right)\frac{h}{cR}. \tag{3.17.34}$$

Notice the $\left(p^2\right)_{\min}$ is indeed a large number, as $R/h\gg 1$.

ii) $q=1$. In this case, the equation becomes

$$\left(1-\upsilon^2\right)p^4+\frac{h^2}{12R^2}p^4\left(p^2+1\right)^2+\lambda_x p^2\left[\left(p^2+1\right)^2+1\right]$$
$$+\lambda_y\left(p^4-p^2\right)=0. \tag{3.17.35}$$

Because $\lambda_y \ll 1$, the term p^4 in its coefficient may be neglected compared to the term $(1-\upsilon^2)p^4$. We must now distinguish various possibilities:

a) $p \gg 1$: In this case, we recover the results for $q = 0$.
b) $p \ll 1$: The equation can now be reduced to

$$(1-\upsilon^2)p^2 + 2\lambda_x - \lambda_y = 0, \tag{3.17.36}$$

which yields (with $k = 1$)

$$2\lambda_x - \lambda_y = -(1-\upsilon^2)p^2 = -(1-\upsilon^2)\frac{\pi^2 R^2}{\ell^2}. \tag{3.17.37}$$

This is the case of Euler buckling of the cylinder, which can easily be seen by setting $\lambda_y = 0$. In this case, using (3.17.8) we have

$$F = -2\pi R N_{xx} = \frac{\pi^2 E}{\ell^2}\pi R^3 h, \tag{3.17.38}$$

where $\pi R^3 h$ is the moment of inertia of the cylinder. When $\lambda_x = 0$, λ_y is positive, namely,

$$\lambda_y = (1-\upsilon^2)\frac{\pi^2 R^2}{\ell^2}. \tag{3.17.39}$$

This is a case of buckling under internal uniform pressure P.

Using (3.17.8), we can write

$$F = -P\pi R^2 = \frac{\pi^2 E}{\ell^2}\pi R^3 h, \tag{3.17.40}$$

which is again the Euler load.

c) $p = O(1)$: In this case, λ_x is of order unity, which is not possible in the elastic range. Hence, values of $p = O(1)$ must be excluded.

Before proceeding with the general discussion of (3.17.30), we notice that this is the equation of a straight line ℓ in the λ_x, λ_y plane. Let p_i, q_i $(i \in \mathbb{N}^+)$ denote minimizing values of p and q, respectively. The corresponding values $(\lambda_x)_{cr}$, $(\lambda_y)_{cr}$ then lie on a straight line ℓ_i, which is the boundary between a stable and an unstable region. Let $\mathcal{D}_i^s$ be the stable region. Let $\mathcal{D}^s = \cap_{i=1\ldots n}\mathcal{D}_i^s$; then $\mathcal{D}^s$ is convex (i.e., if $x, y \in \mathcal{D}^s$, then $\mu x + (1-\mu)y \in \mathcal{D}^s$, $0 \le \mu \le 1$), if the fundamental state is linear. (The origin $(\lambda_x, \lambda_y) = (0, 0) \in \mathcal{D}^s$.)

To prove this property we proceed as follows: Let $S_{ij}^{(k)}$ be the stress due to a unit load in the loading case k. The total stress under n loads can then be written

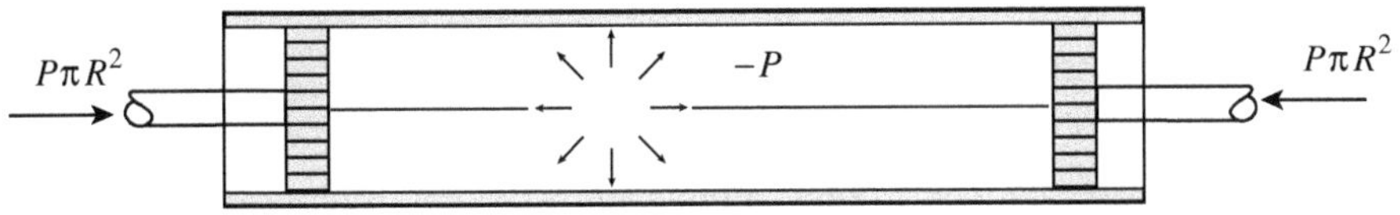

Figure 3.17.1

as $S_{ij} = \sum_{k=1}^{n} \lambda_k S_{ij}^{(k)}$ (linearity). Substitution of this expression into the functional $P_2[\mathbf{u}]$ yields

$$P_2[\mathbf{u}] = \int_V \left[\sum_{k=1}^{n} \lambda_k S_{ij}^{(k)} u_{h,i} u_{h,j} + \frac{1}{2} E_{ijkl} \theta_{ij} \theta_{k\ell} \right] dV. \tag{3.17.41}$$

The necessary condition for stability is that

$$P_2[\mathbf{u}] \geq 0 \quad \forall \mathbf{u} \text{ that are kinemafically admissible,} \tag{3.17.42}$$

where the equality sign holds on $\partial \mathcal{D}^s$. Let $\lambda_k^{(1)}$ and $\lambda_k^{(2)}$ be two load factors $\in \mathcal{D}^s + \partial \mathcal{D}^s$; then

$$\mu P_2\left[\mathbf{u}; \lambda_k^{(1)}\right] + (1-\mu) P_2\left[\mathbf{u}; \lambda_k^{(2)}\right] \geq 0, \quad 0 \leq \mu \leq 1, \tag{3.17.43}$$

and hence

$$\int_V \left\{ \sum_{k=1}^{n} \left[\mu \lambda_k^{(1)} + (1-\mu) \lambda_k^{(2)} \right] S_{ij}^{(k)} + \frac{1}{2} E_{ijkl} \theta_{ij} \theta_{k\ell} \right\} dV \geq 0 \tag{3.17.44}$$

so that $\mu \lambda_k^{(1)} + (1-\mu) \lambda_k^{(2)} \in \mathcal{D}^s + \partial \mathcal{D}^s$, q.e.d.

Let us now return to the buckling equation (3.17.30) and let us first consider the case that there is no axial loading, $\lambda_x = 0$. In this case, λ_y is given by

$$\lambda_y = -\frac{(1-\upsilon^2) p^4 + \frac{h^2}{12R^2} (p^2+q^2)^2 (p^2+q^2-1)^2}{q^2 \left[(p^2+q^2)^2 - (3p^2+q^2) \right]}. \tag{3.17.45}$$

The critical value of λ_y is obtained by minimizing (3.17.45) with respect to p^2 and q^2. We shall not carry out these calculations, but we notice that λ_y will attain its minimum value for the smallest value of p, i.e., $p^2 = \pi^2 R^2 / \ell^2$.

Let us now consider the special case of a very long cylinder, $R/\ell \to 0$. In this case, λ_y is given by

$$\lambda_y = -\frac{h^2}{12R^2} (q^2 - 1). \tag{3.17.46}$$

The critical value is attained for the smallest possible value of q. However, $q = 0$ cannot be taken into account because then the limit process is not valid. The value $q = 1$, the Euler buckling, also can not be taken, so that the minimum value for q is $q = 2$,

$$\lambda_{y_{cr}} = -\frac{h^2}{4R^2}. \tag{3.17.47}$$

This is a long cylinder (or a ring) under uniform external pressure (see Figure 3.17.2).

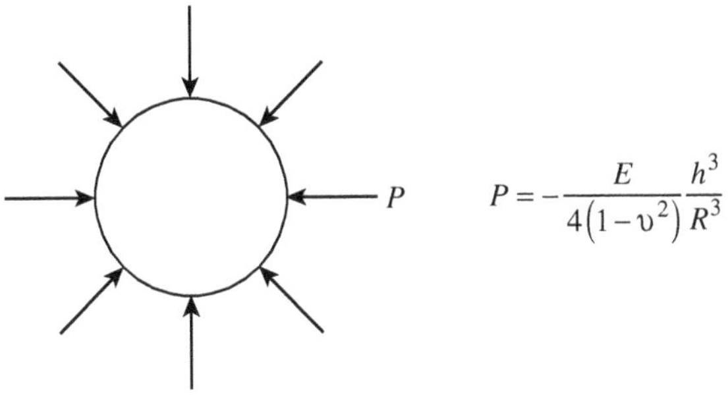

Figure 3.17.2

Let us now consider the case that $\lambda_y = 0$, i.e., we have only axial loading. We then have

$$-\lambda_x = \frac{(1-\upsilon^2)\,p^2}{(p^2+q^2)^2+q^2} + \frac{h^2}{12R^2}\frac{(p^2+q^2)^2\,(p^2+q^2-1)^2}{p^2\left[(p^2+q^2)^2+q^2\right]}. \tag{3.17.48}$$

Let us now consider the case $q = O(1)$ but $q \neq 1$. Then, to stay in the elastic range, we must have either $p^2 \ll 1$ or $p^2 \gg 1$.

Let us first consider the case $p^2 \ll 1$. In this case, (3.17.48) can be simplified to yield

$$-\lambda_x = \frac{(1-\upsilon^2)\,p^2}{q^2\,(q^2+1)} + \frac{h^2}{12R^2}\frac{q^2\,(q^2-1)^2}{p^2\,(q^2+1)}. \tag{3.17.49}$$

Minimizing this expression with respect to p^2, we obtain

$$\frac{1-\upsilon^2}{q^2\,(q^2+1)} - \frac{h^2}{12R^2}\frac{q^2\,(q^2-1)}{p^4\,(q^2+1)} = 0,$$

from which

$$p^2 = \frac{h}{2cR}q^2\,(q^2-1)\,. \tag{3.17.50}$$

Because $p = k\pi R/\ell \ll 1$, we must have

$$\ell = k\pi R\sqrt{\frac{2cR}{h}}\frac{1}{q\sqrt{q^2-1}} \gg 1, \tag{3.17.51}$$

i.e., a sufficiently long cylinder. The corresponding critical load is then

$$(-\lambda_x)_{cr} = (1-\upsilon^2)\,\frac{h}{cR}\frac{q^2-1}{q^2+1}. \tag{3.17.52}$$

For $q = 2$, we have

$$(-\lambda_x)_{cr} = 0.6\,(1-\upsilon^2)\,\frac{h}{cR}. \tag{3.17.53}$$

Let us next consider the case $p^2 \gg 1$, $q = O(1)$, $q \neq 1$. In this case, (3.17.48) can be simplified to yield

$$-\lambda_x = \frac{1-\upsilon^2}{p^2} + \frac{h^2}{12R^2}p^2. \tag{3.17.54}$$

Minimizing this expression with respect to p^2, we obtain

$$(p^2)_{\min} = \frac{2cR}{h}, \quad (-\lambda_x)_{cr} = (1-\upsilon^2)\frac{h}{cR}. \tag{3.17.55}$$

Notice that this critical value is always larger than the value given in (3.17.48), which means that for $q = O(1)$ the critical load is given by (3.17.52), and this result holds for long cylinders.

Let us now consider the case $q \gg 1$, i.e., a short wavelength (shallow shell theory). Then (3.17.48) can be simplified to yield

$$-\lambda_x = (1-\upsilon^2)\frac{p^2}{(p^2+q^2)} + \frac{h^2}{12R^2}\frac{(p^2+q^2)^2}{p^2}. \tag{3.17.56}$$

Minimizing this expression with respect to $p^2/(p^2+q^2)^2$, we find

$$\left[\frac{(p^2+q^2)^2}{p^2}\right]_{\min} = \frac{2cR}{h}, \tag{3.17.57}$$

so that

$$(-\lambda_x)_{cr} = (1-\upsilon^2)\frac{h}{cR}, \quad \text{or} \quad (-\sigma_x)_{cr} = -\frac{Eh}{cR}. \tag{3.17.58}$$

This result holds for cylinders of moderate length. As we are only interested in positive values of p and q, (3.17.57) can be rewritten as

$$p^2 + q^2 = \sqrt{2cR/h}\, p. \tag{3.17.59}$$

For $p, q \in \mathbb{R}^+$, this is the equation of a circle with radius $\frac{1}{2}\sqrt{2cR/h}$ in the p, q plane (see Figure 3.17.3).

However, because $q \in \mathbb{N}^+$, we have only discrete values, and to a given value of q there corresponds two values of p, say, p_{q_1} and p_{q_2}. This means that for the same

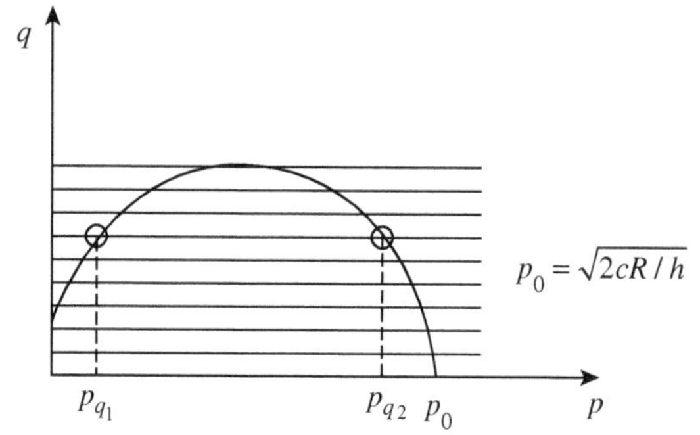

Figure 3.17.3

buckling load, we have buckling modes of the form

$$\begin{pmatrix}\cos\\ \sin\end{pmatrix}\begin{pmatrix}p_{q_1}\\ \quad x/R\\ p_{q_2}\end{pmatrix}\begin{pmatrix}\cos\\ \sin\end{pmatrix}(q\,y/R). \tag{3.17.60}$$

Let us now consider a particular combination of buckling modes represented by

$$\begin{aligned} w = h\,\{&b_0\cos(p_0x/R) + \sum_q [c_{q_1}\cos(p_{q_1}\,x/R) + c_{q_2}\cos(p_{q_2}\,x/R)]\cos(qy/R) \\ &+ c_m\cos(mx/R)\cos(my/R)\}, \end{aligned} \tag{3.17.61}$$

where the last term must only be taken into account when $m = \frac{1}{2}p_0 = \sqrt{cR/2h}$. The corresponding in-plane displacements are then

$$\begin{aligned} u = h\Bigg\{&-\frac{\upsilon}{p_0}\,b_0\sin(p_0x/R) - \sum_q \frac{p_{q_1}\left(\upsilon p_{q_1}^2 - q^2\right)}{\left(p_{q_1}^2 + q^2\right)^2}\,c_{q_1}\sin(p_{q_1}x/R)\cos(qy/R) \\ &- \sum_q \frac{p_{q_2}\left(\upsilon p_{q_2}^2 - q^2\right)}{\left(p_{q_2}^2 + q^2\right)^2}\,c_{q_2}\sin(p_{q_2}x/R)\cos(qy/R) \\ &+ \frac{1-\upsilon}{4m}c_m\sin(mx/R)\cos(my/R)\Bigg\} \end{aligned} \tag{3.17.62}$$

$$\begin{aligned} v = h\Bigg\{&-\sum_q \frac{q\left[(2+\upsilon)p_{q_1}^2 + q^2\right]}{\left(p_{q_1}^2 + q^2\right)^2}\,c_{q_1}\cos(p_{q_1}x/R)\sin(qy/R) \\ &- \sum_q \frac{q\left[(2+\upsilon)p_{q_2}^2 + q^2\right]}{\left(p_{q_2}^2 + q^2\right)^2}\,c_{q_2}\cos(p_{q_2}x/R)\sin(qy/R) \\ &- \frac{3+\upsilon}{4m}c_m\cos(mx/R)\sin(my/R)\Bigg\}. \end{aligned} \tag{3.17.63}$$

A necessary condition for stability at the critical load is that the third variation vanishes. Because we are dealing with the case $q \gg 1$ or $p \gg 1$, we can use (3.15.12), which is valid for shallow shell theory. However, we must note that in the present case we have chosen W positive in the outward direction, so that in the notation adopted in this section, we have

$$\begin{aligned} P_3[\mathbf{u}] = \frac{Eh^3}{12\left(1-\upsilon^2\right)R^3}\int_A &\{[u' + \upsilon(v^{\cdot} + w)]\,w'^2 \\ &+ (\upsilon u' + v' + w)\,w^{\cdot 2} + (1-\upsilon)\,(u^{\cdot} + v')\,w'w^{\cdot}\}\,dA. \end{aligned} \tag{3.17.64}$$

For the evaluation of this integral, one must deal with integrands of the form

$$\begin{aligned} &\begin{pmatrix}\cos\\ \sin\end{pmatrix}(q_1y/R)\begin{pmatrix}\cos\\ \sin\end{pmatrix}(q_2y/R)\begin{pmatrix}\cos\\ \sin\end{pmatrix}(q_3y/R)\begin{pmatrix}\cos\\ \sin\end{pmatrix}(p_1x/R) \\ &\times\begin{pmatrix}\cos\\ \sin\end{pmatrix}(p_2x/R)\begin{pmatrix}\cos\\ \sin\end{pmatrix}(p_3y/R), \end{aligned} \tag{3.17.65}$$

which can be reduced to

$$\begin{pmatrix}\cos\\ \sin\end{pmatrix}(p_1 \pm p_2 \pm p_3)\frac{x}{R}, \quad \begin{pmatrix}\cos\\ \sin\end{pmatrix}(q_1 \pm q_2 \pm q_3)\frac{y}{R}. \tag{3.17.66}$$

For a sufficiently long cylinder, the only non-vanishing terms (after integration) are the cosine terms with an argument of zero; i.e., we must only consider the combinations

$$p_1 \pm p_2 \pm p_3 = 0, \qquad q_1 \pm q_2 \pm q_3 = 0, \tag{3.17.67}$$

under the side condition that (p_i, q_i) $(i = 1, 2, 3)$ lies on the semicircle (3.17.59). The only combination that satisfies these conditions is

$$\begin{aligned} p_{1,2} &= \frac{1}{2}p_0 \pm \frac{1}{2}\sqrt{p_0^2 - 4q^2}, \quad 0/q \le \frac{1}{2}p_0 \\ p_3 &= p_0, \quad q_3 = 0. \end{aligned} \tag{3.17.68}$$

The result of a fairly lengthy but elementary calculation is

$$P_3[\mathbf{u}] = \frac{3\pi}{4}\frac{Eh^4\ell}{R^2}\left(\sum_q q^2 b_0 c_{q_1} c_{q_2} + \frac{1}{2}mb_0 c_m^2\right) \neq 0, \tag{3.17.69}$$

which means that the equilibrium at the critical load is unstable.

Let us now consider the post-buckling behavior. Because $P_3[\mathbf{u}] \not\equiv 0$, the post-buckling behavior is determined by the functional

$$P_2[a_h\mathbf{u}_h; \lambda] + P_3[a_h u_h], \tag{3.17.70}$$

and in both terms, shallow shell theory is used. For the evaluation of the first term, we notice that at the critical load,

$$\int_A \frac{1}{2}\sigma_{x_{cr}}\frac{h}{R^2}w'^2\, dA + \int_A \frac{1}{2}hE_{\alpha\beta\lambda\mu}\left(\theta_{\alpha\beta}\theta_{\lambda\mu} + \frac{h^2}{12}\rho_{\alpha\beta}\rho_{\lambda\mu}\right) dA = 0, \tag{3.17.71}$$

so that

$$P_2[a_h\mathbf{u}_h; \lambda] = \frac{1}{2}(1-\lambda)\frac{Eh^2}{cR^3}\int_A w'^2\, dA, \tag{3.17.72}$$

where $\lambda = -\sigma_x cR/Eh$. Using (3.17.61) and (3.17.69), we obtain

$$\begin{aligned} P_2[a_h\mathbf{u}_h; \lambda] + P_3[a_h\mathbf{u}_h] &= \frac{\pi}{4c}\frac{Eh^4\ell}{R^2}\Bigg\{(1-\lambda)\Big[2p_0^2\, b_0^2 \\ &+ \sum_q \left(p_{q_1}^2 c_{q_1}^2 + p_{q_2}^2 c_{q_2}^2\right) + m^2 c_m^2\Big] + 3c\left[\sum_q q^2 b_0 c_{q_1} c_{q_2} + \frac{1}{2}m^2 b_0 c_m^2\right]\Bigg\} \end{aligned} \tag{3.17.73}$$

The equations of equilibrium are obtained by taking the variations with respect to b_0, c_{q_1}, c_{q_2}, and c_m, respectively, which yields

$$4(1-\lambda)p_0^2 b_0 + 3c\left(\sum_q q^2 c_{q_1} c_{q_2} + \frac{1}{2}m^2 c_m^2\right) = 0$$

$$2(1-\lambda)p_{q_1}^2 c_{q_1} + 3cq^2 b_0 c_{q_2} = 0$$
$$2(1-\lambda)p_{q_2}^2 c_{q_2} + 3cq^2 b_0 c_{q_1} = 0 \quad (3.17.74)$$
$$2(1-\lambda)m^2 c_m + 3cm^2 b_0 c_m = 0.$$

The second and third of these equations contain only c_{q_1} and c_{q_2}, and have a non-trivial solution when

$$4(1-\lambda)^2 p_{q_1}^2 p_{q_2}^2 - 9c^2 q^4 b_0^2 = 0. \quad (3.17.75)$$

Bearing in mind that p_{q_1} and p_{q_2} lie on the semicircle and satisfy the relation $p_{q_1} p_{q_2} = q^2$ (see Figure 3.17.4), we can simplify (3.17.75) to yield

$$4(1-\lambda)^2 - 9cb_0^2 = 0, \quad (3.17.76)$$

from which

$$3cb_0 = \pm 2(1-\lambda). \quad (3.17.77)$$

The last of the equations (3.17.74) has the solution

$$3cb_0 = -2(1-\lambda) \quad \vee \quad c_m = 0. \quad (3.17.78)$$

When the first of these solutions is chosen, compatible with (3.17.77), c_m remains undetermined. This result is due to the fact that we have not taken into account the higher order terms $P_4[a_h \mathbf{u}_h]$. With this choice for b_0, we find from the second and third of the equations of (3.17.75),

$$\frac{c_{q_2}}{c_{q_1}} = -\frac{2(1-\lambda)p_{q_1}^2}{3cq^2[-2(1-\lambda)/3c]} = \frac{p_{q_1}}{p_{q_2}} \quad (317.79)$$

so that $p_{q_1} c_{q_1} = p_{q_2} c_{q_2}$.

The first of the equations of (3.17.74) can be written as

$$\sum_q q^2 c_{q_1} c_{q_2} + \frac{1}{2} m^2 c_m^2 = \frac{8(1-\lambda)^2}{9c^2} p_0^2, \quad (3.17.80)$$

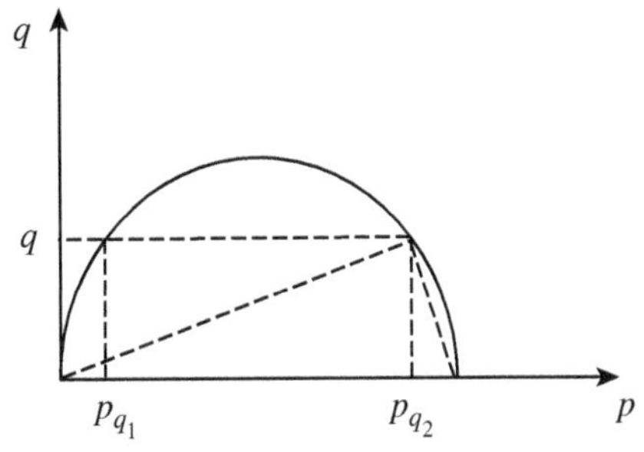

Figure 3.17.4

or, using (3.17.79) and the relations $p_{q_1}p_{q_2} = q^2$ and $p_0^2 = 2cR/h$,

$$\sum_q p_{q_1}^2 c_{q_1}^2 + \frac{1}{2} m^2 c_m^2 = \frac{16\,(1-\lambda)^2}{9c} \frac{R}{h}. \tag{3.17.81}$$

Now b_0, the amplitude of the axisymmetric deflection is small, namely,

$$b_0 = -\frac{2\,(1-\lambda)}{3c} \approx -0.4\,(1-\lambda) \quad \text{for } c = 5/3, \tag{3.17.82}$$

i.e., 40% of the shell thickness, when $\lambda = 0$. c_m is the amplitude of the non-symmetric buckling mode, and for $c_{q_1} = 0$ is given by

$$c_m^2 = \frac{32\,(1-\lambda)^2}{9c} \frac{R}{h} \frac{1}{m^2}, \tag{3.17.83}$$

which with $m = \frac{1}{2} p_0 = \sqrt{cR/2h}$ can be rewritten in the form

$$c_m = \pm \frac{8\,(1-\lambda)}{3c}, \tag{3.17.84}$$

i.e., $|c_m| = 4b_0$, twice the shell thickness, much larger than the amplitude of the axisymmetric buckling mode.

Although the amplitudes of the buckling modes are indeterminate, the shortening of the shell is determined. From (2.4.61), the general formula for the additional shortening is given by

$$\Delta\varepsilon\ell = -\frac{\partial P\,[\mathbf{u}^{\text{eq}}; N]}{\partial N}. \tag{3.17.85}$$

The compression force N is given by

$$N = 2\pi R h \sigma = \frac{2\pi}{c} E h^2 \lambda, \tag{3.17.86}$$

and the elastic energy is given by (3.17.73). Using this relation, we find

$$\begin{aligned} \Delta\varepsilon\ell &= -\frac{\partial P}{\partial N} = -\frac{\partial P}{\partial \lambda} \frac{c}{2\pi E h^2} \\ &= \frac{1}{8} \frac{h^2 \ell}{R^2} \left\{ 2p_0^2 b_0^2 + \sum_q \left(p_{q_1}^2 c_{q_1}^2 + p_{q_2}^2 c_{q_2}^2\right) + m^2 c_m^2 \right\}, \end{aligned} \tag{3.17.87}$$

which with (3.17.79) and (3.17.81) can be rewritten to yield

$$\Delta\varepsilon\ell = \frac{2}{3} \frac{h\ell}{cR} (1-\lambda)^2 = \frac{2}{3} \varepsilon_{cr} \ell\,(1-\lambda)^2, \tag{3.17.88}$$

so that

$$\frac{\Delta\varepsilon}{\varepsilon_{cr}} = \frac{2}{3} (1-\lambda)^2. \tag{3.17.89}$$

Notice that this formula is valid for all post-buckling paths. This relation is shown in the load-generalized displacement diagram (see Figure 3.17.5).

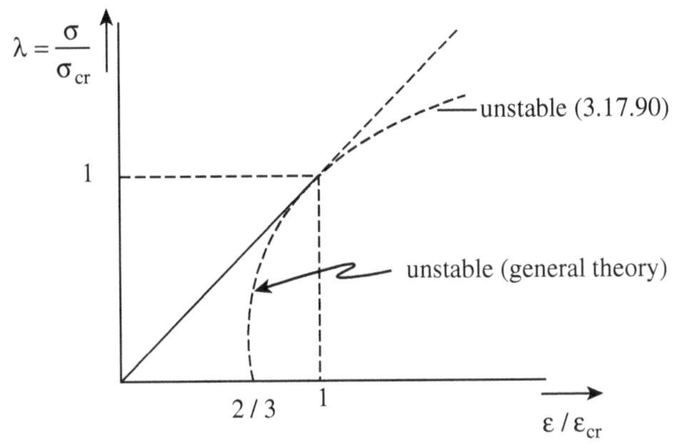

Figure 3.17.5

Because

$$\frac{\partial^2 P}{\partial b_0^2} = \frac{\pi}{c}\frac{Eh^4\ell}{R^2}p_0^2(1-\lambda) < 0 \quad \text{for } \lambda > 1, \tag{3.17.90}$$

it follows that for $\lambda > 1$ the equilibrium path is unstable. Because there are no stable equilibrium configurations in the vicinity of the bifurcation point, the shell will collapse explosively.

Because the post-buckling behavior of the perfect cylinder is unstable, the cylinder is an imperfection-sensitive structure. To get quantitative results for the imperfection sensitivity, we consider first the case of an axisymmetric imperfection in the direction of the buckling mode. Let this imperfection be given by

$$w_0 = \kappa h \cos p_0 x/R. \tag{3.17.91}$$

According to the general theory, the term $\frac{1}{2}\sigma h w_{,x}^2$ in the energy functional must be replaced by

$$\sigma h\left(\frac{1}{2}w_{,x}^2 + w_{0x}w_{,x}\right) \tag{3.17.92}$$

so that in the presence of the axisymmetric imperfection, the energy functional is given by

$$\begin{aligned} P_2[\mathbf{u},\mathbf{u}_0;\lambda] + P_3[\mathbf{u}] &= P_2[\mathbf{u};\lambda] + P_3[\mathbf{u}] \\ &- \frac{\pi}{c}\frac{Eh^4\ell}{R^2}p_0^2 b_0\kappa\lambda. \end{aligned} \tag{3.17.93}$$

The extra term contains only the amplitude b_0, so that only the first of the equilibrium equations (3.17.76) is affected. This equation must be replaced by

$$4(1-\lambda)p_0^2 b_0 + 3c\left(\sum_q q^2 c_{q_1}c_{q_2} + \frac{1}{2}m^2 c_m^2\right) - 4\lambda p_0^2\kappa = 0. \tag{3.17.94}$$

Because prior to buckling c_{q_1}, c_{q_2}, and c_m are zero, we readily obtain

$$b_0 = \frac{\lambda}{1-\lambda}\kappa. \tag{3.17.95}$$

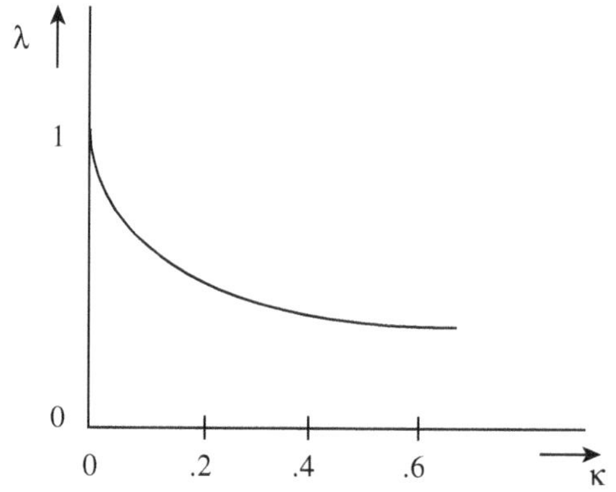

Figure 3.17.6

The values of b_0 for a non-trivial solution for c_{q_1}, c_{q_2} of the other equilibrium equations follow from (3.17.76), so we must have

$$b_0^2 = \frac{4\,(1-\lambda)^2}{9c} = \frac{\lambda^2}{(1-\lambda)^2}\kappa^2, \tag{3.17.96}$$

from which follows

$$(1-\lambda)^2 = \frac{3}{2}c\lambda\,|\kappa| \approx \frac{5}{2}\lambda\,|\kappa| \quad \text{for } c = \frac{5}{2}. \tag{3.17.97}$$

In this case, c_{q_1} and c_{q_2} are nonzero, which means that branching from the axisymmetric mode to a non-axisymmetric buckling mode will occur. It follows that small imperfections reduce the buckling load considerably, e.g., a reduction of the buckling load by 50% is obtained when $\kappa = 1/5$, say, for an amplitude of the imperfection of 20% of the shell thickness. This effect is shown in Figure 3.17.6.

Let us now consider a more general imperfection again in the direction of the buckling mode,

$$w_0 = \kappa h\left[\cos(p_0 x/R) + 4\cos(mx/R)\cos(my/R)\right], \tag{3.17.98}$$

where we have used the fact that $c_m = 4b_0$; i.e., the amplitude of the non-symmetric mode is four times that of the symmetric mode. In this case, the energy functional is given by

$$\begin{aligned} P_2[\mathbf{u}, \mathbf{u}_0; \lambda] + P_3[\mathbf{u}] &= P_2[\mathbf{u}; \lambda] + P_3[\mathbf{u}] \\ &- \frac{\pi}{c}\frac{Eh^4\ell}{R^2}\left(p_0^2 b_0 + 2m^2 c_m\right) u\lambda. \end{aligned} \tag{3.17.99}$$

The equilibrium equations now read

$$\begin{aligned} &4\,(1-\lambda)\,p_0^2 b_0 + 3c\left(\sum_q q^2 c_{q_1} c_{q_2} + \frac{1}{2} m^2 c_m^2\right) - 4\lambda p_0^2 \kappa = 0 \\ &2\,(1-\lambda)\,p_{q_1}^2 c_{q_1} + 3cq^2 b_0 c_{q_2} = 0 \\ &2\,(1-\lambda)\,p_{q_2}^2 c_{q_2} + 3cq^2 b_0 c_{q_1} = 0 \\ &2\,(1-\lambda)\,m^2 c_m + 3cm^2 b_0 c_m - 8\lambda m^2 \kappa = 0. \end{aligned} \tag{3.17.100}$$

For $\lambda = 0$, these equations yield the null solution. For $\lambda < \lambda_{cr}$, c_{q_1} and c_{q_2} are equal to zero, and the first and last of these equations are simplified to read

$$
\begin{aligned}
&4(1-\lambda)\,b_0 + \frac{3}{8}cc_m^2 - 4\lambda\kappa = 0 \\
&2(1-\lambda)\,c_m + 3cb_0c_m - 8\lambda\kappa = 0
\end{aligned}
\tag{3.17.101}
$$

and have a solution for $c_m = 4b_0$. The resulting equation for b_0 then reads as

$$
b_0^2 + \frac{2}{3c}(1-\lambda)\,b_0 - \frac{2}{3c}\lambda\kappa = 0, \tag{3.17.102}
$$

which yields

$$
b_0 = -\frac{1-\lambda}{3c} \pm \sqrt{\frac{(1-\lambda)^2}{9c^2} + \frac{2}{3c}\lambda\kappa}. \tag{3.17.103}
$$

We must choose the positive solution because the result of the negative solution does not refer to the undeformed state.

The expression under the square-root sign is always positive when $\kappa > 0$, and then the imperfection has a stabilizing effect. However, for $\kappa < 0$ the expression may become negative, and the stability limit is reached when

$$
(1-\lambda)^2 = -6c\lambda\kappa, \quad \kappa < 0. \tag{3.17.104}
$$

Notice that now $(1-\lambda)^2$ is four times larger than in the case of an axisymmetric imperfection. In this case, we do not have a branch point but a limit point. The behavior is shown in Figure 3.17.7. Notice that an imperfection of 5% of the shell thickness ($\kappa = 0.05$) yields a reduction of the critical load by 50%.

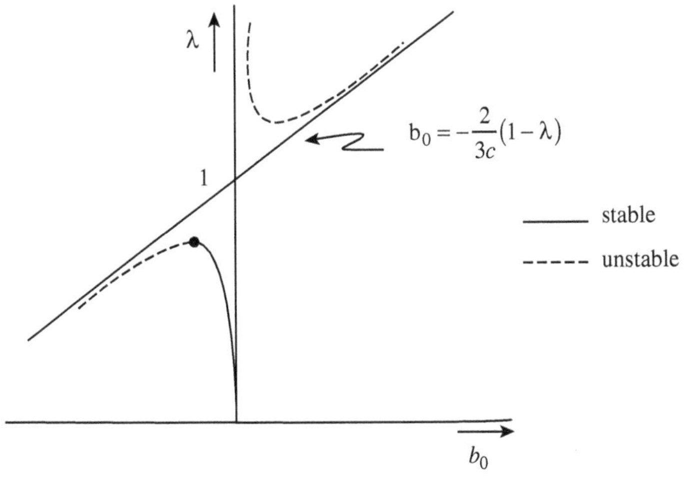

Figure 3.17.7

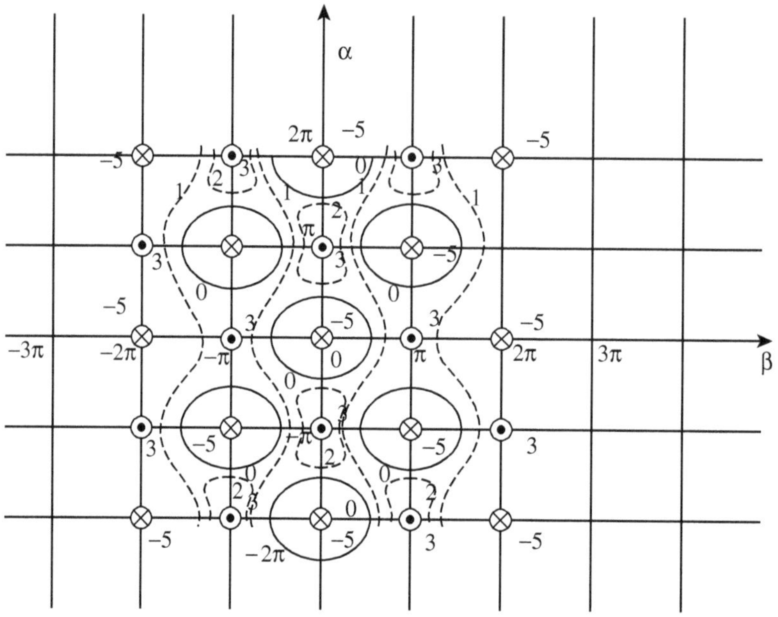

Figure 3.17.8

To get an impression of the pattern of the imperfection, we consider the function

$$f(\alpha, \beta) = \cos 2\alpha + 4\cos\alpha\cos\beta. \tag{3.17.105}$$

For

$$\begin{Bmatrix} \alpha = (2j+1)\pi \\ \beta = (2k+1)\pi \end{Bmatrix} \quad \text{and} \quad \begin{Bmatrix} \alpha = 2j\pi \\ \beta = 2k\pi \end{Bmatrix}, \tag{3.17.106}$$

we have $f(\alpha, \beta) = 5$, i.e., a maximum inward deflection. The maximum outward deflection is $f(\alpha, \beta) = -3$ and is obtained for

$$\begin{Bmatrix} \alpha = (2j+1)\pi \\ \beta = 2k\pi \end{Bmatrix} \quad \text{and} \quad \begin{Bmatrix} \alpha = 2j\pi \\ \beta = (2k+1)\pi \end{Bmatrix}. \tag{3.17.107}$$

Additional information is obtained by considering equipotential lines$f(\alpha, \beta) =$ const. The results are shown in Figure 3.17.8.

These results are in agreement with experimental results obtained at Lockheed.† These experiments were carried out with a mandrel with a radius slightly smaller than that of the cylinder, inside the cylinder. The tests were filmed with a high- speed camera (8,000 images per second) to record the dynamic behavior during buckling.

† B. O. Almroh, A. M. C. Holmes, and D. O. Brush, An experimental study of the buckling of cylinders under axial compression. Rep. 6-90-63-104 of Lockheed Missiles and Space Company, Palo Alto, California (1963).

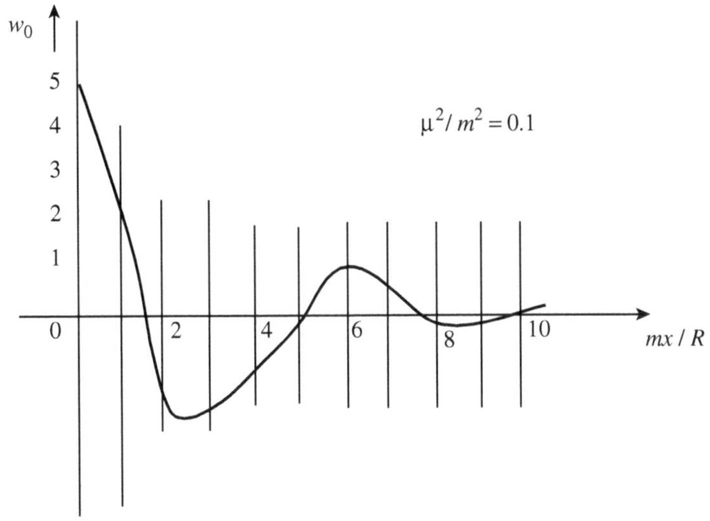

Figure 3.18.1

3.18 The influence of more-or-less localized short-wave imperfections on the buckling of circular cylindrical shells under axial compression†

So far, we have only considered overall imperfections. In practice, one also often encounters localized imperfections (dimples), which are usually inward deflections. To investigate the influence of these local imperfections, we consider an imperfection of the form

$$w_0 = \kappa k[\cos(p_0 x/R) + 4\cos(mx/R)\cos(my/R)]\, e^{-\frac{1}{2}\mu^2(x^2+y^2)/R^2} \quad p_0 = 2m, \quad (3.18.1)$$

i.e., an imperfection in the direction of the buckling mode that is rapidly decaying with increasing values of x and y (even for small values of μ^2/m^2). To get a feeling for the behavior of this imperfection, we have drawn the graph for $\mu^2/m^2 = 0.1$ (see Figure 3.18.1).

To solve this problem, we shall employ the Rayleigh-Ritz method, and we assume deflections of the form

$$\begin{aligned} w &= hb_0\,[\cos(p_0 x/R) + 4\cos(mx/R)\cos(my/R)]\, e^{-\frac{1}{2}\mu^2(x^2+y^2)/R^2} \\ u &= hb_0\left[-\frac{\upsilon}{2m}\sin(p_0 x/R) + \frac{1-\upsilon}{m}\sin(mx/R)\cos(my/R)\right] e^{-\frac{1}{2}\mu^2(x^2+y^2)/R^2} \quad (3.18.2) \\ v &= -hb_0\frac{3+\upsilon}{m}\cos(mx/R)\sin(my/R)\, e^{-\frac{1}{2}\mu^2(x^2+y^2)/R^2}. \end{aligned}$$

Let us now consider the term

$$\int_{-\ell/2}^{\ell/2}\int_0^{2\pi} -\frac{1}{2}\sigma h w_{,x}^2\, dx\, dy, \quad (3.18.3)$$

† Cf. KOITER's paper under the same title, Rep. Lab. For *Appl. Mech. Delft* 534, I. N. Vekua Anniversary volume (Moscow, 1978), pp. 242–244.

which can be evaluated to yield

$$-\frac{1}{2}\lambda\frac{Eh^2}{cR}h^2b_0^2\int_{-\ell/2}^{\ell/2}\int_0^{2\pi}\Big[-\frac{p_0}{R}\sin(p_0x/R)-\frac{4m}{R}\sin(mx/R)\cos(my/R)$$

$$-\frac{\mu^2x}{R^2}\left\{\cos(p_0x/R)+4\cos(mx/R)\cos(my/R)\right\}\Big]^2 e^{-\mu^2(x^2+y^2)/R^2}dx\,dy$$

$$\begin{aligned}=-\frac{1}{2}\lambda\frac{Eh^4}{cR^3}b_0^2\int_{-\ell/2}^{\ell/2}\int_0^{2\pi}&\Big\{[p_0\sin(p_0x/R)+4m\sin(mx/R)\cos(my/R)]^2\\ &+2\frac{\mu^2x}{R}[p_0\sin(p_0x/R)+4m\sin(mx/R)\cos(my/R)]\\ &\times[\cos(p_0x/R)+4\cos(mx/R)\cos(my/R)]\\ &+\frac{\mu^4x^2}{R^2}[\cos(p_0x/R)+4\cos(mx/R)\cos(my/R)]^2\Big\}e^{-\mu^2(x^2+y^2)/R^2}dx\,dy.\end{aligned} \tag{3.18.4}$$

Let us now first consider integration in the x-direction. First, we notice that the first term in the integrand leads to integrals in the x-direction of the form

$$\int_{-\ell/2}^{\ell/2}\cos\,(nx/R)\,e^{-\mu^2x^2/R^2}\,dx. \tag{3.18.5}$$

The second term leads to integrals of the form

$$\int_{-\ell/2}^{\ell/2}\frac{x}{R}\sin\,(nx/R)\,e^{-\mu^2x^2/R^2}\,dx, \tag{3.18.6}$$

and the third term to

$$\int_{-\ell/2}^{\ell/2}\frac{x^2}{R^2}\cos\,(nx/R)\,e^{-\mu^2x^2/R^2}\,dx. \tag{3.18.7}$$

Because the integrants are rapidly decaying functions, we can replace the boundaries of the integrals by $-\infty$ to ∞, so that we must deal with the following integrals,

$$\begin{aligned}&\int_{-\infty}^{\infty}\cos(nx/R)\,e^{-\mu^2x^2/R^2}\,dx,\qquad \int_{-\infty}^{\infty}\frac{x}{R}\sin\,(nx/R)\,e^{-\mu^2x^2/R^2}\,dx,\\ &\int_{-\infty}^{\infty}\frac{x^2}{R^2}\cos\,(nx/R)\,e^{-\mu^2x^2/R^2}\,dx.\end{aligned} \tag{3.18.8}$$

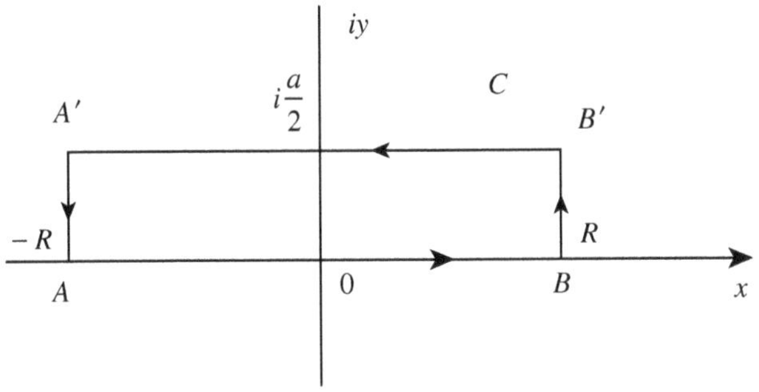

Figure 3.18.2

Let us consider the first of these integrals. It is convenient to consider the integral

$$\oint_C e^{-z^2}\, dz \tag{3.18.9}$$

in the complex plane, where the contour C is shown in Figure 3.18.2.

By Cauchy's theorem, the value of the integral is zero. On the vertical sides, we have

$$\left|e^{-z^2}\right| = \left|e^{-(x^2-y^2)}e^{-2ixy}\right| = e^{-R^2}e^{y^2} < e^{-R^2}e^{a^2/4},$$

and this expression tends uniformly to zero as R tends to infinity. Thus, the portions of the whole integral that arise from the vertical sides tend to zero, and if we carry out the passage to the limit $R \to \infty$ and note that $dz = d(x + \frac{1}{2}ia) = dx$, on A′B′ we can express the result of Cauchy's theorem as follows,

$$\int_{-\infty}^{\infty} e^{-(x+\frac{1}{2}ia)^2}\, dx - \int_{-\infty}^{\infty} e^{-x^2}\, dx = 0.$$

The second integral is easily evaluated by noticing that

$$\left(\int_{-\infty}^{\infty} e^{-x^2}\, dx\right)^2 = \int_{-\infty}^{\infty}\int_{-\infty}^{\infty} e^{-(x^2+y^2)}\, dx\, dy = \int_{-\infty}^{\infty}\int_{0}^{2\pi} e^{-r^2} r\, dr\, d\theta = \pi,$$

so that we have

$$e^{a^2/4}\int_{-\infty}^{\infty} e^{-x^2}(\cos ax - i\sin ax)\, dx = \sqrt{\pi},$$

and because $\sin ax$ is an odd function, its contribution in the integral vanishes so that

$$\int_{-\infty}^{\infty} \cos ax\, e^{-x^2}\, dx = \sqrt{\pi}\, e^{-a^2/4}. \tag{3.18.10}$$

The result for the first of the integrals in (3.18.8) is then

$$\int_{-\infty}^{\infty} \cos(nx/R)\, e^{-\mu^2 x^2/R}\, dx = \frac{R}{\mu}\sqrt{\pi}\, e^{-\frac{n^2}{4\mu^2}}. \tag{3.18.11}$$

Noting that n is either zero or a multiple of m ($m \gg 1$), it follows that only the case $n = 0$ must be taken into account.

Let us now consider the other integrals in (3.18.8). Differentiating (3.18.11) with respect to n, we obtain

$$\int_{-\infty}^{\infty} \frac{x}{R} \sin(nx/R)\, e^{-\mu^2 x^2/R^2}\, dx = \frac{R\sqrt{\pi}}{2\mu^3} n e^{-\frac{n^2}{4\mu^2}}, \tag{3.18.12}$$

which is zero for $n = 0$ and is exponentially small for large values of n, and may thus be neglected. Differentiating (3.18.12) with respect to n, we obtain

$$\int_{-\infty}^{\infty} \frac{x^2}{R^2} \cos(nx/R)\, e^{-\mu^2 x^2/R^2}\, dx = \frac{R\sqrt{\pi}}{2\mu^3}\left(1 - \frac{n}{2\mu^2}\right) e^{-\frac{n^2}{4\mu^2}}. \tag{3.18.13}$$

For large values of n this integral vanishes, and for $n = 0$ the contribution is $R\sqrt{\pi}/2\mu^3$. Summarizing our results, we find that the contribution of the first term in (3.18.4) is of order $m^2 R\sqrt{\pi}/\mu$, whereas the contribution of the third term in (3.18.4) is of order $(1/4)\mu^2 R\sqrt{\pi}/\mu$. For small values of μ^2/m^2, we may neglect the contribution of the third term compared with the first term, making a relative error in the energy of order μ^2/m^2.

Let us now evaluate the first term using the fact that the integrand is also exponentially decaying in the y-direction, so that the boundaries of integration 0 to 2π can be replaced by $-\infty$ to ∞,

$$\begin{aligned}
&\int_{-\infty}^{\infty}\int_{-\infty}^{\infty} [2m\sin(2mx/R) + 4m\sin(mx/R)\cos(my/R)]^2 e^{-\mu^2(x^2+y^2)/R^2}\, dx\, dy \\
&= 4m^2 \int_{-\infty}^{\infty}\int_{-\infty}^{\infty} [\sin^2(2mx/R) + 4\sin(2mx/R)\sin(mx/R)\cos(my/R) \\
&\quad + 4\sin^2(mx/R)\cos^2(my/R)] e^{-\mu^2(x^2+y^2)/R^2}\, dx\, dy \\
&= 4m^2 \int_{-\infty}^{\infty}\int_{-\infty}^{\infty} \left\{\frac{3}{2} - \cos(2mx/R) - \frac{1}{2}\cos(4mx/R) - \cos(2my/R)\right. \\
&\quad + 2[\cos(mx/R) - \cos(3mx/R)]\cos(my/R) \\
&\quad \left. + \cos(2mx/R)\cos(2my/R)\right\} e^{-\mu^2(x^2+y^2)/R^2}\, dx\, dy \\
&= 4m^2 \left[\frac{3}{2}\frac{R^2}{\mu^2}\pi + \text{exponentially small terms}\right].
\end{aligned} \tag{3.18.14}$$

The corresponding term for an imperfection of the form (3.17.98) is $4m^2\pi\ell R$, which means that we can use the results from the previous section if we replace ℓ by $R/2\mu^2$. By similar arguments, we find that for the cubic terms ℓ must be replaced by $R/3\mu^2$, so the energy expression becomes

$$F^{**} = \frac{3\pi}{2}\frac{Eh^3}{\mu^2}\left[(1-\lambda)\, b_0^2 + \frac{2}{3}cb_0^3 - 2\lambda\kappa b_0\right]. \qquad (3.18.15)$$

The equilibrium equation now reads

$$2\,(1-\lambda)\, b_0 + 2cb_0^2 - 2\lambda\kappa = 0, \qquad (3.18.16)$$

from which

$$b_0 = -\frac{1-\lambda}{2c} \pm \sqrt{\frac{(1-\lambda)^2}{4c^2} + \frac{\lambda\kappa}{c}}. \qquad (3.18.17)$$

For positive values of κ, the term under the square-root sign is always positive, which means that an imperfection in the outward direction is stabilizing. However, for negative values of κ a limit point will occur, i.e., when

$$(1-\lambda^*)^2 = -4\lambda^* c\kappa, \qquad \kappa < 0. \qquad (3.18.18)$$

When we compare this expression with (3.17.104), we see that the factor 6 is replaced by a factor 4; i.e., a local imperfection is equally harmful as an imperfection of the corresponding periodic type, with an amplitude reduced by a factor 2/3.

To investigate the accuracy of the present first-order approximation, GRISTCHAK,† who worked as a research fellow in our laboratory, calculated a second approximation with an error of order μ^4/m^4. His improved approximation yields the following equation for the critical load:

$$\left[1 + (d-\varepsilon)\frac{\mu^2}{m^2} - \lambda^*\right]^2 = -4\lambda^* c\kappa\left[1 + (e-\varepsilon)\frac{\mu^2}{m^2}\right], \qquad (3.18.19)$$

where

$$\varepsilon = \frac{3}{8}, \qquad e = \frac{7+5\upsilon-6\upsilon^2}{12\,(1-\upsilon^2)},$$
$$d = \frac{1}{9b\,(1-\upsilon^2)}\{(3-\upsilon)\,[\upsilon^2 + 2\,(1-\upsilon)^2 + 2\,(3+\upsilon)^2] + 80\,(1-\upsilon^2)\}, \qquad (3.18.20)$$

For $\upsilon = 0.3$, we have

$$\varepsilon = 0.375, \quad e = 0.729, \quad d = 1.540, \qquad (3.18.21)$$

and the equation for the critical load then reads

$$\left(1 + 1.165\frac{\mu^2}{m^2} - \lambda^*\right)^2 = -4\lambda^* c\kappa\left(1 + 0.354\frac{\mu^2}{m^2}\right). \qquad (3.18.22)$$

In the graph of Figure 3.18.3, the relation between λ^* and $c\kappa$ is given for $\upsilon = 0.3$.

† V. Z. GRISTCHAK, Asymptotic formula for the buckling stress of axially compressed circular cylindrical shells with more-or-less localized shortwave imperfections. W.T.H.D., Rep. 88.

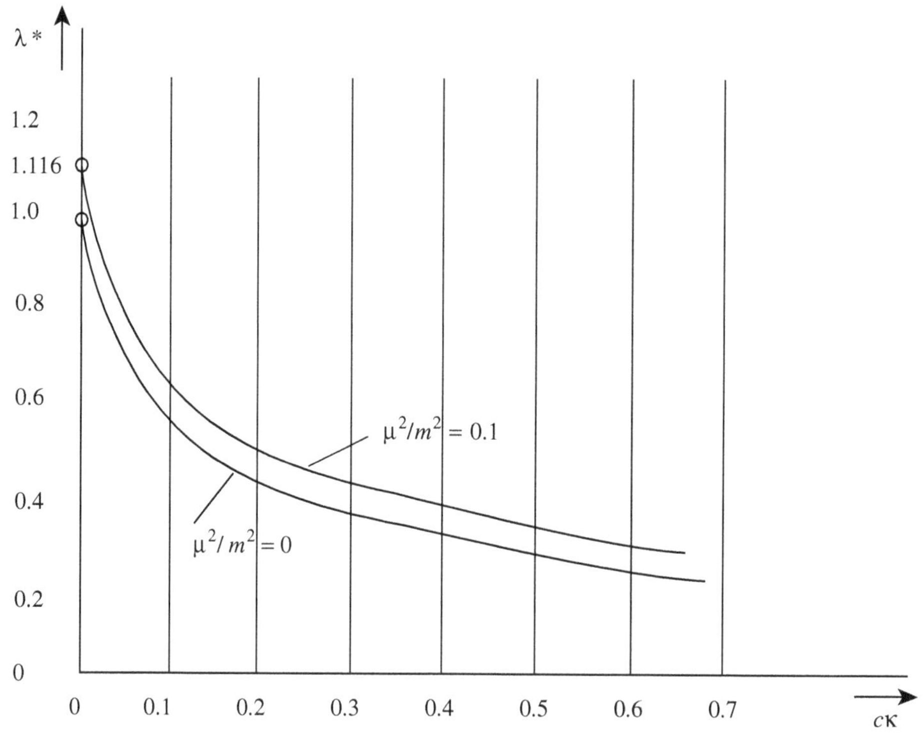

Figure 3.18.3

These results show that the first approximation is indeed a good approximation. In our analysis, we have employed the Rayleigh-Ritz method, which in itself yields an upper bound for the critical load. However, we have also neglected terms in powers of μ/m, so it is not certain that the present results yield an upper bound.

Let us finally remark that actual shells usually have local imperfections due to damage, and these imperfections are in the inward direction, and are therefore strongly destabilizing. It is further interesting to note that even the carefully manufactured cylindrical shells used for the Lockheed tests† had local imperfections.

† Cf. footnote at the end of Section 3.17.

Selected Publications of W. T. Koiter on Elastic Stability Theory

Over de stabiliteit van het elastisch evenwicht (On the stability of elastic equilibrium). Proefschrift T. H. Delft (1945).

On the stability of elastic equilibrium (Washington, 1967), 6 + 202 pp. (NASA Technical Translation F-10, 833, Clearinghouse US Dept. of Commerce/Nat. Bur. of Standards N67-25033).

The stability of elastic equilibrium (Palo Alto, Cal., 1970), 13 + 306 pp. (Stanford Univ., Dept. of Aeronaut. and Astronaut. Report AD 704124/Air Force Flight Dynamics Lab., Report TR-70-25).

"Buckling and post buckling behaviour of a cylindrical panel under axial compression," *Reports and Transactions National Aeronautical Research Institute*, **20** (1956) pp. 71–84 (Nat. Aero. Res. Inst. Report No. S476).

"A consistent first approximation in the general theory of thin elastic shells." *Proc. I. U. T. A. M. Symposium on the Theory of Thin Elastic Shells* (Deflt, August 1959). North-Holland Publishing Cy., Amsterdam, 12 (1960).

"A sysematic simplification of the general equations in the linear theory of thin shells." *Proc. Kon. Ned. Ak. Wet.*, **B64**, 612–619 (1961).

Introduction to the post-buckling behaviour of flat plates. Actes du Colloque sur le Comportement Postcritique des Plaques Utilisées en Construction Métallique, Liège (1963).

"Elastic stability and post-buckling behaviour." *Proc. Symp. Nonlinear Problems*, University of Wisconsin Press, Madison, 257–275 (1963).

"The concept of stability of equilibrium for continuous bodies." *Proc. Kon. Ned. Ak. Wet.*, **B66**, 173–177 (1963).

"The effect of axisymmetric imperfections on the buckling of cylindrical shells under axial compression. *Proc. Kon. Ned. Ak. Wet.*, **B66**, 265–279 (1963).

"Knik van schalen (buckling of shells)." *De Ingenieur*, **76**, 033–041 (1964).

"On the instability of equilibrium in the absence of a minimum of the potential energy." *Proc. Kon. Ned. Wet.*, **B68**, 107–113 (1965).

"The energy criterion of stability for continuous elastic bodies." *Proc. Kon. Ned. Ak. Wet.*, **B68**, 178–202 (1965).

"Post buckling analysis of a simple two-bar frame." *Recent Progress in Applied Mechanics (The Folke Odqvist Volume)*, Almqvist & Wiksell, Stockholm (1967) 337–354.

"On the nonlinear theory of thin elastic shells" *Proc. Kon. Ned. Ak. Wet.*, **B69**, 1–54.

"General equations of elastic stablity for thin shells."*Proc. Symposium on the Theory of Shells*, April 4–6, 1066, Houston, Texas, U. S. A. (Houston, Texas U. S. A., 1967) pp. 187–227.

"A sufficient condition for the stability of shallow shells." *Proc. Kon. Ned. Ak. Wet.*, **B70**, 367–375 (1967).

"The nonlinear buckling problem of a complete spherical shell under uniform external pressure." *Proc. Kon. Ned. Ak. Wet.*, **B72**, 40–123 (1969).
"Postbuckling theory. Jointly with J. W. Hutchinson." *App. Mech. Reviews*, **23**, 1353–1366 (1970).
"Thermodynamics of elastic stablity." *Proc. 3rd Canadian Congress of Applied Mechanics*, Calgary, May 1971, 29–37.
"An alternative approach to the interaction of local and overall buckling in stiffened panels. Jointly with M. Pignataro. *Proc. IUTAM Symposium on Buckling of Structures*, Harvard University, June 1974, Springer-Verlag (1976), 133–148.
"A basic open problem in the theory of elastic stability." Joint IUTAM/IMU Symposium on Applications of Methods of Functional Analysis to Problems of Mechanics. Résumés des conférences, September 1 to 6, 1075, Marseille. Part XXIX (Marseille, 1975).
"Buckling of a flexible shaft under torque loads transmitted by cardan joints." *Ingenieur-Archiv*, **49**, 369–373 (1980).

Index

For EU product safety concerns, contact us at Calle de José Abascal, 56–1°, 28003 Madrid, Spain or eugpsr@cambridge.org.

www.ingramcontent.com/pod-product-compliance
Ingram Content Group UK Ltd.
Pitfield, Milton Keynes, MK11 3LW, UK
UKHW050110230326
469255UK00020B/478
9781107407015